AF793011

Structural Aspects in the Theory of Probability

A Primer in Probabilities on Algebraic-Topological Structures

SERIES ON MULTIVARIATE ANALYSIS

Editor: M M Rao

Published

Vol. 1: Martingales and Stochastic Analysis
J. Yeh

Vol. 2: Multidimensional Second Order Stochastic Processes
Y. Kakihara

Vol. 3: Mathematical Methods in Sample Surveys
H. G. Tucker

Vol. 4: Abstract Methods in Information Theory
Y. Kakihara

Vol. 5: Topics in Circular Statistics
S. R. Jammalamadaka and A. SenGupta

Vol. 6: Linear Models: An Integrated Approach
D. Sengupta and S. R. Jammalamadaka

Vol. 7: Structural Aspects in the Theory of Probability
A Primer in Probabilities on Algebraic-Topological Structures
H. Heyer

Series on
Multivariate
Analysis
Vol. 7

Structural Aspects in the Theory of Probability

A Primer in Probabilities on Algebraic-Topological Structures

Herbert Heyer

Mathematisches Institut
Universität Tübingen
Germany

World Scientific

NEW JERSEY • LONDON • SINGAPORE • BEIJING • SHANGHAI • HONG KONG • TAIPEI • CHENNAI

Published by

World Scientific Publishing Co. Pte. Ltd.

5 Toh Tuck Link, Singapore 596224

USA office: Suite 202, 1060 Main Street, River Edge, NJ 07661

UK office: 57 Shelton Street, Covent Garden, London WC2H 9HE

Library of Congress Cataloging-in-Publication Data

Heyer, Herbert.

Structural aspects in the theory of probability : a primer in probabilities on algebraic-topological structures / Herbert Heyer.

p. cm. -- (Series on multivariate analysis ; v. 7)

Includes bibliographical references and index.

ISBN 981-238-937-7 (alk. paper)

1. Probabilities. 2. Topological groups. 3. Banach spaces. 4. Probability measures. 5. Abelian groups. I. Title. II. Series.

QA273.43 H53 2004

519.2--dc22 2004043061

British Library Cataloguing-in-Publication Data

A catalogue record for this book is available from the British Library.

Copyright © 2004 by World Scientific Publishing Co. Pte. Ltd.

All rights reserved. This book, or parts thereof, may not be reproduced in any form or by any means, electronic or mechanical, including photocopying, recording or any information storage and retrieval system now known or to be invented, without written permission from the Publisher.

Cover design by Katrin Heyer.

For photocopying of material in this volume, please pay a copying fee through the Copyright Clearance Center, Inc., 222 Rosewood Drive, Danvers, MA 01923, USA. In this case permission to photocopy is not required from the publisher.

Printed in Singapore by World Scientific Printers (S) Pte Ltd

Preface

The present book has been written for mathematically prepared readers who like to look beyond the boundary of a single topic in order to discover the interrelations with others. More concretely the author's idea is to direct the attention of probabilists to the applicability of the enlightening notion of a group to probability theory.

The interplay between probability theory and group theory is as old as the early investigations on translation invariant probability distributions and stochastic processes and has become an increasingly important field of research which meanwhile reached a certain state of maturity.

While the traditional approach to the basic theorems of probability theory often overshadows part of the structure of the problems, the awareness of group - theoretical concepts leads to a quick detection of common features of apparently unrelated situations. In other words, the perception of algebraic-topological structures in the state space of stochastic processes does not only yield interesting and applicable generalizations of known results but also sets a limit to such generalizations by describing their domains of validity within the general framework. In practice this approach helps to provide at least more transparent proofs of well-established theorems including Lévy's continuity theorem, the Lévy-Khintchine representation of infinitely divisible probability measures, transience criteria for convolution semi-groups and characterizations of recurrent or transient random walks.

This primer in probabilities on Abelian topological groups with emphasis on separable Banach spaces and on locally compact Abelian groups is by its very conception an elementary introduction to the structural access to probability theory, no text book in the habitual understanding and by now means a monograph. It should be studied by graduate students along with the course work and will make interesting accompanying reading for their lecturers. At the same time the book provides information beyond the particular topic and lays bare the possibility of incorporating certain problems of probability theory into a wider setting which may be chosen according to the actual aims of study.

Since the pioneering work of Grenander and Parthasarathy going back to the early 1960's structural aspects of probability theory have been stressed in various monographs. For probabilities on locally compact groups we mention the books by Berg and Forst and by Revuz, both of 1975, as well as the author's book of 1977. There is also an extensive literature on probabilities on linear spaces. We just cite the books by Araujo and Giné of 1980, by Linde of 1986 and by Vakhania, Tarieladze and Chobanyan of 1987. Our selection of topics from these sources has at least two motives: to stress the significance of the problems within the development of the theory, and to choose an approach to their solutions which at the same time is as direct and informative as possible. Clearly these aims can hardly be achieved without reference to some basic notions and facts from topological groups, topological vector spaces and commutative Banach algebras. Appendices at the end of the book are offered as desirable aids.

In the first part of the book (Chapters 1 to 3) we start by collecting the necessary measure theory on metric spaces including the Riesz and Prohorov theorems. It follows a detailed analysis of the Fourier transform for separable Banach spaces. The main focus of the subsequent discussion is the arithmetic of probability measures on such spaces, in particular the study of infinitely divisible probability measures. We establish the embedding of infinitely divisible probability measures into continuous convolution semigroups and then examine Gauss and Poisson measures. The Ito-Nishio theorem is applied to a construction of Brownian motion. The proof of the Lévy-Khintchine representation is prepared by a detailed discussion of Lévy measures and generalized Poisson measures. It is clear that the theory exposed for general separable Banach spaces covers the case of Euclidean space and also various cases of function spaces.

The second half of the book (Chapters 4 to 6) begins with the notion of convolution of Radon measures on a locally compact group. The exposition continues by developing the duality theory of locally compact Abelian groups including positive definite functions and measures. Then negative definite functions on such groups are studied, their duality with positive definite functions and their correspondence in the sense of Schoenberg with convolution semigroups. The construction of Lévy functions for any locally compact Abelian group is the basic step towards a Lévy-Khintchine representation of negative definite functions. The concluding chapter contains a discussion of transient convolution semigroups and random walks. A measure-theoretic proof of the Port-Stone transience criterion precedes the characterization of groups admitting recurrent random walks and the classification of transient random walks which solves the problem of renewal of random walks on a locally compact Abelian group. The theory developed in this part of the book can be easily specialized to the Euclidean case, but moreover to infinite dimensional lattices and tori.

Now the methodical framework of the book becomes visible. For separable Banach spaces as well as for locally compact Abelian groups dual objects and Fourier transforms of measures as functions on these dual objects are employed in order to determine the structure of infinitely divisible probability measures and convolution semigroups. For Banach spaces only restricted versions of the Lévy continuity theorem can be proved. In fact, by the lack of an appropriate Bochner theorem for positive definite functions harmonic analysis soon reaches its limits. In the case of locally compact Abelian groups, however, the Pontryagin duality provides a far more elaborate harmonic analysis which can be applied to obtain not only strong versions of the Lévy continuity theorem but also deep results on the potential theory of stochastic processes with stationary independent increments and random walks in the group.

To write a primer in probabilities on algebraic-topological structures became a matter of concern during the author's lecturing over about three decades, mostly at the University of Tübingen in Germany. Along with his research work at the interface between probability theory and harmonic analysis he taught on probability measures on Banach spaces, locally compact groups and homogeneous spaces. It turned out that graduate students majoring in probability theory or in analysis took those courses which led to seminars on "Stochastics and Analysis" in which central limit theorems for generalized random variables, stochastic processes in and random fields over general algebraic-topological structures were discussed. In recent years also analogs of these probabilistic objects for generalized convolution structures as Jacobi and Sturm-Liouville translation structures were considered. For the harmonic analysis of these structures the presentation of the case of a locally compact Abelian group provides the appropriate basis. Consequently, the present book may also be used as a preparatory text for the study of probability measures on hypergroups and hypercomplex systems.

In conceiving his book the author received encouragement from many colleagues and friends spread over the globe. Various scientific agencies like the German and the Japanese Research Societies made it possible to test preliminary versions of the manuscript in workshops and crash courses during research stays and sabbaticals at universities in Australia (Perth), Japan (Tokyo) and the US (San Diego). Acknowledgement of prime importance goes to Christian Berg and Gunnar Forst, to Werner Linde and to Daniel Revuz for their excellent monographs the contents of which reaches far beyond our exposition. Several people have read drafts of the text. Especially valuable was Gyula Pap's constructive criticism for which the author is most thankful. There were also capable secretaries who did a great job in preparing the typescript: Kerstin Behrends and Erika Gugl deserve praise for their skillful work. Last but not least I am grateful to M.M. Rao from the University of California at Riverside who invited the book into the series on Multivariate Analysis with World Scientific.

The author expresses his expectation that all obscurities contained in the text will be communicated to him and that despite of such inevitable deficiencies the book may serve its modest purpose. There is no doubt that the following statement due to Pablo Picasso also applies to an author in mathematics

"Ce que je fais aujourd'hui est déjà vieux pour demain."

Tübingen, March 2004 Herbert Heyer

Contents

1

Probability Measures on Metric Spaces

1.1 Tight measures

Let (E, d) denote a metric space, $\mathcal{O}(E), \mathcal{A}(E), \mathcal{K}(E)$ the systems of open, closed and compact subsets of E respectively. On (E, d) we have the notions of the *Borel σ-algebra*

$$\mathfrak{B}(E) := \sigma(\mathcal{O}(E)) = \sigma(\mathcal{A}(E))$$

of E and of a (Borel) measure on E, i.e. a non-negative extended real-valued σ-additive set function μ on $\mathfrak{B}(E)$ with the properties that $\mu(\emptyset) = 0$ and $\mu(K) < \infty$ for all $K \in \mathcal{K}(E)$.

Definition 1.1.1 *A finite measure μ on E is called*

(a) ***regular*** *if for every $B \in \mathfrak{B}(E)$ and for every $\varepsilon > 0$ there exist $A \in \mathcal{A}(E)$ and $O \in \mathcal{O}(E)$ such that $A \subset B \subset O$ and $\mu(O) - \mu(A) < \varepsilon$, and*

(b) ***tight*** *if*

$$\mu(E) = \sup\{\mu(K) : K \in \mathcal{K}(E)\}.$$

Theorem 1.1.2 *Let μ be a finite measure on E. Then*

(i) μ is regular.

(ii) If μ is tight then it must be ***inner-regular*** *in the sense that for each $B \in \mathfrak{B}(E)$*

$$\mu(B) = \sup\{\mu(K) : K \in \mathcal{K}(E),\ \ K \subset B\}\,.$$

In particular, for finite measures the notions of tightness and inner-regularity coincide.

Proof. *(i)* Let $\mathfrak{D} := \mathfrak{D}_\mu$ be the system of all $B \in \mathfrak{B}(E)$ with respect to which μ is regular. Then $\mathfrak{D}$ is a Dynkin system in the sense that $E \in \mathfrak{D}$, $B \in \mathfrak{D}$ implies that $E \setminus B \in \mathfrak{D}$, and whenever $(B_n)_{n\geq 1}$ is a disjoint sequence in $\mathfrak{D}$ then $B := \bigcup_{n\geq 1} B_n \in \mathfrak{D}$.

The proof of the first property is clear, and for the second one we observe that if $A \in \mathcal{A}(E)$ and $O \in \mathcal{O}(E)$ are chosen as in Definition 1.1.1(a) then $E\setminus O \subset E\setminus B \subset E\setminus A$ and, noting that $E\setminus O \in \mathcal{A}(E)$ and $E \setminus A \in \mathcal{O}(E)$, we have

$$\mu(E\setminus A) - \mu(E\setminus O) = \mu(O) - \mu(A) < \varepsilon\,.$$

As for the third property, given $\varepsilon > 0$ we can find $A_n \in \mathcal{A}(E)$ and $O_n \in \mathcal{O}(E)$ with $A_n \subset B_n \subset O_n$ and

$$\mu(O_n) - \mu(A_n) < \frac{1}{2^{n+2}}\ \varepsilon$$

for all $n \in \mathbf{N}$. Let $O := \bigcup_{n\geq 1} O_n$, choose n_0 with $\mu\big(\bigcup_{n>n_0} A_n\big) < \varepsilon/4$ and put $A := \bigcup_{n=1}^{n_0} A_n$. Then $A \in \mathcal{A}(E)$, $O \in \mathcal{O}(E)$, $A \subset B \subset O$ and

$$\begin{aligned}\mu(O\setminus A) &\leq \sum_{n\geq 1} \mu(O_n\setminus A) \leq \sum_{n=1}^{n_0} \mu(O_n\setminus A) + \sum_{n>n_0} \mu(O_n)\\ &\leq \frac{\varepsilon}{4} + \sum_{n>n_0}\left(\mu(A_n) + \frac{1}{2^{n+2}}\varepsilon\right) \leq \frac{3}{4}\varepsilon < \varepsilon\,.\end{aligned}$$

Furthermore $\mathcal{A}(E) \subset \mathfrak{D}$. Indeed, given $A \in \mathcal{A}(E)$ for each $n \in \mathbf{N}$ we observe that

$$A^{\frac{1}{n}} := \left\{x \in E : d(x,A) < \frac{1}{n}\right\}$$

is open, and from $A^{\frac{1}{n}} \downarrow A$ (which holds as E is metric) it follows that $\mu(A^{\frac{1}{n}}) \downarrow \mu(A)$.

Now $\mathcal{A}(E)$ is $\cap$-stable, and therefore

$$\mathfrak{B}(E) = \sigma(\mathcal{A}(E)) = \mathfrak{D}(\mathcal{A}(E)) \subset \mathfrak{D} \subset \mathfrak{B}(E),$$

whence $\mathfrak{D} = \mathfrak{B}(E)$. Here $\mathfrak{D}(\mathcal{A}(E))$ denotes the Dynkin hull of $\mathcal{A}(E)$.

(ii) Let $B \in \mathfrak{B}(E)$ and $\varepsilon > 0$. Using *(i)* there exists $A \in \mathcal{A}(E)$ with $A \subset B$ such that $\mu(B) - \mu(A) < \frac{\varepsilon}{2}$, and also $K \in \mathcal{K}(E)$ with $\mu(E) - \mu(K) < \frac{\varepsilon}{2}$. Then $A \cap K$ is a compact subset of B, and

$$\mu(B) - \mu(A \cap K) \leq \mu(B \setminus A) + \mu(E \setminus K) < \frac{\varepsilon}{2} + \frac{\varepsilon}{2} = \varepsilon. \qquad \blacksquare$$

Corollary 1.1.3 *If μ is tight, then for every downward filtered family $(A_\iota)_{\iota \in I}$ in $\mathcal{A}(E)$*

$$\mu\left(\bigcap_{\iota \in I} A_\iota\right) = \inf_{\iota \in I} \mu(A_\iota)\,.$$

Proof. From $A := \bigcap_{\iota \in I} A_\iota \subset A_\kappa$ we have $\mu(A) \leq \mu(A_\kappa)$ for all $\kappa \in I$, and hence $\mu(A) \leq \inf_{\iota \in I} \mu(A_\iota)$.

In the reverse direction, appealing to Theorem 1.1.2*(ii)*, to each $\varepsilon > 0$ there exists $K \in \mathcal{K}(E)$ with $K \subset E \setminus A$ such that

$$\mu(E \setminus K) - \mu(A) = \mu(E \setminus A) - \mu(K) < \varepsilon\,.$$

Now $K \subset \bigcup_{\iota \in I}(E \setminus A_\iota)$, and hence by compactness there exist $\iota_1, \iota_2, ..., \iota_n \in I$ with $K \subset \bigcup_{i=1}^{n}(E \setminus A_{\iota_i})$. Also $(E \setminus A_\iota)_{\iota \in I}$ is an upward filtered family, and hence there exists $\iota_0 \in I$ such that $K \subset E \setminus A_{\iota_0}$. From $\mu(E \setminus K) - \mu(A) < \varepsilon$ it follows that

$$\mu(A_{\iota_0}) \leq \mu(E \setminus K) < \mu(A) + \varepsilon$$

and so ε being arbitrary we obtain $\inf_{\iota \in I} \mu(A_\iota) \leq \mu(A)$. $\blacksquare$

Theorem 1.1.4 *Let μ be a tight measure on E.*

(i) There exists a smallest closed subset A_0 *of* E *with* $\mu(A_0) = \mu(E)$.

(ii) A_0 *is separable.*

(iii) $A_0 = \{x \in E : \mu(U) > 0$ *for all open neighborhoods* U *of* $x\}$.

Proof. *(i)*. The family

$$\{A \in \mathcal{A}(E) : \mu(A) = \mu(E)\}$$

is downward filtered, even $\cap$-stable. The result now follows from Corollary 1.1.3.

(ii). By Theorem 1.1.2 *(ii)* there exists a sequence $(K_n)_{n\geq 1}$ of compact and hence separable subsets of A_0 with $\mu(A_0) = \sup_{n\geq 1} \mu(K_n)$. Thus $A := (\bigcup_{n\geq 1} K_n)^-$ is separable and closed with $A \subset A_0$, from which it follows that

$$\mu(A) = \mu(A_0) = \mu(E)$$

and by *(i)*, $A = A_0$ so that A_0 must be separable.

(iii). Write

$$B_0 := \{x \in E : \mu(U) > 0 \text{ for all open neighborhoods } U \text{ of } x\}.$$

Given $x \in E \setminus A_0$ then $E \setminus A_0$ is an open neighborhood of x with $\mu(E \setminus A_0) = 0$, and hence $x \in E \setminus B_0$. In the reverse direction given $x \in E \setminus B_0$ there exists an open neighborhood U of x with $\mu(U) = 0$. Hence $\mu(E \setminus U) = \mu(E)$. Thus $A_0 \subset E \setminus U$ and hence $x \in E \setminus A_0$. ■

Definition 1.1.5 *The set* A_0 *in Theorem* 1.1.4 *is called the* ***support*** *of* μ, *and will be denoted by* supp (μ).

Theorem 1.1.6 *Let* (E, d) *be a separable complete metric space. Then every finite measure* μ *on* E *is tight.*

Proof. Let $\{x_k : k \in \mathbf{N}\}$ be a dense subset of E. Then for each $n \in \mathbf{N}$

$$\bigcup_{k\geq 1} B(x_k, \frac{1}{n})^- = E$$

where

$$B(x,\delta) := \{y \in E : d(x,y) < \delta\}$$

is the open ball of radius $\delta > 0$ with centre x. Choose $\varepsilon > 0$. Then to each $n \in \mathbf{N}$ there exists $k_n \in \mathbf{N}$ satisfying

$$\mu\left(E \setminus \bigcup_{k=1}^{k_n} B(x_k, \frac{1}{n})^-\right) \leq \frac{\varepsilon}{2^n}.$$

The set

$$K := \bigcap_{n\geq 1} \bigcup_{k=1}^{k_n} B(x_k, \frac{1}{n})^-$$

is closed and totally bounded. From the completeness of E it follows that K is compact. Finally

$$\mu(E \setminus K) = \mu\left(\bigcup_{n\geq 1}\left(E \setminus \bigcup_{k=1}^{k_n} B(x_k, \frac{1}{n})^-\right)\right) \leq \sum_{n\geq 1} \frac{\varepsilon}{2^n} = \varepsilon. \qquad \blacksquare$$

Theorem 1.1.7 *Let E, F be metric spaces, and $\varphi : E \to F$ a continuous mapping. If μ is a tight measure on E then the image measure $\varphi(\mu)$ of μ under φ is tight on F.*

Proof. Since φ is a continuous mapping it must be $\mathfrak{B}(E) - \mathfrak{B}(F)$-measurable, and hence $\varphi(\mu)$ is a finite measure on F. Given $\varepsilon > 0$ there exists a compact subset K of E with $\mu(E \setminus K) < \varepsilon$. Also $\varphi(K)$ is a compact subset of F, and

$$\begin{aligned}\varphi(\mu)(F \setminus \varphi(K)) &= \mu(\varphi^{-1}(F \setminus \varphi(K))) \\ &= \mu(E \setminus \varphi^{-1}(\varphi(K))) \\ &\leq \mu(E \setminus K) < \varepsilon.\end{aligned} \qquad \blacksquare$$

1.2 The topology of weak convergence

Although in the following discussion the set $M^b(E)$ of all tight (finite Borel) measures on E and its subset $M^1(E) := \{\mu \in M^b(E) : \mu(E) = 1\}$ of probability measures will remain the basic measure-theoretic objects, for some technical arguments we need a few facts on regular normed contents on E and related integrals. A **content** on E is a non-negative extended real-valued (finitely) additive set function μ on the algebra $\mathfrak{A}(\mathcal{O}(E))$ generated by $\mathcal{O}(E)$ satisfying $\mu(\emptyset) = 0$. Regular (finite) contents and probability contents on E are introduced in analogy to regular (finite) and probability measures on E.

Given a regular finite content μ on E, the μ-integral of a bounded real-valued function f on E is defined as follows. Let $\mathcal{P}$ be a partition of E consisting of finitely many pairwise disjoint sets $E_1, ..., E_n \in \mathfrak{A}(\mathcal{O}(E))$. We put

$$S_{\mathcal{P}} := \sum_{j=1}^{n} M_j \mu(E_j)$$

and

$$s_{\mathcal{P}} := \sum_{j=1}^{n} m_j \mu(E_j),$$

where $M_j := \sup\{f(x) : x \in E_j\}$ and $m_j := \inf\{f(x) : x \in E_j\}$ for $j = 1, ..., n$. f is said to be ***μ-integrable*** if

$$\inf_{\mathcal{P}} S_{\mathcal{P}} = \sup_{\mathcal{P}} s_{\mathcal{P}},$$

and in this case

$$\int f d\mu := \inf_{\mathcal{P}} S_{\mathcal{P}}$$

is the ***μ-integral*** of f. Obviously every bounded continuous function f is μ-integrable, and

$$f \mapsto \int f d\mu$$

defines a normed positive linear functional on the vector space $C^b(E)$ of bounded continuous functions on E. Moreover, we have

Theorem 1.2.1 *(F. Riesz) There is a one-to-one correspondence*

$$\mu \longmapsto L_\mu$$

between the set of regular finite (probability) contents μ on E and the set of bounded (normed) positive linear functionals L_μ on $C^b(E)$ given by

$$L_\mu(f) := \int f d\mu$$

for all $f \in C^b(E)$.

The Proof will be carried out only for the case in parentheses.
1. Let L be a normed positive linear functional on $C^b(E)$, and let

$$\lambda_L(A) := \inf\{L(f) : f \in C^b(E),\ \ f \geq 1_A\}$$

for every $A \in \mathcal{A}(E)$. $\lambda_L : \mathcal{A}(E) \to [0,1]$ is a smooth probability content in the sense of the following four properties

(a) $\lambda_L(\emptyset) = 0,\ \ \lambda_L(E) = 1$.

(b) $\lambda_L(A_1) \leq \lambda_L(A_2)$ for all $A_1, A_2 \in \mathcal{A}(E)$ with $A_1 \subset A_2$.

(c) $\lambda_L(A_1 \cup A_2) \leq \lambda_L(A_1) + \lambda_L(A_2)$ for all $A_1, A_2 \in \mathcal{A}(E)$, where equality holds whenever $A_1 \cap A_2 = \emptyset$.

(d) For all $A \in \mathcal{A}(E)$

$$\lambda_L(A) = \inf\{\lambda_L(O^-) : O \in \mathcal{O}(E),\ \ O \supset A\}.$$

2. Now, λ_L can be uniquely extended to a regular probability content $\mu_L : \mathfrak{A}(\mathcal{O}(E)) \to [0,1]$, and it turns out that

$$L(f) = \int f d\mu_L$$

for all $f \in C^b(E)$.

In order to verify this identity we pick $f \in C^b(E)$ with $0 \le f \le 1$ and introduce the sets

$$G_i := \left\{x \in E : f(x) > \frac{i}{n}\right\} \in \mathcal{O}(E)$$

for all $i = 0, 1, \ldots, n$, $n \ge 1$. Clearly, $G_0 \supset G_1 \supset \ldots \supset G_n = \emptyset$. Now we define functions $\alpha_i \in C\big([0,1]\big)$ by

$$\alpha_i := \begin{cases} \equiv 0 & \text{on } \left[0, \dfrac{i-1}{n}\right] \\ \text{linear} & \text{on } \left[\dfrac{i-1}{n}, \dfrac{i}{n}\right] \\ \equiv 1 & \text{on } \left[\dfrac{i}{n}, 1\right] \end{cases}$$

and functions f_i on E by

$$f_i := \alpha_i \circ f$$

for $i = 0, 1, \ldots, n, n \ge 1$. Then

$$\frac{1}{n}\sum_{i=1}^{n} \alpha_i(t) = t$$

for all $t \in [0,1]$, hence

$$\frac{1}{n}\sum_{i=1}^{n} f_i = f \qquad \text{and} \qquad \frac{1}{n}\sum_{i=1}^{n} L(f_i) = L(f)\,.$$

Since $f_i \ge 1_{G_i}$, and for any $A \in \mathcal{A}(E)$, $A \subset G_i$, $1_{G_i} \ge 1_A$ we obtain that $f_i \ge 1_A$ and hence that

$$L(f_i) \ \ge \lambda_L(A) = \mu_L(A)\,.$$

From the regularity of μ_L we infer that

$$L(f_i) \ge \mu_L(G_i)$$

and thus

$$\begin{aligned}
L(f) &\geq \frac{1}{n}\sum_{i=1}^{n} \mu_L(G_i) = \sum_{i=1}^{n}\left(\frac{i}{n} - \frac{i-1}{n}\right)\mu_L(G_i) \\
&= \sum_{i=1}^{n-1} \frac{i}{n}\left(\mu_L(G_i) - \mu_L(G_{i+1})\right) \\
&= \left(\sum_{i=1}^{n-1} \frac{i+1}{n}\mu_L(G_i \backslash G_{i+1})\right) - \frac{1}{n}\mu_L(G_1) \\
&\geq \left(\sum_{i=1}^{n-1} \int_{G_i \backslash G_{i+1}} f\, d\mu_L\right) - \frac{1}{n}\mu_L(G_1) \\
&= \int_{G_1} f d\mu_L - \frac{1}{n}\mu_L(G_1) \geq \int_E f\, d\mu_L - \frac{1}{n}.
\end{aligned}$$

For $n \to \infty$ we obtain that

$$L(f) \geq \int f d\mu_L$$

whenever $f \in C^b(E)$ with $0 \leq f \leq 1$.

But since $f \in C_+(E)$ there exists a constant $c > 0$ such that $0 \leq cf \leq 1$, hence

$$L(f) = \frac{1}{c}L(cf) \geq \frac{1}{c}\int cf\, d\mu_L = \int f\, d\mu_L\,.$$

Moreover, if $f \in C^b(E)$, there exists a constant $c_1 > 0$ satisfying $f + c_1 \geq 0$, hence

$$L(f) = L(f + c_1) - c_1 \geq \int (f + c_1)\, d\mu_L - c_1 = \int f\, d\mu_L\,.$$

Thus we have

$$L(f) \geq \int f\, d\mu_L$$

for all $f \in C^b(E)$. Replacing f by $-f$ yields the assertion.

3. The injectivity of the correspondence $\mu \mapsto L_\mu$ can be seen as follows: Let μ, ν be regular probability contents of E satisfying

$$\int f d\mu = \int f d\nu$$

for all $f \in C^b(E)$, and let $A \in \mathcal{A}(E)$. There exist decreasing sequences $(G_n)_{n\geq 1}$ and $(H_n)_{n\geq 1}$ in $\mathcal{O}(E)$ with $G_n \supset A$ and $H_n \supset A$ for all $n \geq 1$ such that

$$\lim_{n\to\infty} \mu(G_n) = \mu(A)$$

and

$$\lim_{n\to\infty} \nu(H_n) = \nu(A)\,.$$

But then $V_n := G_n \cap H_n \downarrow A$ and

$$\lim_{n\to\infty} \mu(V_n) = \mu(A)$$

as well as

$$\lim_{n\to\infty} \nu(V_n) = \nu(A)\,.$$

Choosing for every $n \geq 1$ a function $f_n \in C^b(E)$ with the properties $0 \leq f_n \leq 1$, $f_n(A) = \{1\}$ and $f_n(V_n^c) = \{0\}$ (the existence of which follows from $A \cap V_n^c = \emptyset$ for all $n \geq 1$) we obtain

$$\int f_n d\mu \;=\; \int_A f_n d\mu + \int_{V_n \setminus A} f_n d\mu \;=\; \mu(A) + \int_{V_n \setminus A} f_n d\mu$$

and

$$\int_{V_n \setminus A} f_n d\mu \;\leq\; \mu(V_n \setminus A) \;=\; \mu(V_n) - \mu(A)\,,$$

hence

$$\lim_{n\to\infty} \int_{V_n \setminus A} f_n d\mu = 0\,.$$

Therefore

$$\lim_{n\to\infty} \int f_n d\mu = \mu(A)$$

and

$$\lim_{n\to\infty} \int f_n d\nu = \nu(A),$$

thus

$$\mu(A) = \nu(A) \text{ for all } A \in \mathcal{A}(E)$$

and by the regularity of μ, ν also

$$\mu(B) = \nu(B) \text{ for all } B \in \mathfrak{A}(\mathcal{A}(E)) = \mathfrak{A}(\mathcal{O}(E))$$

which implies that $\mu = \nu$. ■

At a later stage we will apply the following consequences of the theorem.

Corollary 1.2.2 *If for measures $\mu, \nu \in M^b(E)$*

$$\int f d\mu = \int f d\nu$$

holds whenever $f \in C^b(E)$ then $\mu = \nu$ (on $\mathfrak{B}(E)$).

Corollary 1.2.3 *Let (E, d) be a compact metric space. There is a one-to-one correspondence*

$$\mu \mapsto L_\mu$$

between the set $M^b(E)$ and the set $L^1_+(C(E))$ of positive normed linear functionals on $C(E)$ given by

$$L_\mu(f) = \int f d\mu$$

for all $f \in C(E) = C^b(E)$.

The **Proof** follows directly from Theorem 1.2.1 by applying the fact that for compact E every regular finite content on E is in fact σ-additive and hence uniquely extendable to a measure in $M^b(E)$. For the latter property see Theorem 1.3.1. ■

We proceed to introducing a topology in $M^b(E)$.

Definition 1.2.4 *Given* $\mu \in M^b(E)$, $n \geq 1$, $f_1, f_2, ..., f_n \in C^b(E)$ *and* $\varepsilon > 0$, *define*

$$V(\mu; f_1, f_2, ..., f_n; \varepsilon) :$$
$$= \left\{\nu \in M^b(E) : \left| \int f_i d\mu - \int f_i d\nu \right| < \varepsilon \text{ for all } i = 1, 2, ..., n \right\}.$$

The ***weak topology*** τ_w *on* $M^b(E)$ *is the uniquely determined topology for which*

$$\{V(\mu; f_1, f_2, ..., f_n; \varepsilon) : n \geq 1, \ f_1, f_2, ..., f_n \in C^b(E), \ \varepsilon > 0\}$$

is a neighborhood system of μ *for each* $\mu \in M^b(E)$.

Remark 1.2.5

(a) The weak topology on $M^b(E)$ *is Hausdorff due to Corollary 1.2.2.*

(b) A net $(\mu_\iota)_{\iota \in I}$ *in* $M^b(E)$ ***converges weakly*** (τ_w) *to* $\mu \in M^b(E)$ *whenever*

$$\lim_{\iota \in I} \int f d\mu_\iota = \int f d\mu$$

for all $f \in C^b(E)$*; we write* $\tau_w - \lim_\iota \mu_\iota = \mu$.

(c) In the functional-analytic context of Appendix B 10 one introduces for the dual pair $(C^b(E)', C^b(E))$ *of topological vector spaces the weak topology on* $C^b(E)'$. *If* E *is compact, then Corollary 1.2.3 yields the homeomorphism*

$$C^b(E)'_+ \cong M^b(E)$$

and consequently the coincidence of the weak topology restricted to $C^b(E)'_+$ *with the weak topology* τ_w *on* $M^b(E)$.

Definition 1.2.6 *Let* $\mu \in M^b(E)$. *A set* $B \in \mathfrak{B}(E)$ *is called a* μ-***continuity*** *(*μ*-null boundary) set if* $\mu(\partial B) = 0$ *where* $\partial B := B^- \backslash B^0$ $(\in \mathcal{A}(E))$.

Theorem 1.2.7 *(Portemanteau) Let* $(\mu_\iota)_{\iota \in I}$ *be a net in* $M^b(E)$ *and* $\mu \in M^b(E)$. *The following statements are equivalent:*

(i) $\tau_w - \lim_\iota \mu_\iota = \mu$.

(ii) $\lim_{\iota \in I} \mu_\iota(E) = \mu(E)$ *and* $\limsup_{\iota \in I} \mu_\iota(A) \leq \mu(A)$ *for all* $A \in \mathcal{A}(E)$.

(iii) $\lim_{\iota \in I} \mu_\iota(E) = \mu(E)$ *and* $\liminf_{\iota \in I} \mu_\iota(O) \geq \mu(O)$ *for all* $O \in \mathcal{O}(E)$.

(iv) $\lim_{\iota \in I} \mu_\iota(B) = \mu(B)$ *for all* μ*-continuity sets* B.

Proof.

(i) $\Rightarrow$ *(ii)*. As $\mathbf{1}_E \in C^b(E)$ we have $\lim_{\iota \in I} \mu_\iota(E) = \mu(E)$. Now consider $A \in \mathcal{A}(E)$. Then, as $A^{\frac{1}{n}} \downarrow A$ as $n \to \infty$, to each $\varepsilon > 0$ we can find $n \in \mathbf{N}$ with $\mu(A^{\frac{1}{n}}) - \mu(A) < \varepsilon$. Choose $f \in C^b(E)$ with $0 \leq f \leq 1, f(A) = \{1\}$ and $f(E \setminus A^{\frac{1}{n}}) = \{0\}$. Then

$$\limsup_{\iota \in I} \mu_\iota(A) \leq \limsup_{\iota \in I} \int f d\mu_\iota \leq \mu(A^{\frac{1}{n}}) < \mu(A) + \varepsilon$$

and hence

$$\limsup_{\iota \in I} \mu_\iota(A) \leq \mu(A)\,.$$

(ii) $\Leftrightarrow$ *(iii)*. This follows by considering complements.

(ii),(iii) $\Rightarrow$ *(iv)*. Let B be a μ-continuity set. Then

$$\begin{aligned}\limsup_{\iota \in I} \mu_\iota(B) &\leq \limsup_{\iota \in I} \mu_\iota(B^-) \leq \mu(B^-) \\ &= \mu(B^0) \leq \liminf_{\iota \in I} \mu_\iota(B^0) \leq \liminf_{\iota \in I} \mu_\iota(B)\end{aligned}$$

and this yields the result.

(iv) $\Rightarrow$ *(i)*. Let $f \in C^b(E)$. Since $f(\mu)$ contains at most countably many atoms, to each $\varepsilon > 0$ there exists a strictly increasing sequence $(t_i)_{i=0,1,\dots,k}$ in $\mathbf{R}$ with $f(\mu)(\{t_i\}) = 0$ for all $i = 0, 1, ..., k$, $t_i - t_{i-1} \leq \varepsilon$ for all $i = 1, 2, ..., k$, and $f(E) \subset [t_0, t_k[$. For each $i = 1, 2, ..., k$ put $B_i := f^{-1}([t_{i-1}, t_i[)$. Then $B_i \in \mathfrak{B}(E)$ and, since

$$\partial B_i \subset f^{-1}(\partial[t_{i-1}, t_i[) = f^{-1}(\{t_{i-1}, t_i\}),$$

we see that B_i is a μ-continuity set. We now define

$$g := \sum_{i=1}^{k} t_{i-1} \mathbf{1}_{B_i} \text{ and } h := \sum_{i=1}^{k} t_i \mathbf{1}_{B_i} .$$

Then

$$g \leq f \leq g + \varepsilon \text{ and } h - \varepsilon \leq f \leq h$$

and

$$\begin{aligned}\limsup_{\iota \in I} \int f d\mu_\iota &\leq \limsup_{\iota \in I} \int g d\mu_\iota + \varepsilon\mu(E) \\ &= \int g d\mu + \varepsilon\mu(E) \leq \int f d\mu + \varepsilon\mu(E)\end{aligned}$$

and

$$\begin{aligned}\liminf_{\iota \in I} \int f d\mu_\iota &\geq \liminf_{\iota \in I} \int h d\mu_\iota - \varepsilon\mu(E) \\ &= \int h d\mu - \varepsilon\mu(E) \geq \int f d\mu - \varepsilon\mu(E)\end{aligned}$$

as

$$\int g d\nu = \sum_{i=1}^{k} t_{i-1}\nu(B_i) \text{ and } \int h d\nu = \sum_{i=1}^{k} t_i\nu(B_i)$$

for all $\nu \in M^b(E)$, and E is a μ-continuity set as $\partial E = \emptyset$. It now follows that

$$\limsup_{\iota \in I} \int f d\mu_\iota \leq \int f d\mu \leq \liminf_{\iota \in I} \int f d\mu_\iota$$

and this gives the desired equality. ■

Corollary 1.2.8 *Let $\mu \in M^b(E)$. Then each of the following is a τ_w-neighborhood basis of μ.*

(a) $\{\nu \in M^b(E) : |\nu(E) - \mu(E)| < \varepsilon$ *and* $\nu(A_i) < \mu(A_i) + \varepsilon$ *for all* $i = 1, 2, ..., n\}$, *where* $A_1, A_2, ..., A_n \in \mathcal{A}(E)$, $n \in \mathbf{N}$ *and* $\varepsilon > 0$.

(b) $\{\nu \in M^b(E) : |\nu(E) - \mu(E)| < \varepsilon$ *and* $\nu(O_i) > \mu(O_i) - \varepsilon$ *for all* $i = 1, 2, ..., n\}$, *where* $O_1, O_2, ..., O_n \in \mathcal{O}(E)$, $n \in \mathbf{N}$ *and* $\varepsilon > 0$.

(c) $\{\nu \in M^b(E) : |\nu(B_i) - \mu(B_i)| < \varepsilon$ *for all* $i = 1, 2, ..., n\}$, *where* $B_1, B_2, ..., B_n \in \mathfrak{B}(E)$ *are μ-continuity sets with* $n \in \mathbf{N}$ *and* $\varepsilon > 0$.

Theorem 1.2.9 *Let $(\mu_\iota)_{\iota \in I}$ be a net in $M^b(E)$ with $\tau_w - \lim_\iota \mu_\iota = \mu \in M^b(E)$. Furthermore let f be a bounded Borel-measurable real-valued function on E. If the set D_f of discontinuity points of f is a μ-null set, then*

$$\lim_{\iota \in I} \int f d\mu_\iota = \int f d\mu .$$

Proof. Let $A \in \mathcal{A}(\mathbf{R})$. Since $f^{-1}(A)^- \subset D_f \cup f^{-1}(A)$ we can apply Theorem 1.2.7 to obtain

$$\begin{aligned}\limsup_{\iota \in I} f(\mu_\iota)(A) &= \limsup_{\iota \in I} \mu_\iota(f^{-1}(A)) \leq \limsup_{\iota \in I} \mu_\iota(f^{-1}(A)^-) \\ &\leq \mu(f^{-1}(A)^-) \leq \mu(D_f \cup f^{-1}(A)) = \mu(f^{-1}(A)) = f(\mu)(A).\end{aligned}$$

In addition

$$\lim_{\iota \in I} f(\mu_\iota)(\mathbf{R}) = \lim_{\iota \in I} \mu_\iota(E) = \mu(E) = f(\mu)(\mathbf{R}) .$$

A second application of Theorem 1.2.7 gives $\tau_w - \lim_\iota f(\mu_\iota) = f(\mu)$. Now consider $\varphi \in C^b(\mathbf{R})$ such that $\mathrm{Res}_B \varphi = \mathrm{id}_B$, where B is any bounded interval containing the bounded set $f(B)$. Since $\varphi \circ f = f$ we have

$$\lim_{\iota \in I} \int f d\mu_\iota = \lim_{\iota \in I} \int \varphi d f(\mu_\iota) = \int \varphi d f(\mu) = \int f d\mu . \qquad \blacksquare$$

Corollary 1.2.10 *Let $(\mu_\iota)_{\iota\in I}$ be a net in $M^b(E)$ satisfying $\tau_w - \lim_\iota \mu_\iota = \mu \in M^b(E)$, and let $B \in \mathfrak{B}(E)$ be a μ-continuity set. Then for the corresponding measures induced on B we have $\tau_w - \lim_\iota (\mu_\iota)_B = \mu_B$.*

Proof. Let $f \in C^b(E)$. Then

$$\int f d\nu_B = \int f\mathbf{1}_B d\nu$$

for all $\nu \in M^b(E)$. From $D_{f\mathbf{1}_B} \subset \partial B$ we see that $D_{f\mathbf{1}_B}$ is a μ-null set. In addition $f1_B$ is bounded and Borel measurable. Referring to Theorem 1.2.9 it follows that

$$\lim_{\iota\in I} \int f d(\mu_\iota)_B = \lim_{\iota\in I} \int f\mathbf{1}_B d\mu_\iota = \int f d\mu_B . \qquad \blacksquare$$

Theorem 1.2.11 *The set $D(E) := \{\varepsilon_x : x \in E\}$ of Dirac measures on E is τ_w-closed in $M^b(E)$, and $x \mapsto \varepsilon_x$ is a homeomorphism of E onto $D(E)$.*

Proof. Let $(x_\iota)_{\iota\in I}$ be a net in E such that $\lim_{\iota\in I} x_\iota = x \in E$. Then

$$\lim_{\iota\in I} \int f d\varepsilon_{x_\iota} = \lim_{\iota\in I} f(x_\iota) = f(x) = \int f d\varepsilon_x$$

for all $f \in C^b(E)$, and thus $\tau_w - \lim_\iota \varepsilon_{x_\iota} = \varepsilon_x$ which shows that $x \mapsto \varepsilon_x$ is a continuous mapping $E \to D(E)$. In addition $x \mapsto \varepsilon_x$ is injective, which is easily seen by simply choosing $B \in \mathfrak{B}(E)$ with $x \in B$ and $y \notin B$, and indeed $B = \{x\}$ will suffice.

Now suppose that $\tau_w - \lim_\iota \varepsilon_{x_\iota} = \mu \in M^b(E)$. From $\lim_{\iota\in I} \varepsilon_{x_\iota}(E) = \mu(E)$ we see that $\mu(E) = 1$, and hence $\operatorname{supp}(\mu) \neq \emptyset$. Choose $x \in \operatorname{supp}(\mu)$ and an open neighborhood U of x. It follows from the properties of $\operatorname{supp}(\mu)$ and Theorem 1.2.7 that

$$\liminf_{\iota\in I} \varepsilon_{x_\iota}(U) \geq \mu(U) > 0$$

so that there exists $\iota_U \in I$ with $\varepsilon_{x_{\iota_U}}(U) > 0$, and hence $x_\iota \in U$ for all $\iota > \iota_U$. This shows that $\lim_{\iota \in I} x_\iota = x$, and in particular that $\varepsilon_x \mapsto x$ is continuous. It follows that $\mu = \varepsilon_x$, since τ_w is Hausdorff, and therefore $D(E)$ is τ_w-closed. ∎

Theorem 1.2.12 *Let E, F be metric spaces, and $\varphi : E \to F$ a continuous mapping. Then $\mu \mapsto \varphi(\mu)$ is a τ_w-continuous mapping from $M^b(E)$ into $M^b(F)$.*

Proof. According to Theorem 1.1.7 we see that $\varphi(\mu) \in M^b(F)$ for all $\mu \in M^b(E)$. Let $(\mu_\iota)_{\iota \in I}$ be a net in $M^b(E)$ with $\tau_w - \lim_\iota \mu_\iota = \mu \in M^b(E)$. For each $f \in C^b(F)$ we have $f \circ \varphi \in C^b(E)$, and this implies that

$$\lim_{\iota \in I} \int f d\varphi(\mu_\iota) = \lim_{\iota \in I} \int f \circ \varphi d\mu_\iota = \int f \circ \varphi d\mu = \int f d\varphi(\mu) . \qquad \blacksquare$$

In the following we will show that we can restrict the study of weak convergence to bounded sequences. For this purpose we consider the metrizability of the space $M^b(E)$ for an arbitrary metric space (E, d).

Lemma 1.2.13 *The mapping $\rho : M^b(E) \times M^b(E) \to \mathbf{R}_+$ given by*

$$\varrho(\mu, \nu) = \inf \{\varepsilon > 0 : \mu(B) \leq \nu(B^\varepsilon) + \varepsilon \text{ and } \nu(B) \leq \mu(B^\varepsilon) + \varepsilon \text{ for all } B \in \mathfrak{B}(E)\}$$

for all $\mu, \nu \in M^b(E)$ is a metric on $M^b(E)$, called the ***Prohorov metric*** *(It should be noted that the existence of numbers $\varepsilon > 0$ satisfying the above is guaranteed by the boundedness of μ, ν).*

Proof. It is clear that $\rho(\mu, \nu) \geq 0, \rho(\mu, \nu) = \rho(\nu, \mu)$ and $\rho(\mu, \mu) = 0$ for all $\mu, \nu \in M^b(E)$. Now suppose $\rho(\mu, \nu) = 0$. For each $A \in \mathcal{A}(E)$ we know that $A^{\frac{1}{n}} \downarrow A$. Then $\mu(A) = \nu(A)$ and hence by Theorem 1.1.2(i), $\mu = \nu$.

Finally, to prove that ρ satisfies the triangle inequality, consider $\lambda, \mu, \nu \in M^b(E)$ and $\alpha, \beta > 0$ with $\rho(\lambda, \mu) < \alpha$ and $\rho(\mu, \nu) < \beta$. By definition of $\rho(\mu, \nu)$ we have

$$\mu(B) \leq \nu(B^\beta) + \beta$$

for all $B \in \mathfrak{B}(E)$. From

$$(B^\alpha)^\beta \cup (B^\beta)^\alpha \subset B^{\alpha+\beta}$$

it follows that

$$\lambda(B) \le \mu(B^\alpha) + \alpha \le \nu((B^\alpha)^\beta) + \beta + \alpha \le \nu(B^{\alpha+\beta}) + (\alpha + \beta)$$

and correspondingly

$$\nu(B) \le \lambda(B^{\alpha+\beta}) + (\alpha + \beta) .$$

Thus $\rho(\lambda, \nu) \le \alpha + \beta$. Now since $\alpha > \rho(\lambda, \mu)$ and $\beta > \rho(\mu, \nu)$ were chosen arbitrarily it follows that

$$\rho(\lambda, \nu) \le \rho(\lambda, \mu) + \rho(\mu, \nu) .$$

■

Theorem 1.2.14 *The Prohorov metric induces the weak topology τ_w on $M^b(E)$.*

Proof. In each τ_w-neighborhood U of $\mu \in M^b(E)$ there is an open ρ-ball centered in μ. Without loss of generality we may assume U to be chosen as in Corollary 1.2.8(a). Now since μ is continuous from above and $A_j^\delta \downarrow A_j$ as $\delta \downarrow 0$ there exists $\delta \in]0, \frac{\varepsilon}{2}[$ such that

$$\mu(A_j^\delta) < \mu(A_j) + \frac{\varepsilon}{2} \text{ for all } j = 1, 2, ..., n .$$

Choose $\nu \in M^b(E)$ satisfying $\rho(\mu, \nu) < \delta$. Since $E^\delta = E$ it follows that

$$|\nu(E) - \mu(E)| \le \delta < \varepsilon .$$

Furthermore

$$\nu(A_j) \le \mu(A_j^\delta) + \delta < \mu(A_j) + \frac{\varepsilon}{2} + \delta < \mu(A_j) + \varepsilon$$

for all $j = 1, 2, ..., n$, and this implies that $\nu \in U$.

For the reverse direction it is to be shown that each open ρ-ball centered on μ with radius $\varepsilon > 0$ contains a τ_w-neighborhood U. Let $\delta \in]0, \frac{\varepsilon}{4}[$. Since μ is tight, by Theorem 1.1.4 there exists a σ-compact set $G \subset E$ with $\mu(G) = \mu(E)$. To each $x \in G$ there corresponds $\delta(x) \in]0, \frac{\delta}{2}[$ with $\mu(\partial B(x, \delta(x))) = 0$. Note that

$$\partial B(x, \eta) \subset \{y \in E : d(x, y) = \eta\}\,.$$

For each $n \in \mathbf{N}$ there exist finitely many $\eta > 0$ such that $\mu(\partial B(x, \eta)) \geq \frac{1}{n}$. Thus there exist only countably many η with $\mu(\partial B(x, \eta)) \neq 0$. Since $G \subset \bigcup_{x \in G} B(x, \delta(x))$, the σ-compactness of G yields the existence of a sequence $(x_n)_{n \geq 1}$ with

$$G \subset \bigcup_{n \geq 1} B(x_n, \delta(x_n))\,.$$

For each $n \in \mathbf{N}$ put $G_n := B(x_n, \delta(x_n))$. Then $(G_n)_{n \geq 1}$ is a sequence of μ-continuity sets with diameter less than δ, and

$$G \subset \bigcup_{n \geq 1} G_n\,.$$

Hence there exists $k \in \mathbf{N}$ such that

$$\mu\left(\bigcup_{n=1}^{k} G_n\right) > \mu(E) - \delta\,.$$

Put

$$\mathfrak{C} := \left\{\bigcup_{j \in J} G_j : J \subset \{1, 2, ..., k\}\right\}.$$

Now $\mathfrak{C}$ is a finite system of μ-continuity sets since

$$\partial\left(\bigcup_{j \in J} G_j\right) \subset \bigcup_{j \in J} G_j^- \setminus \bigcup_{j \in J} G_j \subset \bigcup_{j \in J} \partial G_j\,.$$

It follows from Corollary 1.2.8 that

$$U := \{\nu \in M^b(E) : |\nu(E) - \mu(E)| < \delta, |\nu(C) - \mu(C)| < \delta \quad \text{for all } C \in \mathfrak{C}\}$$

is a τ_w-neighborhood of μ.

We wish to show that for each $\nu \in U$ we have $\rho(\mu, \nu) < \varepsilon$. Put $C_0 := \bigcup_{n=1}^{k} G_n$. Since $C_0 \in \mathfrak{C}$ we have

$$\nu(C_0) > \mu(C_0) - \delta > \mu(E) - 2\delta > \nu(E) - 3\delta\,.$$

Now choose $B \in \mathfrak{B}(E)$. Then

$$C := \bigcup_{\{j \le k: G_j \cap B \neq \emptyset\}} G_j \in \mathfrak{C},$$

$C \subset B^\delta$ (as diam $G_j < \delta$) and $B \subset C \cup (E \setminus C_0)$ (where $C \cap (E \setminus C_0) = \emptyset$). Then

$$\mu(B) \le \mu(C) + \mu(E \setminus C_0) < \nu(C) + \delta + \delta \le \nu(B^\delta) + 2\delta < \nu(B^{4\delta}) + 4\delta$$

and analogously

$$\nu(B) \le \nu(C) + \nu(E \setminus C_0) < \mu(C) + \delta + 3\delta \le \mu(B^{4\delta}) + 4\delta\,.$$

Thus

$$\rho(\mu, \nu) \le 4\delta < \varepsilon\,.$$

■

Application 1.2.15 *Let $(X_n)_{n \ge 0}$ be a sequence of E-valued random variables on a probability space $(\Omega, \mathfrak{A}, \mathbf{P})$ with distributions $\mathbf{P}_{X_n} \in M^1(E)$ for $n \ge 0$. Then*

$X_n \longrightarrow X_0$ $\mathbf{P}$*-stochastically implies*

$\mathbf{P}_{X_n} \longrightarrow \mathbf{P}_{X_0}$ *weakly* *as* $n \to \infty$.

In fact, given $\varepsilon > 0$, there exists $n_\varepsilon \ge 1$ such that

$$\mathbf{P}([d(X_n, X_0) \ge \varepsilon]) < \varepsilon \text{ for all } n \ge n_\varepsilon\,.$$

Let $B \in \mathfrak{B}(E)$. Then

$$\begin{aligned}[X_n \in B] &\subset [d(X_n, X_0) \ge \varepsilon] \cup \Big([d(X_n, X_0) < \varepsilon] \cap [X_0 \in B^\varepsilon]\Big) \\ &\subset [d(X_n, X_0) \ge \varepsilon] \cup [X_0 \in B^\varepsilon]\end{aligned}$$

and analogously,

$$[X_0 \in B] \subset [d(X_n, X_0) \geq \varepsilon] \cup [X_n \in B^\varepsilon].$$

For $n \geq n_\varepsilon$ follows

$$\mathbf{P}_{X_n}(B) \leq \varepsilon + \mathbf{P}_{X_0}(B^\varepsilon)$$

as well as

$$\mathbf{P}_{X_0}(B) \leq \varepsilon + \mathbf{P}_{X_n}(B^\varepsilon)$$

whenever $B \in \mathfrak{B}(E)$. But this implies

$$\rho(\mathbf{P}_{X_n}, \mathbf{P}_{X_0}) < \varepsilon,$$

thus Theorem 1.2.14 yields the assertion.

1.3 The Prohorov theorem

As before (E, d) is a metric space, and the space $M^b(E)$ is given the topology τ_w, or equivalently the Prohorov metric ρ. Hence $(M^b(E), \rho)$ is a metric space.

A finite content μ on E is called ***inner-regular*** if

$$\mu(B) = \sup\{\mu(K) : K \in \mathcal{K}(E), K \subset B\}$$

for all $B \in \mathfrak{B}(E)$.

Theorem 1.3.1 *Every inner-regular content μ on E is σ-additive, so that $\mu \in M^b(E)$.*

Proof. All that needs to be shown is that μ is $\emptyset$-continuous. Let $(B_n)_{n\geq 1}$ be a sequence in $\mathfrak{B}(E)$ with $B_n \downarrow \emptyset$, and ε be given. For each $n \in \mathbf{N}$ there exists a compact set $K_n \subset B_n$ such that

$$\mu(B_n) - \mu(K_n) < \frac{1}{2^n}\varepsilon\,.$$

Put $L_n := \bigcap_{i=1}^n K_i$. Then

$$\mu(B_n) - \mu(L_n) \leq \sum_{i=1}^{n}(\mu(B_i) - \mu(K_i)) < \varepsilon$$

for all $n \in \mathbf{N}$, and $L_n \downarrow \emptyset$. From the finite intersection property there exists $n_0 \in \mathbf{N}$ such that $L_n = \emptyset$ for all $n \geq n_0$, and hence $\mu(L_n) = 0$ for all $n \geq n_0$. Then $\mu(B_n) < \varepsilon$ for all $n \geq n_0$, and we have shown that $\lim_{n\to\infty} \mu(B_n) = 0$. ■

Corollary 1.3.2 *Let $(\mu_n)_{n\geq 1}$ be an increasing sequence in $M^b(E)$ satisfying*

$$\sup_{n\geq 1} \mu_n(E) < \infty\,.$$

Then $\sup_{n\geq 1} \mu_n \in M^b(E)$.

Proof. Write $\mu(B) := \sup_{n \geq 1} \mu_n(B)$ for all $B \in \mathfrak{B}(E)$. Clearly μ is a finite content on $\mathfrak{B}(E)$. From Theorem 1.1.2 we have that μ_n is inner-regular, and so μ is an inner-regular content, and the result follows from Theorem 1.3.1. ■

Theorem 1.3.3 *Let (E, d) be a compact metric space. Then for each $a > 0$*

$$M^{(a)}(E) := \{\mu \in M^b(E) : \mu(E) \leq a\}$$

is τ_w-compact.

Proof. According to Appendix B 14.1 (Alaoglu, Bourbaki), the set

$$V^{(a)} := \{\varphi \in C^b(E)' : \|\varphi\| \leq a\}$$

is weakly-compact, i.e compact with respect to the topology $\sigma(C^b(E)', C^b(E))$. Hence

$$V_+^{(a)} := \{\varphi \in V^{(a)} : \varphi \geq 0\} = \bigcap_{f \in C_+^b(E)} \{\varphi \in V^{(a)} : \varphi(f) \geq 0\}$$

is weakly compact. Furthermore by Corollary 1.2.3 the mapping

$$\mu \longmapsto \left(f \longmapsto \int f d\mu\right)$$

is a bijection from $M^{(a)}(E)$ onto $V_+^{(a)}$. From the definition of τ_w it follows that it is a homeomorphism, and hence that $M^{(a)}(E)$ is τ_w-compact. ■

Definition 1.3.4 *A set $H \subset M^b(E)$ is called* ***uniformly tight*** *if*

(a) $\sup\{\mu(E) : \mu \in H\} < \infty$*;*

(b) to each $\varepsilon > 0$ there exists a compact set $K \subset E$ such that

$$\mu(E \setminus K) < \varepsilon \ \textit{for all} \ \mu \in H.$$

Theorem 1.3.5 *Suppose (E, d) is separable complete. For each $H \subset M^b(E)$ the following statements are equivalent:*

(i) H is uniformly tight.

(ii) (a) $\sup\{\mu(E) : \mu \in H\} < \infty$;

(b) For all $\varepsilon > 0$ and $n \in \mathbf{N}$ there exist $x_1, x_2, ..., x_k \in E$ such that

$$\mu(E \setminus B_n) < \varepsilon \ \textit{for all} \ \mu \in H$$

where $B_n := \bigcup_{j=1}^k B(x_j, \frac{1}{n})$.

Proof. *(i)⇒(ii).* To each $\varepsilon > 0$ there exists a compact set $K \subset E$ such that $\mu(E \setminus K) < \varepsilon$ for all $\mu \in H$. Also to each $n \in \mathbf{N}$ there exist $x_1, x_2, ..., x_k \in E$ such that $K \subset B_n$, so that

$$\mu(E \setminus B_n) < \varepsilon$$

for all $\mu \in H$.

(ii)⇒(i). Let $\varepsilon > 0$. From the assumption for each $n \in \mathbf{N}$ there exists $B_n \subset E$ such that B_n is the finite union of $\frac{1}{n}$-balls and

$$\mu(E \setminus B_n) < \frac{1}{2^n}\varepsilon$$

for all $\mu \in H$. Put $L := \bigcap_{n \geq 1} B_n$. Then $\mu(E \setminus L) < \varepsilon$ for all $\mu \in H$. Now L is totally bounded, and hence so is L^-. As E is complete it follows that L^- is compact. The result now follows from the fact that

$$\mu(E \setminus L^-) \leq \mu(E \setminus L) < \varepsilon \ \text{for all} \ \mu \in H . \qquad \blacksquare$$

Lemma 1.3.6 *Let $A \in \mathcal{A}(E)$, $\mu_\iota \in M^b(E)$ for all $\iota \in I$, $\mu \in M^b(E)$, $\mu_A \in M^b(A)$. If $\tau_w - \lim_\iota \mu_\iota = \mu$ and $\tau_w - \lim_\iota \mathrm{Res}_A \mu_\iota = \mu_A$* then *$\mu_A \leq \mathrm{Res}_A \mu$.*

Proof. We can use Theorem 1.1.2(*ii*) to deduce that $\mathrm{Res}_A\mu \in M^b(A)$. Consider a continuous function $g : A \to [0,1]$. By Tietze's extension theorem there exists a continuous function $f : E \to [0,1]$ with $\mathrm{Res}_A f = g$. Then

$$\int g\, d\mu_A = \lim_{\iota \in I} \int g\, d(\mathrm{Res}_A \mu_\iota) = \lim_{\iota \in I} \int_A f d\mu_\iota$$

and

$$\int g\, d(\mathrm{Res}_A \mu) = \int_A f d\mu \,.$$

But in slight modification of Corollary 1.2.2 one obtains that for measures $\mu, \nu \in M^b(E)$ satisfying

$$\int f d\mu \le \int f d\nu \text{ for all } f \in C^b(E) \text{ with } 0 \le f \le 1$$

$\mu \le \nu$ holds. Therefore we need only prove that

$$\lim_{\iota \in I} \int_A f d\mu_\iota \le \int_A f d\mu \,.$$

However this inequality follows immediately from the Portemanteau theorem 1.2.7, as clearly $\tau_w - \lim_\iota f(\mu_\iota) = f\mu$. ∎

Theorem 1.3.7 *(Prohorov) Consider $H \subset M^b(E)$.*

(i) If H is uniformly tight then H is τ_w-relatively compact.

(ii) If E is separable complete then any τ_w-relatively compact set H is uniformly tight.

Proof. *(i)*. Let $(K_n)_{n\ge1}$ be an increasing sequence of compact subsets of E satisfying $\mu(E \setminus K_n) < \frac{1}{n}$ for all $\mu \in H$. Let $(\mu_k)_{k\ge1}$ be a sequence in H. We have to show that $(\mu_k)_{k\ge1}$ possesses a τ_w-convergent subsequence. From

$$a := \sup\{\mu(K_n) : \mu \in H, n \in \mathbf{N}\} \le \sup\{\mu(E) : \mu \in H\} < \infty$$

appealing to Theorem 1.3.3 we have that $\{\text{Res}_{K_n}\mu : \mu \in H\}(\subset M^{(a)}(K_n))$ is τ_w-relatively compact in $M^b(K_n)$. By the metrizability of $M^b(E)$, which is the content of Theorem 1.2.14, a diagonal argument provides for each $n \in \mathbf{N}$ a measure $\nu'_n \in M^b(K_n)$ with

$$\tau_w - \lim_k \text{Res}_{K_n}\mu_k = \nu'_n .$$

Put

$$\nu_n(B) := \nu'_n(B \cap K_n) \ \text{ for all } \ B \in \mathfrak{B}(E) .$$

Then $\nu_n \in M^b(E)$ for all $n \in \mathbf{N}$. Also from Lemma 1.3.6, $\nu'_n \leq \text{Res}_{K_n}\nu'_{n+1}$ and (recall that $K_n \subset K_{n+1}$)

$$\nu_n(B) \leq \nu'_{n+1}(B \cap K_n) \leq \nu_{n+1}(B)$$

for all $B \in \mathfrak{B}(E)$ which says that $(\nu_n)_{n\geq 1}$ is an increasing sequence. Moreover,

$$\nu_n(E) = \nu'_n(K_n) = \lim_{k\to\infty} \mu_k(K_n) \leq \liminf_{k\to\infty} \mu_k(E) < \infty$$

for all $n \in \mathbf{N}$. Now appeal to Corollary 1.3.2 to obtain

$$\nu := \sup_{n\geq 1} \nu_n \in M^b(E)$$

and moreover

$$\nu(E) = \sup_{n\geq 1} \nu_n(E) \leq \liminf_{k\to\infty} \mu_k(E) .$$

Let $A \in \mathcal{A}(E)$. Applying the Portemanteau theorem 1.2.7 we get

$$\begin{aligned}\nu(A) \geq \nu_n(A) = \nu'_n(A \cap K_n) &\geq \limsup_{k\geq 1} \mu_k(A \cap K_n) \\ &\geq \limsup_{k>1} \mu_k(A) - \frac{1}{n}\end{aligned}$$

for all $n \in \mathbf{N}$, where for the second inequality we have used the fact that $A \cap K_n$ is closed, and for the third that

$$\mu(A \cap K_n) \geq \mu(A) - \frac{1}{n}$$

for all $\mu \in H$. Hence

$$\nu(A) \geq \limsup_{k \geq 1} \mu_k(A)$$

and

$$\nu(E) = \lim_{k \to \infty} \mu_k(E).$$

A further application of the Portemanteau theorem 1.2.7 gives $\tau_w - \lim_k \mu_k = \nu$. Thus every sequence in H has a τ_w-convergent subsequence, and this just says that H is τ_w-relatively compact.

(ii). Assume that H is τ_w-relatively compact and at the same time fails to be uniformly tight. The mapping

$$\mu \longmapsto \mu(E) = \int \mathbf{1}_E d\mu$$

is continuous, and this implies that

$$\sup\{\mu(E) : \mu \in H\} < \infty.$$

Let $\mathcal{F} = \{F \subset E : |F| < \infty\}$. Theorem 1.3.5 implies that there exist $\varepsilon > 0$ and $n \in \mathbf{N}$ such that to each $F \in \mathcal{F}$ there is $\mu_F \in H$ with

$$\mu_F\left(E \setminus \left(\bigcup_{x \in F} B(x, \frac{1}{n})\right)\right) \geq \varepsilon.$$

The net $(\mu_F)_{F \in \mathcal{F}}$ contains a τ_w-convergent subnet $(\mu_{F_\iota})_{\iota \in I}$ with limit μ say. For each $\iota \in I$ define

$$B_\iota = \bigcup_{x \in F_\iota} B(x, \frac{1}{n}).$$

Now $(E \setminus B_\iota)_{\iota \in I}$ is a downward filtered family in $\mathcal{A}(E)$ with $\bigcap_{\iota \in I} E \setminus B_\iota = \emptyset$. It follows from Corollary 1.1.3 that

$$\inf_{\iota \in I} \mu(E \setminus B_\iota) = 0.$$

On the other hand applying the Portemanteau theorem we have

$$\mu(E \setminus B_\kappa) \geq \limsup_{\iota \in I} \mu_{F_\iota}(E \setminus B_\kappa) \geq \limsup_{\iota \in I} \mu_{F_\iota}(E \setminus B_\iota) \geq \varepsilon$$

for all $\kappa \in I$, and this is a contradiction. ■

1.4 Convolution of measures

In this section (E, d) will denote a separable complete metric Abelian group which means that E is an Abelian group (with binary operation denoted by addition and 0 as neutral element), (E, d) is a separable and complete metric space with distance function d, and the mapping $(x, y) \mapsto x - y$ from $E \times E$ into E is continuous. Along with $(E, d), (E \times E, d \times d)$ is also a separable complete metric Abelian group. Here the metric $d \times d$ is defined by $d \times d((x, y), (u, v)) := \max\{d(x, u), d(y, v)\}$ for all $(x, y), (u, v) \in E$. A prominent example of a separable complete metric Abelian group is a separable Banach space $(E, \|\cdot\|)$ over $\mathbf{R}$, where the distance function is given by $d(x, y) := \|x - y\|$ for all $x, y \in E$.

For separable complete metric groups E we have that each finite measure on E is tight (Theorem 1.1.6) and that the notions "uniform tightness" and "τ_w-relative compactness" are equivalent (Prohorov's theorem 1.3.7).

Theorem 1.4.1 *Let (E, d) be a separable complete metric Abelian group.*

(i) $\mathfrak{B}(E \times E) = \mathfrak{B}(E) \otimes \mathfrak{B}(E)$.

(ii) The mapping $(x, y) \mapsto m(x, y) := x + y$ from $E \times E$ into E is $(\mathfrak{B}(E) \otimes \mathfrak{B}(E), \mathfrak{B}(E))$-measurable.

Proof. *(i).* $\mathcal{O}(E)$ is a generator of $\mathfrak{B}(E)$ with the exhaustion property which says that there exists a sequence $(O_n)_{n \geq 1}$ in $\mathcal{O}(E)$ such that $O_n \uparrow E$. Thus $\mathcal{O}(E) \times \mathcal{O}(E)$ is a generator of $\mathfrak{B}(E) \otimes \mathfrak{B}(E)$. Furthermore $\mathcal{O}(E) \times \mathcal{O}(E) \subset \mathcal{O}(E \times E)$ from which it follows that $\mathfrak{B}(E) \otimes \mathfrak{B}(E) \subset \mathfrak{B}(E \times E)$. Now choose a countable dense subset D of E, so that $D \times D$ is a countable dense subset of $E \times E$. The open balls in E and $E \times E$ centered at points in D and $D \times D$ respectively and with rational radii make up countable bases $\mathcal{B}$ and $\mathcal{C}$ in $\mathcal{O}(E)$ and $\mathcal{O}(E \times E)$ respectively. Now

$$B((x, y), r) = B(x, r) \times B(y, r)$$

for the corresponding balls, and hence $\mathcal{B} \times \mathcal{B} \supset \mathcal{C}$. Now $\mathcal{B}$ and $\mathcal{C}$ are generators of $\mathfrak{B}(E)$ and $\mathfrak{B}(E \times E)$ respectively with the exhaustion property, which implies that $\mathfrak{B}(E) \otimes \mathfrak{B}(E) \supset \mathfrak{B}(E \times E)$.

(ii). The mapping m from $E \times E$ into E is continuous, and hence $(\mathfrak{B}(E \times E), \mathfrak{B}(E))$-measurable. Now apply (i). ■

Application 1.4.2 *Let X, Y be E-valued random variables on a probability space $(\Omega, \mathfrak{A}, \mathbf{P})$. Then by Theorem 1.4.1(ii) the mapping $\omega \mapsto (X+Y)(\omega) = X(\omega) + Y(\omega)$ from Ω into E is also an E-valued random variable on $(\Omega, \mathfrak{A}, \mathbf{P})$, since $X + Y = m \circ (X, Y)$.*

Definition 1.4.3 *For $\mu, \nu \in M^b(E)$ we refer to the measure $\mu * \nu := m(\mu \otimes \nu)$ on E as the* ***convolution*** *of μ and ν.*

We have the following properties of the convolution mapping.

Properties 1.4.4

1.4.4.1 *For all $f \in C^b(E)$*

$$\int f d(\mu * \nu) = \int \left(\int f(x+y) \mu(dx) \right) \nu(dy)$$
$$= \int \left(\int f(x+y) \nu(dy) \right) \mu(dx)$$

which follows from Fubini's theorem.

1.4.4.2 *In particular, for all $B \in \mathfrak{B}(E)$*

$$(\mu * \nu)(B) = \int \mu(B - y) \nu(dy) = \int \nu(B - x) \mu(dx) .$$

1.4.4.3 *For all $B, C \in \mathfrak{B}(E)$ we have $B + C \in \mathfrak{B}(E)$ and*

$$(\mu * \nu)(B + C) \geq \mu(B) \nu(C) .$$

The latter can easily be seen as follows:

$$(\mu*\nu)(B+C) = (\mu\otimes\nu)(m^{-1}(B+C)) \geq (\mu\otimes\nu)(B\times C) = \mu(B)\nu(C).$$

1.4.4.4 *The convolution is commutative and associative, that is*

$$\mu * \nu = \nu * \mu$$

and

$$(\lambda * \mu) * \nu = \lambda * (\mu * \nu)$$

for all $\lambda, \mu, \nu \in M^b(E)$.

The proof uses 1.4.4.1 in conjunction with the injectivity of the mapping

$$\mu \mapsto \left(f \mapsto \int f d\mu\right)$$

from $M^b(E)$ into $C^b(E)'_+$.

1.4.4.5 *For each* $x \in E$ and $B \in \mathfrak{B}(E)$

$$(\mu * \varepsilon_x)(B) = \mu(B - x).$$

1.4.4.6 *From* 1.4.4.5 *and* 1.4.4.2 *we have* $\mu*\varepsilon_0 = \mu$ *and* $\varepsilon_x*\varepsilon_y = \varepsilon_{x+y}$ *for all* $x, y \in E$.

Application 1.4.5 *Let* X, Y *be independent* E*-valued random variables on a probability space* $(\Omega, \mathfrak{A}, \mathbf{P})$. *Then* $\mathbf{P}_{X+Y} = \mathbf{P}_X * \mathbf{P}_Y$.

Proof. To show this we note that independence of X, Y gives $\mathbf{P}_{(X,Y)} = \mathbf{P}_X \otimes \mathbf{P}_Y$ which implies that

$$\mathbf{P}_{X+Y} = \mathbf{P}_{m\circ(X,Y)} = m(\mathbf{P}_{(X,Y)}) = m(\mathbf{P}_X \otimes \mathbf{P}_Y) = \mathbf{P}_X * \mathbf{P}_Y. \quad \blacksquare$$

Theorem 1.4.6 *(Support formula) For* $\mu, \nu \in M^b(E)$

$$\operatorname{supp}(\mu * \nu) = (\operatorname{supp}(\mu) + \operatorname{supp}(\nu))^-.$$

Proof. First we note that

$$\begin{aligned}(\mu * \nu)((\mathrm{supp}\,(\mu)&+\mathrm{supp}\,(\nu))^-)\\ &= (\mu \otimes \nu)(m^{-1}((\mathrm{supp}\,(\mu) + \mathrm{supp}\,(\nu))^-))\\ &\geq (\mu \otimes \nu)(\mathrm{supp}\,(\mu) \times \mathrm{supp}\,(\nu)) = \mu(E)\nu(E)\\ &= (\mu \otimes \nu)(E \times E) = (\mu * \nu)(E).\end{aligned}$$

Thus by Theorem 1.1.4(i)

$$\mathrm{supp}\,(\mu * \nu) \subset (\mathrm{supp}\,(\mu) + \mathrm{supp}\,(\nu))^- .$$

To prove the reverse inclusion, consider $x \in \mathrm{supp}\,(\mu), U \in \mathfrak{V}(x), y \in \mathrm{supp}\,(\nu)$ and $V \in \mathfrak{V}(y)$. Then $U + V \in \mathfrak{V}(x + y)$. Now Property 1.4.4.3 together with Theorem 1.1.4(iii) gives

$$(\mu * \nu)(U + V) \geq \mu(U)\mu(V) > 0 .$$

To each $W \in \mathfrak{V}(x + y)$ there exist $U \in \mathfrak{V}(x), V \in \mathfrak{V}(y)$ with $U + V \subset W$. A further application of Theorem 1.1.4(iii) gives $x + y \in \mathrm{supp}\,(\mu + \nu)$. Thus

$$\mathrm{supp}\,(\mu) + \mathrm{supp}\,(\nu) \subset \mathrm{supp}\,(\mu * \nu)$$

and this completes the proof. ■

Corollary 1.4.7 *If $\mu * \nu$ is a Dirac measure then so are each of μ, ν.*

Proof. We just observe that a measure is Dirac precisely when it has a single element support. ■

Theorem 1.4.8 *Let $(\mu_n)_{n\geq 1}, (\nu_n)_{n\geq 1}$ be sequences with $\tau_w - \lim_n \mu_n = \mu$, $\tau_w - \lim_n \nu_n = \nu$ in $M^b(E)$. Then $\tau_w - \lim_n \mu_n \otimes \nu_n = \mu \otimes \nu$.*

Proof. Appealing to Prohorov's theorem 1.3.7 to each $\varepsilon > 0$ there exist $K \in \mathcal{K}(E)$ with

$$\mu_n(E \setminus K) < \varepsilon \text{ and } \nu_n(E \setminus K) < \varepsilon$$

and $\alpha > 0$ such that

$$\mu_n(E) \leq \alpha \text{ and } \nu_n(E) \leq \alpha$$

for all $n \in \mathbf{N}$. Clearly $K \times K \in \mathcal{K}(E \times E)$, and

$$\begin{aligned}(\mu_n \otimes \nu_n)(K \times K) = \mu_n(K)\nu_n(K) &\geq (\mu_n(E) - \varepsilon)(\nu_n(E) - \varepsilon) \\ &> (\mu_n \otimes \nu_n)(E \times E) - \varepsilon(\mu_n(E) + \nu_n(E)).\end{aligned}$$

Therefore

$$(\mu_n \otimes \nu_n)((E \times E) \setminus (K \times K)) < 2\alpha\varepsilon$$

and

$$(\mu_n \otimes \nu_n)(E \times E) = \mu_n(E)\nu_n(E) \leq \alpha^2$$

for all $n \in \mathbf{N}$. Furthermore from Prohorov's theorem 1.3.7 we see that $\{\mu_n \otimes \nu_n : n \in \mathbf{N}\}$ is τ_w-relatively compact.

Let $\lambda \in M^b(E \times E)$ be a cluster point of the sequence $(\mu_n \otimes \nu_n)_{n\geq 1}$, that is, there exists a subsequence $(\mu_{n_k} \otimes \nu_{n_k})_{k\geq 1}$ with limit λ. Now

$$\begin{aligned}\mathcal{E} := \{&B_1 \times B_2 \in \mathfrak{B}(E) \times \mathfrak{B}(E) : \\ &B_1 \times B_2 \text{ is both a } \mu \otimes \nu\text{- and a } \lambda\text{-continuity set}\}\end{aligned}$$

is an $\cap$-stable generator of $\mathfrak{B}(E) \otimes \mathfrak{B}(E) = \mathfrak{B}(E \times E)$. Let $B_1 \times B_2 \in \mathcal{E}$. Then

$$\partial(B_1 \times B_2) = (B_1^- \times \partial B_2) \cup (\partial B_1 \times B_2^-)\,.$$

Therefore we have the following three cases to consider.

(1) B_1 is a μ-continuity set and B_2 is a ν-continuity set.

(2) $\mu(B_1^-) = 0$.

(3) $\nu(B_2^-) = 0$.

With the help of the Portemanteau theorem 1.2.7 we have

$$\begin{aligned}\lambda(B_1 \times B_2) &= \lim_{k\to\infty} (\mu_{n_k} \otimes \nu_{n_k})(B_1 \times B_2) = \lim_{k\to\infty} \mu_{n_k}(B_1)\nu_{n_k}(B_2) \\ &= \mu(B_1)\nu(B_2) = (\mu \otimes \nu)(B_1 \times B_2),\end{aligned}$$

$$\lambda(B_1 \times B_2) \leq \lim_{k\to\infty} \mu_{n_k}(B_1)\alpha = \mu(B_1)\alpha = 0 = (\mu \otimes \nu)(B_1 \times B_2).$$

and similarly in the remaining case.

Thus

$$\lambda(B_1 \times B_2) = (\mu \otimes \nu)(B_1 \times B_2) \text{ for all } B_1 \times B_2 \in \mathcal{E}$$

and hence $\lambda = \mu \otimes \nu$. Hence the τ_w-relatively compact sequence $(\mu_n \otimes \nu_n)_{n\geq 1}$ has a unique cluster point $\mu \otimes \nu$, and it follows that $\tau_w - \lim_n \mu_n \otimes \nu_n = \mu \otimes \nu$. ∎

Theorem 1.4.9 *Let (E, d) be a separable complete metric Abelian group. Then the space $(M^b(E), \tau_w, *)$ with the convolution $*$ defined above is a commutative metric semigroup, in the sense that*

(i) $(M^b(E), \tau_w)$ is a metric space (with the Prohorov metric inducing the topology τ_w),

(ii) $(M^b(E), *)$ is a commutative semigroup with neutral element ε_0, that is, $\varepsilon_0 * \mu = \mu$ for all $\mu \in M^b(E)$, and

(iii) the mapping $(\mu, \nu) \mapsto \mu * \nu$ from $M^b(E) \times M^b(E)$ into $M^b(E)$ is τ_w-continuous.

Moreover, $(M^1(E), \tau_w, *)$ is a sub-semigroup of $(M^b(E), \tau_w, *)$.

Proof. In view of the assumed metrizability it suffices to consider sequences. So given sequences $(\mu_n)_{n\geq 1}, (\nu_n)_{n\geq 1}$ with $\tau_w - \lim_n \mu_n = \mu, \tau_w - \lim_n \nu_n = \nu$ in $M^b(E)$ we have by Theorem 1.4.8 that $\tau_w - \lim_n \mu_n \otimes \nu_n = \mu \otimes \nu$. Then Theorem 1.2.12 gives

$$\tau_w - \lim_n \mu_n * \nu_n = \tau_w - \lim_n m(\mu_n \otimes \nu_n) = m(\mu \otimes \nu) = \mu * \nu.$$

This proves the continuity of $(\mu, \nu) \mapsto \mu * \nu$. The remaining assertions follow immediately from Properties 1.4.4.2 and 1.4.4.4 of the convolution. ∎

2

The Fourier Transform in a Banach Space

2.1 Fourier transforms of probability measures

Let $(E, \|\cdot\|)$ denote a separable Banach space over $\mathbf{R}$. With the metric

$$(x, y) \mapsto \|x - y\|$$

$(E, \|\cdot\|)$ is a complete metric space. We use E' to designate the topological dual of E. For $x \in E$ and any (continuous real-valued) linear functional $a \in E'$ we put $\langle x, a\rangle := a(x)$. It is well-known that E' separates the points of E, in the sense that for every $x \in E \setminus \{0\}$ there exists $a \in E'$ satisfying $\langle x, a\rangle \neq 0$ (See Appendix B).

Now let $\mathcal{N}(E)$ denote the family of all closed sub (vector) spaces E of finite codimension. For every $a \in E'$ we have $\ker a \in \mathcal{N}(E)$, and therefore $\bigcap\{N : N \in \mathcal{N}(E)\} = \{0\}$. Finally, for every $N \in \mathcal{N}(E)$ let p_N denote the canonical mapping from E onto E/N.

Theorem 2.1.1 *For every $K \in \mathcal{K}(E)$ we have*

$$K = \bigcap\{p_N^{-1}(p_N(K)) : N \in \mathcal{N}(E)\}\,.$$

Proof. Clearly

$$K \subset K_1 := \bigcap\{p_N^{-1}(p_N(K)) : N \in \mathcal{N}(E)\}\,.$$

For the reverse inclusion, let $x \in K_1$. Then for every $N \in \mathcal{N}(E)$ the set

$$K_N := p_N^{-1}(p_N(x)) \cap K$$

is compact and non-empty. Moreover, given $M, N \in \mathcal{N}(E)$ with $M \subset N$ we have $K_M \subset K_N$. Indeed, let p_{NM} denote the canonical mapping from E/M onto E/N satisfying $p_{NM} \circ p_M = p_N$. Then for $y \in K_M$,

$$p_N(y) = p_{NM}(p_M(y)) = p_{NM}(p_M(x)) = p_N(x),$$

and thus $y \in K_N$.

Next we observe that $\mathcal{N}(E)$ is $\cap$-stable, and consequently that $(K_N)_{N \in \mathcal{N}(E)}$ is a downward filtered family. Since each K_N is compact, $\bigcap\{K_N : N \in \mathcal{N}(E)\} \neq \emptyset$. But for $y \in \bigcap\{K_N : N \in \mathcal{N}(E)\}$ we have $p_N(y) = p_N(x)$ whenever $N \in \mathcal{N}(E)$. From $\bigcap\{N : N \in \mathcal{N}(E)\} = \{0\}$ we have $x = y \in K$ and consequently $K_1 \subset K$. ∎

Definition 2.1.2 *Let $\mu \in M^b(E)$. The mapping $\hat{\mu} : E' \to \mathbf{C}$ given by*

$$\hat{\mu}(a) := \int e^{i\langle x,a\rangle} \mu(dx)$$

for all $a \in E'$ is called the ***Fourier transform*** *of μ.*

For Banach spaces E, F denote by $L(E, F)$ the set of all continuous linear mappings from E into F, and consider $T \in L(E, F)$. The adjoint T^t of T is continuous linear mapping from F' into E' given by $\langle x, T^t b\rangle = \langle Tx, b\rangle$ whenever $x \in E$, $b \in F'$.

Lemma 2.1.3 *Let $\mu \in M^b(E)$. Then for any $T \in L(E, F)$*

$$T(\mu)^\wedge = \hat{\mu} \circ T^t .$$

Proof. We know already that $T(\mu) \in M^b(F)$ whenever $\mu \in M^b(E)$. Then for every $b \in F'$,

$$\begin{aligned} T(\mu)^\wedge(b) &= \int e^{i\langle x,b\rangle} T(\mu)(dy) = \int e^{i\langle Tx,b\rangle} \mu(dx) \\ &= \int e^{i\langle x,T^t b\rangle} \mu(dx) = \hat{\mu}(T^t b) \\ &= (\hat{\mu} \circ T^t)(b). \end{aligned}$$

∎

Theorem 2.1.4 *(Uniqueness of the Fourier transform). Let $\mu, \nu \in M^b(E)$ with $\hat{\mu} = \hat{\nu}$. Then $\mu = \nu$.*

Proof. From Lemma 2.1.3 we deduce that

$$p_N(\mu)^\wedge = p_N(\nu)^\wedge$$

for every $N \in \mathcal{N}(E)$. But from Appendix B 3 we infer that $E/N \cong \mathbf{R}^p$ with $p := \dim(E/N)$ (in the sense of a topological isomorphism). Applying Property 4.2.16.3 on the uniqueness of the classical Fourier transform of bounded measures on $\mathbf{R}^p$ we obtain that $p_N(\mu) = p_N(\nu)$ for every $N \in \mathcal{N}(E)$. Then

$$(p_N^{-1}(p_N(K)))_{N \in \mathcal{N}(E)}$$

is a downward filtered family in $\mathcal{A}(E)$ satisfying (by Theorem 2.1.1) the equality

$$K = \bigcap \{p_N^{-1}(p_N(K)) : N \in \mathcal{N}(E)\}\,.$$

It follows that

$$\begin{aligned}
\mu(K) &= \inf\{\mu(p_N^{-1}(p_N(K))) : N \in \mathcal{N}(E)\} \\
&= \inf\{p_N(\mu)(p_N(K)) : N \in \mathcal{N}(E)\} \\
&= \inf\{p_N(\nu)(p_N(K)) : N \in \mathcal{N}(E)\} \\
&= \inf\{\nu(p_N^{-1}(p_N(K))) : N \in \mathcal{N}(E)\} \\
&= \nu(K).
\end{aligned}$$

From Theorem 1.1.6 and 1.1.2(*ii*) we now infer that $\mu = \nu$. ■

Corollary 2.1.5 *Every $\mu \in M^b(E)$ is uniquely determined by the family $\{a(\mu) : a \in E'\}$ of its one-dimensional marginal distributions.*

Proof. For $x \in E$, $a \in E'$ and $t \in \mathbf{R}$ we have

$$\langle x, a^t(t)\rangle = \langle a(x), t\rangle = ta(x) = a(tx) = \langle tx, a\rangle = \langle x, ta\rangle$$

and hence $a^t(1) = a$. Then Lemma 2.1.3 yields $a(\mu)^\wedge(1) = \hat{\mu}(a)$, and Theorem 2.1.4 yields the assertion. ■

From now on we shall employ $\mathfrak{S}$-topologies on E'. Prominent choices for $\mathfrak{S}$ are the families $\mathcal{F}(E)$ and $\mathcal{K}(E)$ of finite and compact subsets E respectively. For every $S \in \mathfrak{S}$ let

$$p_S(a) := \sup\{|\langle x, a\rangle| : x \in S\}$$

whenever $a \in E'$. We know from Appendix B 7, B 8 that p_S is a seminorm on E'. The topologies generated in E' by the sets $\{p_S : S \in \mathfrak{S}\}$ for $\mathfrak{S}$ equal to $\mathcal{F}(E)$ and $\mathcal{K}(E)$ of simple and compact convergence will be denoted by $\sigma(E', E)$ and $\tau(E', E)$ respectively. Clearly

$$\sigma(E', E) \succ \tau(E', E) .$$

Properties 2.1.6 *Let $\mu, \nu \in M^b(E)$ and $a, b \in E'$. Then*

2.1.6.1 $|\hat{\mu}(a)| \leq \hat{\mu}(0) = \mu(E)$.

2.1.6.2 $\hat{\mu}(-a) = \overline{\hat{\mu}(a)}$.

2.1.6.3 $|\hat{\mu}(a) - \hat{\mu}(b)|^2 \leq 2\hat{\mu}(0)\{\hat{\mu}(0) - \mathrm{Re}\hat{\mu}(a-b)\}$.

2.1.6.4 *$\hat{\mu}$ is $\tau(E', E)$-continuous.*

2.1.6.5 *If H is a uniformly tight subset of $M^b(E)$ then $\{\hat{\mu} : \mu \in H\}$ is $\tau(E', E)$-equicontinuous.*

2.1.6.6 *Suppose there exists $\delta > 0$ such that $\hat{\mu}(a) = 1$ for all $a \in E'$ with $\|a\| < \delta$. Then $\mu = \varepsilon_0$.*

2.1.6.7 $(\mu * \nu)^\wedge = \hat{\mu}\hat{\nu}$.

To show 2.1.6.1 we just consider the equalities

$$|\hat{\mu}(a)| = \left| \int e^{i\langle x, a\rangle} \mu(dx) \right| \leq \int |e^{i\langle x, a\rangle}| \mu(dx) = \int \mathbf{1}_E d\mu = \mu(E) .$$

Property 2.1.6.2 follows from

$$\hat{\mu}(-a) = \int e^{-i\langle x,a\rangle}\mu(dx) = \int \overline{e^{i\langle x,a\rangle}}\mu(dx) = \overline{\int e^{i\langle x,a\rangle}\mu(dx)} = \overline{\hat{\mu}(a).}$$

As for the proof of 2.1.6.3 we first note that for $\alpha, \beta \in \mathbf{R}$ we have

$$\begin{aligned} |e^{i\alpha} - e^{i\beta}|^2 &= |e^{i\beta}|^2|e^{i(\alpha-\beta)} - 1|^2 \\ &= (e^{i(\alpha-\beta)} - 1)(e^{-i(\alpha-\beta)} - 1) = 2 - 2\cos(\alpha - \beta). \end{aligned}$$

Applying the Cauchy-Schwarz inequality we then obtain

$$\begin{aligned} |\hat{\mu}(a) - \hat{\mu}(b)|^2 &= \left| \int (e^{i\langle x,a\rangle} - e^{i\langle x,b\rangle})\mu(dx) \right|^2 \\ &\leq \int |e^{i\langle x,a\rangle} - e^{i\langle x,b\rangle}|^2 \mu(dx) \int \mathbf{1}_E^2 d\mu \\ &= 2\int (1 - \cos\langle x, a - b\rangle)\mu(dx)\hat{\mu}(0) \\ &= 2\hat{\mu}(0)\Big(\hat{\mu}(0) - \int \text{Re }\, e^{i\langle x,a-b\rangle}\mu(dx)\Big) \\ &= 2\hat{\mu}(0)(\hat{\mu}(0) - \text{Re }\,\hat{\mu}(a - b)). \end{aligned}$$

In order to show 2.1.6.4 given $\varepsilon > 0$ we choose $K \in \mathcal{K}(E)$ such that $\mu(E \setminus K) < \varepsilon/4$. Moreover, there exists $\delta > 0$ such that

$$|e^{is} - e^{it}| \leq \frac{\varepsilon}{2\mu(E)}$$

for all $s, t \in \mathbf{R}$ with $|s - t| < \delta$. Now suppose $a, b \in E'$ satisfy $p^K(a - b) < \delta$. Then

$$\begin{aligned} |\hat{\mu}(a) - \hat{\mu}(b)| &= \left| \int (e^{i\langle x,a\rangle} - e^{i\langle x,b\rangle})\mu(dx) \right| \\ &\leq 2\mu(E \setminus K) + \int_K |e^{i\langle x,a\rangle} - e^{i\langle x,b\rangle}|\mu(dx) \ < \ \varepsilon. \end{aligned}$$

The proof of 2.1.6.5 is analogous to that of 2.1.6.4. As for 2.1.6.6 we apply 2.1.6.3 in order to deduce that

$$|\hat{\mu}(2a) - \hat{\mu}(a)|^2 \leq 2(1 - \text{Re }\,\hat{\mu}(a))$$

for all $a \in E'$. Now for $a \in E'$ and $\delta > 0$ there exists $k \geq 1$ satisfying $\|a/2^k\| \leq \delta$. By assumption $\hat{\mu}(\frac{a}{2^k}) = 1$ and consequently

$$1 = \hat{\mu}(2\frac{a}{2^k}) = ... = \hat{\mu}(2^k\frac{a}{2^k}) = \hat{\mu}(a)$$

and hence $\mu = \varepsilon_0$.

Finally, 2.1.6.7 is immediate from the following chain of equalities:

$$\begin{aligned}(\mu * \nu)^\wedge(a) &= \int e^{i\langle x,a\rangle}(\mu * \nu)(dx) \\ &= \int\int e^{i\langle x+y,a\rangle}\mu(dx)\nu(dy) \\ &= \int e^{i\langle y,a\rangle}\left(\int e^{i\langle x,a\rangle}\mu(dx)\right)\nu(dy) \\ &= \hat{\mu}(a)\hat{\nu}(a).\end{aligned}$$

■

Given $\delta > 0$ we now consider the set

$$V_\delta := \{a \in E' : \|a\| \leq \delta\}$$

which by Appendix B 14.1 is equicontinuous, hence $\sigma(E', E)$ - compact. Moreover,

$$\mathrm{Res}_{V_\delta}\sigma(E', E) = \mathrm{Res}_{V_\delta}\tau(E', E).$$

From Appendix B 14.2 we infer that, if E is assumed to the separable, V_δ is metrizable with respect to the topologies $\sigma(E', E)$ and $\tau(E', E)$.

Let $C(V_\delta) := C(V_\delta, \tau(E', E), \mathbf{C})$ denote the Banach space of all $\tau(E', E)$-continuous complex-valued functions on V_δ, together with the *sup* norm $\|\cdot\|_\infty$. Then for any $\mu \in M^b(E)$ we obtain

Properties 2.1.7 *(of the Fourier transform)*

2.1.7.1 $\mathrm{Res}_{V_\delta}\hat{\mu} \in C(V_\delta)$.

2.1.7.2 $\|\mathrm{Res}_{V_\delta}\hat{\mu}\| \leq \mu(E)$.

In the following we shall identify $\mathrm{Res}_{V_\delta}\hat{\mu}$ with $\hat{\mu}$.

Theorem 2.1.8 *For every τ_w-relatively compact subset H of $M^b(E)$ the set $\hat{H}$ is relatively compact in $C(V_\delta)$.*

Proof. From Prohorov's theorem 1.3.7 we infer that H is uniformly tight. Then Property 2.1.7.2 implies that $\hat{H}$ is bounded in $C(V_\delta)$, and by Property 2.1.6.5, $\hat{H}$ is equicontinuous. But then the Arzelà-Ascoli theorem yields the assertion. ■

Theorem 2.1.9 *(Continuity of the Fourier transform)*

Let $(\mu_n)_{n\geq 1}$ be a sequence of measures in $M^b(E)$ and let $\mu \in M^b(E)$. The following statements are equivalent:

(i) $(\mu_n)_{n\geq 1}$ converges with respect to τ_w.

(ii) $(\mu_n)_{n\geq 1}$ is τ_w-relatively compact, and for every $\delta > 0$ the sequence $(\hat{\mu}_n)_{n\geq 1}$ converges uniformly on V_δ.

(iii) $(\mu_n)_{n\geq 1}$ is τ_w-relatively compact, and for every $a \in E'$ the sequence $(\hat{\mu}_n(a))_{n\geq 1}$ converges in $\mathbf{C}$.

If in (i) we assume in addition that $\tau_w - \lim_{n\to\infty} \mu_n = \mu$ then in (iii) we have

$$\lim_{n\to\infty} \hat{\mu}_n(a) = \hat{\mu}(a)$$

for all $a \in E'$.

Proof. *(i) $\Longrightarrow$ (ii).* From the τ_w-convergence of the sequence $(\mu_n)_{n\geq 1}$ follows its τ_w-relative compactness. Then Theorem 2.1.8 implies that $(\hat{\mu}_n)_{n\geq 1}$ is relatively compact in $C(V_\delta)$ for every $\delta > 0$. But from $\tau_w - \lim_{n\to\infty} \mu_n = \mu$ it follows that

$$\lim_{n\to\infty} \hat{\mu}_n(a) = \hat{\mu}(a)$$

for all $a \in E'$, since $\mathrm{Re}\, e^{i\langle \cdot, a\rangle}$ and $\mathrm{Im}\, e^{i\langle \cdot, a\rangle}$ belong to $C^b(E)$ whenever $a \in E'$. Note that this argument takes care of the last statement of the theorem. We conclude that all the accumulation points of

$(\hat{\mu}_n)_{n\geq 1}$ in $C(V_\delta)$ coincide with $\text{Res}_{V_\delta}\hat{\mu}$, and hence that $\lim_{n\to\infty}\mu_n = \mu$ uniformly on V_δ.

(ii) $\Rightarrow$ *(iii)* is clear.

(iii) $\Rightarrow$ *(i)*. Let

$$\varphi(a) := \lim_{n\to\infty} \hat{\mu}_n(a)$$

for all $a \in E'$. Moreover, let μ_0 be an accumulation point of $(\mu_n)_{n\geq 1}$, which means that τ_w-$\lim_{k\to\infty}\mu_{n_k} = \mu_0$ for some subsequence $(\mu_{n_k})_{k\geq 1}$ of $(\mu_n)_{n\geq 1}$. But then

$$\hat{\mu}_0(a) = \lim_{k\to\infty} \hat{\mu}_{n_k}(a) = \varphi(a)$$

for all $a \in E'$. The uniqueness theorem 2.1.4 implies that $(\mu_n)_{n\geq 1}$ admits only one accumulation point, and consequently $(\mu_n)_{n\geq 1}$ τ_w-converges to it. ■

We are now in a position to study the logarithm of the Fourier transform.

Theorem 2.1.10 *Consider $\mu \in M^1(E)$ satisfying $\hat{\mu}(a) \neq 0$ for all $a \in E'$. Then there exists exactly one complex-valued function h on E' admitting the following properties:*

(i) $h(0) = 0$.

(ii) h is norm-continuous.

(iii) For each $a \in E'$,

$$\hat{\mu}(a) = \exp h(a)$$

In the sequel we shall write $\text{Log}\,\hat{\mu}$ instead of h.

Example 2.1.11 *For every $x \in E$*

$$\text{Log}\,\hat{\varepsilon}_x = i\langle x, \cdot\rangle .$$

Proof of the theorem. Let $\delta > 0$. Since V_δ is compact and $\hat{\mu}$ is $\tau(E', E)$-continuous by Property 2.1.6.4, we have that

$$\alpha := \inf\{|\hat{\mu}(a)| : a \in V_\delta\} > 0 .$$

It follows from Properties 2.1.6.3 and 2.1.6.4 that $\hat{\mu}$ is uniformly continuous on E' (with respect to the norm in E'), and hence there exists $\varepsilon > 0$ such that

$$|\hat{\mu}(a) - \hat{\mu}(b)| < \frac{\alpha}{2}$$

for all $a, b \in E'$ with $\|a - b\| < \varepsilon$. Let $\{t_0, t_1, ..., t_n\}$ be a partition of $[0,1]$ of mesh size less than ε/δ. Then for all $a \in V_\delta$ and $j \in \{1, 2, ..., n\}$ we have

$$\|t_j a - t_{j-1} a\| = (t_j - t_{j-1})\|a\| < \frac{\varepsilon}{\delta}\delta = \varepsilon$$

and therefore

$$|\hat{\mu}(t_j a) - \hat{\mu}(t_{j-1} a)| < \frac{\alpha}{2}\,.$$

Consequently

$$\left|\frac{\hat{\mu}(t_j a)}{\hat{\mu}(t_{j-1} a)} - 1\right| < \frac{\alpha}{2|\hat{\mu}(t_{j-1} a)|} \leq \frac{1}{2}$$

whenever $a \in V_\delta$ and $j \in \{1, 2, ..., n\}$. Now, let Log denote the main branch of the complex logarithm. We define

$$h_\delta(a) := \sum_{j=1}^{n} \operatorname{Log} \frac{\hat{\mu}(t_j a)}{\hat{\mu}(t_{j-1} a)}$$

for all $a \in V_\delta$. Then h_δ is a complex-valued function on V_δ with the following properties:

(*i'*) $h_\delta(0)=0$.

(*ii'*) h_δ is $\tau(E', E)$-continuous, and hence also $\|\cdot\|_\infty$-continuous.

(*iii'*) For all $a \in V_\delta$

$$\hat{\mu}(a) = \exp(h_\delta(a))\,.$$

We shall show that h_δ is uniquely determined by properties (*i'*) to (*iii'*). In fact, let g_δ be another complex-valued function satisfying these three properties. Then for each $a \in V_\delta$ there exists $k(a) \in \mathbf{Z}$ satisfying

$$h_\delta(a) - g_\delta(a) = 2\pi i k(a)$$

(by (iii')). But k is $\|\cdot\|_\infty$-continuous on V_δ (by (ii')) and satisfies $k(0)=0$ (by (i')). Since V_δ is $\|\cdot\|_\infty$-connected, $k(V_\delta)$ is a connected subset of $\mathbf{Z}$. This implies $k = 0$, and hence $h_\delta = g_\delta$.

Now, choose $0 < \delta_1 < \delta_2$. By the above discussion we have $\text{Res}_{V_{\delta_1}} h_{\delta_2} = h_{\delta_1}$, and hence $h(a) := h_\delta(a)$ for all $a \in V_\delta$ defines the desired function on E' which is certainly uniquely determined by properties (i) to (iii). ■

Corollary 2.1.12 *For every $\delta > 0$ Log $\hat{\mu}$ is $\tau(E', E)$-continuous on V_δ. In particular*

$$\sup\{|\text{Log}\hat{\mu}(a)| : \|a\| \leq \delta\} < \infty\,.$$

Proof. The first assertion follows from the representation

$$(\text{Res}_{V_\delta}\hat{\mu})(a) = \sum_{j=1}^{n} \text{Log}\frac{\hat{\mu}(t_j a)}{\hat{\mu}(t_{j-1}a)}$$

valid for all $a \in V_\delta$, together with the $\tau(E', E)$-continuity of $\hat{\mu}$ (by Property 2.1.6.4). The second assertion is a consequence of the $\tau(E', E)$-compactness of V_δ. ■

Theorem 2.1.13 *Let $H \subset M^1(E)$ such that $\hat{\mu}(a) \neq 0$ for all $a \in E'$, $\mu \in H$. Suppose there exist $\alpha, \delta > 0$ satisfying $|\hat{\mu}(a)| \geq \alpha$ for all $a \in V_\delta$, $\mu \in H$, and such that $\hat{H}$ is equicontinuous on V_δ with respect to $\tau(E', E)$. Then*

$$\{\text{Res}_{V_\delta}\text{Log } \hat{\mu} : \mu \in H\}$$

is relatively compact in $C(V_\delta)$.

Proof. From Property 2.1.6.3 together with the assumption, $\hat{H}$ is uniformly equicontinuous on V_δ with respect to the norm $\|\cdot\|_\infty$.

Therefore there exists a partition $\{t_0, t_1, ..., t_n\}$ of $[0,1]$ with sufficiently small mesh such that

$$\left| \frac{\hat{\mu}(t_j a)}{\hat{\mu}(t_{j-1}a)} - 1 \right| \leq \frac{1}{2}$$

for all $j \in \{1, 2, ..., n\}$ and

$$\text{Log } \hat{\mu}(a) = \sum_{j=1}^{n} \text{Log } \frac{\hat{\mu}(t_j a)}{\hat{\mu}(t_{j-1}a)}$$

for all $a \in V_\delta,\ \mu \in H$ (compare the proof of Theorem 2.1.10). But then

$$\{\text{Res}_{V_\delta} \text{Log } \hat{\mu} : \mu \in H\}$$

is bounded and equicontinuous with respect to $\tau(E', E)$, and the Arzelà-Ascoli theorem yields the assertion. ■

Lemma 2.1.14 *Let $\delta > 0$ be given. For every $x \in E$ define $J(x)(a) := \langle x, a \rangle$ whenever $a \in V_\delta$. Then $J(x) \in C(V_\delta)$ for all $x \in E$, and $\frac{1}{\delta}J$ is a linear isometry from E onto a closed subspace of $C(V_\delta)$.*

Proof. We first note that $J(x)$ is $\tau(E', E)$-continuous. Moreover, J is a linear mapping from E into $C(V_\delta)$. By Appendix B 14.3

$$\begin{aligned} \|J(x)\|_\delta &:= \sup\{|\langle x, a \rangle| : a \in V_\delta\} \\ &= \delta \sup\{|\langle x, a \rangle| : \|a\| \leq 1\} = \delta \|x\| \end{aligned}$$

for all $x \in E$. Therefore $\frac{1}{\delta}J$ is a linear isometry from E into $C(V_\delta)$. In particular, $\frac{1}{\delta}J(E) = J(E)$ is closed in $C(V_\delta)$. ■

Corollary 2.1.15 *(to Theorem 2.1.13). Let $M \subset E$ be such that*

$$\{\text{Res}_{V_\delta} \hat{\varepsilon}_x : x \in M\}$$

is relatively compact in $C(V_\delta)$ (for some $\delta > 0$). Then M is relatively compact in E. Furthermore $x \mapsto \text{Res}_{V_\delta} \hat{\varepsilon}_x$ is a homeomorphism from E onto a closed subset of $C(V_\delta)$.

Proof. Firstly $|\hat{\varepsilon}_x(a)| = 1$ for all $x \in M, \ a \in E'$. By assumption $\{\hat{\varepsilon}_x : x \in M\}$ is equicontinuous on V_δ with respect to $\tau(E', E)$. Now we infer from the theorem that

$$\{\mathrm{Res}_{V_\delta} \mathrm{Log}\ \hat{\varepsilon}_x : x \in M\}$$

is relatively compact in $C(V_\delta)$, and hence $\{J(x) \in M\}$ is relatively compact in $C(V_\delta)$. Lemma 2.1.14 yields the assertion. ■

Corollary 2.1.16 *Let* $(\mu_k)_{k \geq 1}$ *be a sequence of measures with* τ_w*-limit* $\mu \in M^1(E)$ *such that* $\hat{\mu}(a) \neq 0, \ \hat{\mu}_k(a) \neq 0$ *for all* $a \in E, \ k \geq 1$. *Then*

$$\lim_{k \to \infty} \mathrm{Log}\ \hat{\mu}_k = \mathrm{Log}\ \hat{\mu}$$

uniformly on bounded subsets of E'.

Proof. Let $\delta > 0$ be given. As in the proof of Theorem 2.1.10 we see that

$$\alpha := \frac{1}{2} \inf \left\{ \hat{\mu}(a) : a \in V_\delta \right\} > 0 \,.$$

Applying the continuity theorem 2.1.9 we obtain the existence of $k_0 \in \mathbf{N}$ such that $|\hat{\mu}_k(a)| \geq \alpha$ for all $a \in V_\delta, \ k \geq k_0$. We now infer from Prohorov's theorem 1.3.7 that $\{\mu_k : k \geq k_0\}$ is uniformly tight, and from Property 2.1.6.5 that $\{\hat{\mu}_k : k \geq k_0\}$ is equicontinuous with respect to $\tau(E', E)$. But Theorem 2.1.13 implies that

$$\{\mathrm{Res}_{V_\delta} \mathrm{Log}\ \hat{\mu}_k : k \geq k_0\}$$

is relatively compact in $C(V_\delta)$. On the other hand the representations of Log $\hat{\mu}_k$ and Log $\hat{\mu}$ on V_δ (see the proof of Theorem 2.1.13) together with the continuity theorem 2.1.9 yield the limit relationship

$$\lim_{k \to \infty} \mathrm{Log}\ \hat{\mu}_k(a) = \mathrm{Log}\ \hat{\mu}(a)$$

for all $a \in V_\delta$. Since $\{\mathrm{Res}_{V_\delta} \mathrm{Log}\ \hat{\mu}_k : k \geq k_0\}$ is relatively compact, this convergence is uniform on V_δ. ■

2.2 Shift compact sets of probability measures

We start with two useful results, following from properties of the Fourier transform $\mu \mapsto \hat{\mu}$ on the commutative topological semigroup $(M^1(E), *, \tau_w)$. Here E is a given separable Banach space so that, by Theorem 1.4.9, $(M^1(E), *, \tau_w)$ is a metric semigroup.

Lemma 2.2.1 *Suppose $\mu \in M^1(E)$ has the property that $(\mu^n)_{n\geq 1}$ is τ_w-relatively compact. Then $\mu = \varepsilon_0$.*

Proof. From Prohorov's theorem 1.3.7 together with Property 2.1.6.5 (of the Fourier transform) we conclude that the set $\{(\mu^n)^\wedge : n \in \mathbf{N}\}$ is equicontinuous with respect to $\tau(E', E)$. But we have $(\mu^n)^\wedge = (\hat{\mu})^n$ for all $n \in \mathbf{N}$. Therefore there exists $\delta > 0$ such that

$$\left|1 - (\hat{\mu}(a))^n\right| \leq \frac{1}{2}$$

whenever $a \in V_\delta,\ n \in \mathbf{N}$. Consequently $\hat{\mu}(a) = 1$ for all $a \in V_\delta$ as we shall see now. Indeed, let $z \in \mathbf{C}$ with $|1 - z^n| \leq 1/2$ for all $n \in \mathbf{N}$. Then Re $z^n \geq 1/2$ for all $n \in \mathbf{N}$, and

$$\begin{aligned}\frac{1}{2} &\geq |1 - z^n| = |1 - z|\left|1 + z + \ldots + z^{n-1}\right| \\ &\geq |1 - z|\left(1 + \mathrm{Re}\ z + \ldots + \mathrm{Re}\ z^{n-1}\right) \geq |1 - z|\frac{n}{2},\end{aligned}$$

so that $|1 - z| \leq 1/n$ for all $n \in \mathbf{N}$, and consequently $z = 1$.

The assertion now follows from $\hat{\mu}(a) = 1$ for all $a \in V_\delta$ with the help of Property 2.1.6.6 of the Fourier transform. ∎

Lemma 2.2.2 *Let $\mu, \nu \in M^1(E)$ such that $\mu * \nu = \nu$. Then $\nu = \varepsilon_0$.*

Proof. Properties 2.1.6.1 and 2.1.6.4 of the Fourier transform give the existence of $\delta > 0$ satisfying $\hat{\mu}(a) \neq 0$ for all $a \in V_\delta$. Moreover, $\hat{\mu}(a)\hat{\nu}(a) = \hat{\mu}(a)$ and hence $\hat{\nu}(a) = 1$ for all $a \in V_\delta$. We now apply Property 2.1.6.6 of the Fourier transform in order to obtain $\nu = \varepsilon_0$. ∎

Theorem 2.2.3 *Let $(\mu_n)_{n\geq 1}$, $(\nu_n)_{n\geq 1}$ be sequences in $M^1(E)$ such that $\{\mu_n * \nu_n : n \geq 1\}$ and $\{\mu_n : n \geq 1\}$ are both τ_w-relatively compact. Then $\{\nu_n : n \geq 1\}$ is also τ_w-relatively compact.*

Proof. Let $\varepsilon > 0$. By Prohorov's theorem 1.3.7 there exists $K \in \mathcal{K}(E)$ such that for all $n \geq 1$

$$\mu_n * \nu_n(K) \geq 1 - \varepsilon \text{ and } \mu_n(K) \geq 1 - \varepsilon\,.$$

But

$$\begin{aligned} 1 - \varepsilon \leq \mu_n * \nu_n(K) &= \int \nu_n(K - x)\mu_n(dx) \\ &\leq \int_K \nu_n(K - x)\mu_n(dx) + \varepsilon \leq \nu_n(K - K) + \varepsilon, \end{aligned}$$

so that $\nu_n(K - K) \geq 1 - 2\varepsilon$ whenever $n \geq 1$. One more application of Prohorov's theorem 1.3.7 gives that $\{\nu_n : n \geq 1\}$ is τ_w-relatively compact. ■

Corollary 2.2.4 *Suppose that*

$$\tau_w - \lim_{n\to\infty} \mu_n = \mu \in M^1(E)$$

and

$$\tau_w - \lim_{n\to\infty} \mu_n * \nu_n = \lambda \in M^1(E)\,.$$

Moreover, assume $\hat{\mu}(a) \neq 0$ for all $a \in E'$. Then there exists $\nu \in M^1(E)$ such that

$$\tau_w - \lim_{n\to\infty} \nu_n = \nu$$

and $\mu * \nu = \lambda$.

Proof. Theorem 2.2.3 implies that $\{\nu_n : n \geq 1\}$ is τ_w-relatively compact. Furthermore the continuity theorem 2.1.9 implies that for all $a \in E'$

$$\lim_{n\to\infty} \hat{\nu}_n(a) = \lim_{n\to\infty} \frac{(\mu_n * \nu_n)^\wedge(a)}{\hat{\mu}_n(a)} = \frac{\hat{\lambda}(a)}{\hat{\mu}(a)}\,.$$

Another application of the continuity theorem 2.1.9 yields

$$\tau_w - \lim_{n\to\infty} \nu_n = \nu \in M^1(E)\,.$$

But then

$$\hat{\nu}(a)\hat{\mu}(a) = \hat{\lambda}(a)$$

or

$$(\mu * \nu)^\wedge(a) = \hat{\lambda}(a)$$

for all $a \in E'$ which, by the uniqueness theorem 2.1.4, gives the desired result. ■

Remark 2.2.5 *The assumption made in the corollary that $\hat{\mu}(a) \neq 0$ for all $a \in E'$ cannot be dropped (without replacement). In fact, from Lukacs [29], Theorem 5.1.1, we know that there exist $\mu, \nu_1, \nu_2 \in M^1(\mathbf{R})$ with $\nu_1 \neq \nu_2$ such that $\mu * \nu_1 = \mu * \nu_2$. The sequence $\{\mu * \nu_1, \mu * \nu_2, \mu * \nu_1, ...\}$ obviously converges with respect to τ_w, but the sequence $\{\nu_1, \nu_2, \nu_1, ...\}$ does not.*

Definition 2.2.6 *We refer to $H \subset M^1(E)$ as being* ***relatively shift compact*** *if for every $\mu \in H$ there exists $x_\mu \in E$ such that $\{\mu * \varepsilon_{x_\mu} : \mu \in H\}$ is τ_w-relatively compact.*

We note by Prohorov's theorem 1.3.7 that relative shift compactness of a subset H is equivalent to the property that for every $\varepsilon > 0$ there exists $K \in \mathcal{K}(E)$ such that

$$\mu(K - x_\mu) = \mu * \varepsilon_{x_\mu}(K) \geq 1 - \varepsilon$$

for all $\mu \in H$, which motivates the use of the alternative terminology *shift tightness*.

Let $(\mu_n)_{n\geq 1}$ be a relatively shift compact sequence in $M^1(E)$, so that there exists a sequence $(x_n)_{n\geq 1}$ in E such that

$$\{\mu_n * \varepsilon_{x_n} : n \in \mathbf{N}\}$$

is τ_w-relatively compact. We refer to $(x_n)_{n\geq 1}$ as a ***centralizing sequence*** for $(\mu_n)_{n\geq 1}$ (equivalently, $(x_n)_{n\geq 1}$ ***centralizes*** $(\mu_n)_{n\geq 1}$).

Theorem 2.2.7 *Let $(\mu_n)_{n\geq 1}$, $(\nu_n)_{n\geq 1}$ be sequences in $M^1(E)$ such that $(\mu_n * \nu_n)_{n\geq 1}$ is τ_w-relatively compact. Then both $(\mu_n)_{n\geq 1}$ and $(\nu_n)_{n\geq 1}$ are relatively shift compact. Moreover, if $(x_n)_{n\geq 1}$ is a centralizing sequence for $(\mu_n)_{n\geq 1}$, then $(-x_n)_{n\geq 1}$ is a centralizing sequence for $(\nu_n)_{n\geq 1}$.*

Proof. Choose $(\delta_k)_{k\geq 1}$, $(\varepsilon_k)_{k\geq 1} \subset]0,\infty[$ with $\delta_1 < 1$, $\delta_k \downarrow 0$ and

$$\sum_{k\geq 1} \frac{\varepsilon_k}{\delta_k} \leq \frac{1}{2}\,.$$

By assumption for every $k \in \mathbf{N}$ there exists $K_k \in \mathcal{K}(E)$ satisfying

$$\mu_n * \nu_n(K_k) \geq 1 - \varepsilon_k$$

for all $n \in \mathbf{N}$. Let

$$B_{nk} := \{x \in E : \mu_n(K_k - x) > 1 - \delta_k\}$$

and put

$$B_n := \bigcap_{k\geq 1} B_{nk}\,.$$

We have for all $k \geq 1$

$$\begin{aligned}
\varepsilon_k \geq \mu_n * \nu_n(E \setminus K_k) &= \int \mu_n((E \setminus K_k) - x)\nu_n(dx) \\
&= \int \mu_n(E \setminus (K_k - x))\nu_n(dx) \\
&\geq \int_{E\setminus B_{nk}} (1 - \mu_n(K_k - x))\nu_n(dx) \\
&\geq \nu_n(E \setminus B_{nk})(1 - (1 - \delta_k)) = \delta_k\, \nu_n(E \setminus B_{nk})\,,
\end{aligned}$$

so that

$$\nu_n(E \setminus B_n) \leq \sum_{k\geq 1} \nu_n(E \setminus B_{nk}) \leq \sum_{k\geq 1} \frac{\varepsilon_k}{\delta_k} \leq \frac{1}{2}$$

and thus for all $n \geq 1$

$$\nu_n(B_n) \geq \frac{1}{2}\,.$$

In particular $B_n \neq \emptyset$ and so there exists $x_n \in E$ with

$$\mu_n(K_k - x_n) > 1 - \delta_k$$

for all $k \in \mathbf{N}$. This implies the τ_w-relative compactness of $(\mu_n * \varepsilon_{x_n})_{n\geq 1}$.

But since

$$\mu_n * \nu_n = (\mu_n * \varepsilon_{x_n}) * (\nu_n * \varepsilon_{-x_n})$$

for all $n \geq 1$, Theorem 2.2.3 implies the τ_w-relative compactness of $(\nu * \varepsilon_{-x_n})_{n\geq 1}$. ■

Corollary 2.2.8 *The following statements are equivalent:*

(i) $(\mu_n * \nu_n)_{n\geq 1}$ *is shift tight.*

(ii) $(\mu_n)_{n\geq 1}$ *and* $(\nu_n)_{n\geq 1}$ *are shift tight.*

Proof. $(i) \Rightarrow (ii)$. Let $(x_n)_{n\geq 1}$ be a centralizing sequence for $(\mu_n * \nu_n)_{n\geq 1}$. Then Theorem 2.2.7 implies that $(\mu_n)_{n\geq 1}$ and $(\nu_n * \varepsilon_{x_n})_{n\geq 1}$ are shift tight, and consequently $(\nu_n)_{n\geq 1}$ is also shift tight.

$(ii) \Rightarrow (i)$. Let $(x_n)_{n\geq 1}$, $(y_n)_{n\geq 1}$ be centralizing sequences for $(\mu_n)_{n\geq 1}$, $(\nu_n)_{n\geq 1}$ respectively. Since

$$(\mu * \varepsilon_{x_n}) * (\nu * \varepsilon_{y_n}) = (\mu_n * \nu_n) * \varepsilon_{x_n+y_n}$$

for all $n \geq 1$ and since the convolution $(\mu, \nu) \mapsto \mu*\nu$ is continuous on $M^1(E) \times M^1(E)$ the sequence $(x_n + y_n)_{n\geq 1}$ centralizes $(\mu_n * \nu_n)_{n\geq 1}$. ■

Theorem 2.2.9 *For every shift tight sequence* $(\mu_n)_{n\geq 1}$ *in* $M^1(E)$ *the following statements are equivalent:*

(i) $(\mu_n)_{n\geq 1}$ *is* τ_w*-relatively compact.*

(ii) Each centralizing sequence for $(\mu_n)_{n\geq 1}$ *is relatively compact.*

(iii) There is a relatively compact centralizing sequence for $(\mu_n)_{n\geq 1}$.

Proof. *(i)* $\Rightarrow$ *(ii)*. Let $(x_n)_{n\geq 1}$ be a centralizing sequence for $(\mu_n)_{n\geq 1}$. Then the sequences $(\mu * \varepsilon_{x_n})_{n\geq 1}$ and $(\mu_n)_{n\geq 1}$ are both τ_w-relatively compact. By Theorem 2.2.3 it follows that $(\varepsilon_{x_n})_{n\geq 1}$ is τ_w-relatively compact, and hence $(x_n)_{n\geq 1}$ is relatively compact (See Theorem 1.2.11).

(ii) $\Rightarrow$ *(iii)* is clear.

(iii) $\Rightarrow$ *(i)*. Let $(x_n)_{n\geq 1}$ denote a relatively compact centralizing sequence for $(\mu_n)_{n\geq 1}$. Then the sequences $(\mu * \varepsilon_{x_n})_{n\geq 1}$ and $(\varepsilon_{-x_n})_{n\geq 1}$ are τ_w-relatively compact. But

$$\mu_n = (\mu_n * \varepsilon_{x_n}) * \varepsilon_{-x_n}$$

for every $n \geq 1$; this together with the continuity of the convolution in $M^1(E)$ yields the desired conclusion. ■

Corollary 2.2.10 *Given a centralizing sequence $(x_n)_{n\geq 1}$ for $(\mu_n)_{n\geq 1}$ and an arbitrary sequence $(y_n)_{n\geq 1}$ in E the following statements are equivalent:*

(i) $(y_n)_{n\geq 1}$ centralizes $(\mu_n)_{n\geq 1}$.

(ii) $(x_n - y_n)_{n\geq 1}$ is relatively compact.

Proof. For every $n \geq 1$ we have

$$(\mu_n * \varepsilon_{y_n}) * \varepsilon_{x_n - y_n} = \mu_n * \varepsilon_{x_n} .$$

Therefore $(\mu_n * \varepsilon_{y_n})_{n\geq 1}$ is shift tight and $(x_n - y_n)_{n\geq 1}$ centralizes $(\mu_n * \varepsilon_{y_n})_{n\geq 1}$. The equivalence now follows from Theorem 2.2.9. ■

Corollary 2.2.11 *Let $(\mu_n)_{n\geq 1}, (x_n)_{n\geq 1}$ be sequences in $M^1(E)$ and E respectively satisfying*

$$\tau_w - \lim_{n\to\infty} \mu_n = \mu \in M^1(E)$$

and

$$\tau_w - \lim_{n\to\infty} \mu_n * \varepsilon_{x_n} = \nu \in M^1(E) .$$

Then there exists $x := \lim_{n\to\infty} x_n$, *and* $\nu = \mu * \varepsilon_x$.

Proof. From Theorem 2.2.9 we conclude that $(x_n)_{n\geq 1}$ is relatively compact. Let x, y be accumulation points of $(x_n)_{n\geq 1}$. Since the convolution in $M^1(E)$ is continuous,

$$\mu * \varepsilon_x = \nu = \mu * \varepsilon_y \,.$$

Putting $z := x - y$ we obtain $\mu * \varepsilon_z = \mu$. But now Lemma 2.2.2 applies to give $z = 0$, so that $x = y$. Consequently $\lim_{n\to\infty} x_n = x$, and this takes care of the assertion. ■

Theorem 2.2.12 *Let* $(\mu_n)_{n\geq 1}$ *be a shift tight sequence in* $M^1(E)$. *The following statements are equivalent:*

(i) $(\mu_n)_{n\geq 1}$ *is* τ_w*-relatively compact.*

(ii) For some (each) $\delta > 0$ *the sequence* $(\mathrm{Res}_{V_\delta}\hat{\mu}_n)_{n\geq 1}$ *is relatively compact in* $C(V_\delta)$.

(iii) *For some (each)* $\delta > 0$ *the sequence* $(\mathrm{Res}_{V_\delta}\hat{\mu}_n)_{n\geq 1}$ *is equicontinuous in* 0 *with respect to* $\tau(E', E)$.

Proof. *(ii)* $\Leftrightarrow$ *(iii)*. Property 2.1.6.3 of the Fourier transform together with $|\hat{\mu}_n| \leq 1$ for all $n \geq 1$ and the Arzelà-Ascoli theorem imply the assertion.

(i) $\Rightarrow$ *(ii)* follows from Prohorov's theorem 1.3.7 in conjunction with Property 2.1.6.5 of the Fourier transform.

(iii) $\Rightarrow$ *(i)*. From Theorem 2.2.9 we see that it suffices to show the existence of a relatively compact centralizing sequence for $(\mu_n)_{n\geq 1}$. Let therefore $(x_n)_{n\geq 1}$ be a centralizing sequence for $(\mu_n)_{n\geq 1}$. By assumption and Property 2.1.6.5 of the Fourier transform we have, for every $\varepsilon > 0, K \in \mathcal{K}(E)$ and $\eta > 0$ such that

$$|1 - \hat{\mu}_n(a)| \leq \frac{\varepsilon}{2}$$

and

$$\left|1 - \hat{\mu}_n(a) e^{i\langle x_n, a\rangle}\right| \leq \frac{\varepsilon}{2}$$

for all $a \in V_\delta$ with $p_K(a) < \eta$ and all $n \geq 1$. For such a and all $n \geq 1$ we then obtain the inequalities

$$\left|\hat{\mu}_n(a) - \hat{\mu}_n(a)e^{i\langle x_n, a\rangle}\right| \leq \varepsilon$$

and

$$|\hat{\mu}_n(a)| \geq 1 - \frac{\varepsilon}{2},$$

and hence that

$$|1 - e^{i\langle x_n, a\rangle}| \leq \frac{1}{1 - \frac{\varepsilon}{2}}\left|\hat{\mu}_n(a) - \hat{\mu}_n(a)e^{i\langle x_n, a\rangle}\right| \leq \frac{\varepsilon}{1 - \frac{\varepsilon}{2}}\,.$$

Consequently $(\mathrm{Res}_{V_\delta}\hat{\varepsilon}_{x_n})_{n\geq 1}$ is equicontinuous in 0, and hence relatively compact in $C(V_\delta)$ by Property 2.1.6.3 of the Fourier transform. The homeomorphism $x \mapsto \mathrm{Res}_{V_\delta}\hat{\varepsilon}_x$ between E and $C(V_\delta)$ provides the final step of the proof. ■

Corollary 2.2.13 *Let* $(\mu_n)_{n\geq 1}, (x_n)_{n\geq 1}$ *be sequences in* $M^1(E), E$ *respectively satisfying*

$$\tau_w - \lim_{n\to\infty} \mu_n * \varepsilon_{x_n} =: \nu \in M^1(E)\,.$$

Then the following statements are equivalent:

(i) $(\mu_n)_{n\geq 1}$ *is* τ_w*-convergent.*

(ii) *For some (each)* $\delta > 0$, $(\hat{\mu}_n)_{n\geq 1}$ *converges uniformly on* V_δ.

Proof. $(i) \Rightarrow (ii)$. This follows directly from the continuity theorem 2.1.9.

$(ii) \Rightarrow (i)$. By assumption $(\mu_n)_{n\geq 1}$ is shift tight, and by Theorem 2.2.12 $(\mu_n)_{n\geq 1}$ is τ_w-relatively compact. We now employ Theorem 2.2.3 and the homeomorphism $x \mapsto \varepsilon_x$ in order to obtain that $(x_n)_{n\geq 1}$ is relatively compact in E. Let x, y be accumulation points of $(x_n)_{n\geq 1}$. Moreover, let

$$f := \lim_{n\to\infty} \mathrm{Res}_{V_\delta}\hat{\mu}_n \in C(V_\delta)$$

(for some $\delta > 0$). Then by the continuity theorem 2.1.9

$$f(a)e^{i\langle x,a\rangle} = \hat{\nu}(a) = f(a)e^{i\langle y,a\rangle}$$

for all $a \in V_\delta$. But now $\hat{\nu}(0) = 1$, and $\hat{\nu}$ is $\|\cdot\|$-continuous by Property 2.1.6.4 of the Fourier transform, hence there exists $\delta' \in]0, \delta]$ with $\hat{\nu}(a) \neq 0$ for all $a \in V_{\delta'}$. It follows that

$$e^{i\langle x-y,a\rangle} = 1$$

for all $a \in V_{\delta'}$, and hence $x = y$ by Property 2.1.6.6 of the Fourier transform. Thus $(x_n)_{n\geq 1}$ converges in E. From the factorization

$$\mu_n = (\mu_n * \varepsilon_{x_n}) * \varepsilon_{-x_n}$$

together with the continuity of the convolution in $M^1(E)$ we conclude that $(\mu_n)_{n\geq 1}$ τ_w-converges. ■

Corollary 2.2.14 *Let $(\mu_n)_{n\geq 1}$ be a shift tight sequence in $M^1(E)$ such that $(\hat{\mu}_n)_{n\geq 1}$ converges uniformly on bounded subsets of E'. Then $(\mu_n)_{n\geq 1}$ τ_w-converges.*

Proof. Theorem 2.2.12 yields the τ_w-relative compactness of $(\mu_n)_{n\geq 1}$. But then the continuity theorem 2.1.9 implies the assertion. ■

The next topic will be the discussion of symmetrizing measures in $M^1(E)$, which will place some of the preceding results in a more applicable setting.

Definition 2.2.15 *Given measures $\mu, \nu \in M^1(E)$ we call μ a* ***factor*** *of ν if there exists $\lambda \in M^1(E)$ such that $\mu * \lambda = \nu$, in which case we write $\mu \prec \nu$.*

Properties 2.2.16 *(of the factorization).*

2.2.16.1 *(Reflexivity) $\mu \prec \mu$ for each $\mu \in M^1(E)$.*

2.2.16.2 *(Weak symmetry)* $\mu \prec \nu$ *and* $\nu \prec \mu$ *for* $\mu, \nu \in M^1(E)$ *implies* $\mu = \nu * \varepsilon_x$ *for some* $x \in E$.

2.2.16.3 *(Transitivity)* $\mu \prec \nu$ *and* $\nu \prec \kappa$ *for* $\mu, \nu, \kappa \in M^1(E)$ *implies* $\mu \prec \kappa$.

2.2.16.4 *(Permanence under convolution) If* $\mu_1 \prec \nu_1$ *and* $\mu_2 \prec \nu_2$ *for* $\mu_1, \nu_1, \mu_2, \nu_2 \in M^1(E)$ *then* $\mu_1 * \mu_2 \prec \nu_1 * \nu_2$.

2.2.16.5 *Let* $(\mu_n)_{n\geq 1}$ *and* $(\nu_n)_{n\geq 1}$ *be sequences in* $M^1(E)$ *such that* $(\nu_n)_{n\geq 1}$ *is shift tight and* $\mu_n \prec \nu_n$ *for all* $n \in \mathbf{N}$. *Then* $(\mu_n)_{n\geq 1}$ *is also shift tight.*

2.2.16.6 *(Continuity of* $\prec$*) Let* $(\mu_n)_{n\geq 1}$ *and* $(\nu_n)_{n\geq 1}$ *be sequences in* $M^1(E)$ *such that* $\mu_n \prec \nu_n$ *for all* $n \in \mathbf{N}$ *and both*

$$\tau_w - \lim_{n\to\infty} \mu_n = \mu, \quad \tau_w - \lim_{n\to\infty} \nu_n = \nu\,.$$

Then $\mu \prec \nu$.

2.2.16.7 *If* $\mu \prec \nu$ *for* $\mu, \nu \in M^1(E)$ *then*

$$|\hat{\nu}(a)| \leq |\hat{\mu}(a)|$$

$a \in E'$.

Proof. To prove 2.2.16.2, suppose $\mu * \lambda = \nu$ and $\nu * \kappa = \mu$ for $\lambda, \kappa \in M^1(E)$. Then $\mu * \lambda * \kappa = \mu$. From Lemma 2.2.2 we infer that $\lambda * \kappa = \varepsilon_0$, and hence by Corollary 1.4.7 that there exists $x \in E$ such that $\lambda = \varepsilon_{-x}$ and $\kappa = \varepsilon_x$. 2.2.16.5 follows from Theorem 2.2.7.

To show 2.2.16.6, given $\mu_n * \lambda_n = \nu_n$ for $\lambda_n \in M^1(E)$, $n \in \mathbf{N}$, Theorem 2.2.3 yields the τ_w-relative compactness of $(\lambda_n)_{n\geq 1}$. Let λ be an accumulation point of $(\lambda_n)_{n\geq 1}$. The continuity of the convolution in $M^1(E)$ implies that $\mu * \lambda = \nu$, and this is the desired assertion.

And finally, 2.2.16.7 follows from $\hat{\mu}(a)\hat{\lambda}(a) = \hat{\nu}(a)$ for some $\lambda \in M^1(E)$ together with $|\hat{\lambda}(a)| \leq 1$ for all $a \in E'$. ■

Definition 2.2.17 *For every* $\mu \in M^b(E)$ *the* **adjoint** μ^- *of* μ *is defined (as a measure in* $M^b(E)$*) by*

$$\mu^-(B) := \mu(-B)$$

whenever $B \in \mathfrak{B}(E)$. *We refer to* μ *as* ***symmetric*** *if* $\mu^- = \mu$.

Properties 2.2.18 *(of the adjoint). Let* $x \in E, \mu, \nu \in M^b(E)$.

2.2.18.1 $(\varepsilon_x)^- = \varepsilon_{-x}$.

2.2.18.2 $(\mu^-)^\wedge = \bar{\hat{\mu}}$.

2.2.18.3 μ *is symmetric if and only if* $\hat{\mu}$ *is real-valued, and this in turn holds if and only if*

$$\hat{\mu}(a) = \int \cos\langle x, a\rangle \mu(dx)$$

$a \in E'$.

2.2.18.4 *The mapping* $\mu \mapsto \mu^-$ *from* $M^b(E)$ *into itself is* τ_w*-continuous.*

2.2.18.5 $(\mu^-)^- = \mu$.

2.2.18.6 $(\mu * \nu)^- = \mu^- * \nu^-$.

In particular, the convolution of symmetric measures is again symmetric.

Theorem 2.2.19 *For every sequence* $(\mu_n)_{n\geq 1}$ *of symmetric measures in* $M^1(E)$ *the following statements are equivalent:*

(i) $(\mu_n)_{n\geq 1}$ *is* τ_w*-relatively compact.*

(ii) $(\mu_n)_{n\geq 1}$ *is shift tight.*

Proof. It suffices to demonstrate the implication $(ii) \Rightarrow (i)$. Let $(x_n)_{n\geq 1}$ be a centralizing sequence for $(\mu_n)_{n\geq 1}$. For every $n \in \mathbf{N}$ we have that

$$(\mu_n * \varepsilon_{x_n})^- = \mu_n^- * \varepsilon_{-x_n} = \mu_n * \varepsilon_{-x_n} .$$

Since the adjoint is a continuous mapping by Property 2.2.18.4, $(-x_n)_{n\geq 1}$ is also a centralizing sequence for $(\mu_n)_{n\geq 1}$. But then $(2x_n)_{n\geq 1}$ is relatively compact in E by Corollary 2.2.10, and Theorem 2.2.9 gives the result. ■

Theorem 2.2.20 *Let $(\mu_n)_{n\geq 1}$ be a sequence of symmetric measures in $M^1(E)$ and $(x_n)_{n\geq 1}$ any sequence in E such that $(\mu_n * \varepsilon_{x_n})_{n\geq 1}$ τ_w-converges. Then the sequences $(\mu_n)_{n\geq 1}$ and $(x_n)_{n\geq 1}$ converge in $M^1(E)$ and E respectively.*

Proof. From $\tau_w - \lim_{n\to\infty} \mu_n * \varepsilon_{x_n} = \lambda \in M^1(E)$ it follows that

$$\tau_w - \lim_{n\to\infty} \mu_n * \varepsilon_{-x_n} = \lambda^- .$$

On the other hand we have

$$\mu_n * \varepsilon_{x_n} = (\mu_n * \varepsilon_{-x_n}) * \varepsilon_{2x_n}$$

for all $n \in \mathbf{N}$. From Corollary 2.2.11 we infer that $\lim_{n\to\infty} 2x_n =: y \in E$ exists, and hence

$$x := \lim_{n\to\infty} x_n = \frac{1}{2} y .$$

Finally, the continuity of the convolution in $M^1(E)$ yields

$$\tau_w - \lim_{n\to\infty} \mu_n = \tau_w - \lim_{n\to\infty} (\mu_n * \varepsilon_{x_n}) * \varepsilon_{-x_n} = \lambda * \varepsilon_{-x} .$$ ■

Theorem 2.2.21 *Let $(\mu_n)_{n\geq 1}$, $(\nu_n)_{n\geq 1}$ be sequences in $M^1(E)$, where each μ_n is symmetric. If $(\mu_n * \nu_n)_{n\geq 1}$ is a τ_w-relatively compact sequence, then so are $(\mu_n)_{n\geq 1}$ and $(\nu_n)_{n\geq 1}$.*

Proof. Theorem 2.2.7 implies that $(\mu_n)_{n\geq 1}$ is shift tight. Then, by Theorem 2.2.19, $(\mu_n)_{n\geq 1}$ is τ_w-relatively compact and, by Theorem 2.2.3, $(\nu_n)_{n\geq 1}$ is τ_w-relatively compact. ■

Theorem 2.2.22 *For every sequence $(\mu_n)_{n\geq 1}$ in $M^b(E)$ the following are equivalent:*

(i) $(\mu_n)_{n\geq 1}$ is τ_w-relatively compact.

(ii) $(\mu_n + \mu_n^-)_{n\geq 1}$ is τ_w-relatively compact.

Proof. For every $n \geq 1$ we have

$$(\mu_n + \mu_n^-)(E) = \mu_n(E) + \mu_n^-(E) = 2\mu_n(E)$$

and hence

$$\sup_{n\geq 1}(\mu_n + \mu_n^-)(E) = 2 \sup_{n\geq 1} \mu_n(E)\,.$$

Let $\varepsilon > 0$ and $K \in \mathcal{K}(E)$. If $\mu_n(E \setminus K) < \varepsilon$ then

$$\begin{aligned}(\mu_n + \mu_n^-)(E \setminus (K \cup (-K))) &\leq \mu_n(E \setminus K) + \mu_n^-(-(E \setminus K)) = \\ &= 2\mu_n(E \setminus K) < \varepsilon\end{aligned}$$

and $K \cup (-K) \in \mathcal{K}(E)$. On the other hand, if

$$(\mu_n + \mu_n^-)(E \setminus K) < \varepsilon$$

we obtain that $\mu_n(E \setminus K) < \varepsilon$ for every $n \geq 1$. Prohorov's theorem 1.3.7 yields both of the desired implications. ■

Definition 2.2.23 *For every $\mu \in M^1(E)$ the measure*

$$|\mu|^2 := \mu * \mu^- \in M^1(E)$$

is called the ***symmetrization*** *of μ.*

Properties 2.2.24 *(of the symmetrization). Let $\mu, \nu \in M^1(E)$.*

2.2.24.1 $|\mu|^2$ *is symmetric.*

2.2.24.2 $|\mu * \nu|^2 = |\mu|^2 * |\nu|^2$, *and in particular,* $|\mu^n|^2 = (|\mu|^2)^n$ *for all* $n \in \mathbf{N}$.

2.2.24.3 $(|\mu|^2)^\wedge(a) = |\hat{\mu}(a)|^2$ *for all* $a \in E'$.

Proof. 2.2.24.1 is a simple application of Properties 2.2.18.5 and 2.2.18.6. To show 2.2.24.2, it suffices to apply Property 2.2.18.6, and for 2.2.24.3 we just appeal to Properties 2.1.6.7 and 2.2.18.2 (of the Fourier transform). ■

Theorem 2.2.25 *For every sequence* $(\mu_n)_{n\geq 1}$ *in* $M^1(E)$ *the following are equivalent:*

(i) $(\mu_n)_{n\geq 1}$ *is shift tight.*

(ii) $(|\mu_n|^2)_{n\geq 1}$ *is* τ_w*-relatively compact.*

Proof. $(i) \Rightarrow (ii)$. From the τ_w-continuity of the mapping $\mu \mapsto \mu^-$ (Property 2.2.18.4) we obtain that $(\mu_n^-)_{n\geq 1}$ is shift tight. An application of Corollary 2.2.8 yields the shift tightness of $(|\mu_n|^2)_{n\geq 1}$. Finally, we employ Property 2.2.24.1 together with Theorem 2.2.19 in order to see that $(|\mu_n|^2)_{n\geq 1}$ is τ_w-relatively compact.

$(ii) \Rightarrow (i)$. This follows directly from Corollary 2.2.8. ■

Corollary 2.2.26 *For every* $\mu \in M^1(E)$ *the following are equivalent:*

(i) $(\mu^n)_{n\geq 1}$ *is shift tight.*

(ii) μ *is a Dirac measure.*

Proof. It suffices to demonstrate the implication $(i) \Rightarrow (ii)$. Theorem 2.2.25 together with Property 2.2.24.2 implies that $((|\mu|^2)^n)_{n\geq 1}$ is τ_w-relatively compact. But then Lemma 2.2.2 gives $|\mu|^2 = \varepsilon_0$, and μ itself must be a Dirac measure. ■

2.3 Infinitely divisible and embeddable measures

We start by giving the basic

Definition 2.3.1 *A measure $\mu \in M^1(E)$ is called* ***infinitely divisible*** *if for every $n \in \mathbf{N}$ there exists an n-th* ***root*** *of μ i.e. a measure $\mu_n \in M^1(E)$ such that*

$$\mu_n^n = \mu .$$

By $I(E)$ we denote the set of all infinitely divisible measures in $M^1(E)$.

Example 2.3.2 *For every $x \in E$ the Dirac measure ε_x belongs to $I(E)$.*

In fact, for every $n \in \mathbf{N}$

$$\varepsilon_x = (\varepsilon_{\frac{1}{n}x})^n,$$

since E is divisible.

Clearly, $I(E)$ is a (commutative) subsemigroup of $M^1(E)$. One just notes that for $\mu, \nu \in I(E)$ admitting the representations $\mu = \mu_n^n$ and $\nu = \nu_n^n$ with $\mu_n, \nu_n \in M^1(E)$ respectively ($n \in \mathbf{N}$), we have

$$\mu * \nu = (\mu_n * \nu_n)^n .$$

Moreover, the semigroup $I(E)$ has a neutral element ε_0.

Theorem 2.3.3 *For each $\mu \in I(E)$ one has $\hat{\mu}(a) \neq 0$ whenever $a \in E'$.*

Proof. Let $\mu = \mu_n^n$ with $\mu_n \in M^1(E)$ for every $n \in \mathbf{N}$. Then $\mu_n \prec \mu$ for all $n \in \mathbf{N}$, and by Property 2.2.16.5 $(\mu_n)_{n \geq 1}$ is shift tight, hence by Theorem 2.2.25 $(|\mu_n|^2)_{n \geq 1}$ is τ_w-relatively compact. Let ν be an accumulation point of $(|\mu_n|^2)_{n \geq 1}$. Then ν is symmetric. Moreover, for all $n \geq m$ we have that

$$(|\mu_n|^2)^m \prec |\mu|^2,$$

and consequently that

$$\nu^m \prec |\mu|^2$$

for all $m \in \mathbf{N}$, since $\prec$ is a closed relation. But then $(\nu^m)_{m\geq 1}$ is a τ_w-relatively compact sequence, which follows with the help of Property 2.2.16.5 and Theorem 2.2.19. Thus $\nu = \varepsilon_0$ which implies that

$$|\mu_n|^2 \longrightarrow \varepsilon_0$$

as $n \to \infty$ (in the topology τ_w) and therefore by the continuity theorem 2.1.19 together with Property 2.2.24.3 that

$$\lim_{n\to\infty} |\hat{\mu}_n(a)|^2 = 1$$

for all $a \in E'$. But then for each $a \in E'$ there exists an $n \in \mathbf{N}$ such that $\hat{\mu}_n(a) \neq 0$, thus

$$\hat{\mu}(a) = (\hat{\mu}_n(a))^n \neq 0 \,.$$

■

Corollary 2.3.4 *For every* $\mu \in I(E)$

(i) there exists Log $\hat{\mu}$, *and*

(ii) if μ *is symmetric, then* Log $\hat{\mu}$ *is real which implies that* $\hat{\mu}(a) > 0$ *whenever* $a \in E'$.

Proof. (i) follows from Theorem 2.1.10 on the existence of the function h. Concerning

(ii) we first note that $\hat{\mu}(E') \subset \mathbf{R}$, since μ is symmetric and hence $\hat{\mu}$ is real. Next we realize that

$$\hat{\mu}(a) = \exp \ \mathrm{Log} \ \hat{\mu}(a)$$

for all $a \in E'$, hence that

$$(\mathrm{Log} \ \hat{\mu})(E') \subset \mathbf{R} \cup 2\pi i \mathbf{Z}^{\times} \,.$$

As a continuous image of E' the set $(\text{Log } \hat{\mu})(E')$ is connected. But $\text{Log } \hat{\mu}(0) = 0$, thus $(\text{Log } \hat{\mu})(E') \subset \mathbf{R}$ and the remaining assertion follows. ■

Theorem 2.3.5 *(Uniqueness of roots) Let $\mu \in I(E)$.*

For every $n \in \mathbf{N}$ there exists exactly one (n-th root) $\mu_n \in M^1(E)$ such that $\mu_n^n = \mu$, and

$$\hat{\mu}_n(a) = \exp\left(\frac{1}{n}\text{Log } \hat{\mu}(a)\right)$$

whenever $a \in E'$.

If, moreover, μ is symmetric, then also each n-th root μ_n of μ is symmetric ($n \in \mathbf{N}$).

Proof. From Theorem 2.3.3 we conclude that $\hat{\mu}(a) \neq 0$ for all $a \in E'$. But then also $\hat{\mu}_n(a) \neq 0$ for each $n \in \mathbf{N}$ ($a \in E'$). Theorem 2.1.10 yields the existence of $\text{Log } \hat{\mu}_n$ for every $n \in \mathbf{N}$. We also have that

$$\exp(\text{Log } \hat{\mu}(a)) = \hat{\mu}(a) = (\hat{\mu}_n(a))^n = \exp(n \text{ Log } \hat{\mu}_n(a))$$

for each $n \in \mathbf{N}$ ($a \in E'$), hence that

$$\text{Log } \hat{\mu} = n \text{ Log } \hat{\mu}_n$$

or

$$\text{Log } \hat{\mu}_n = \frac{1}{n} \text{ Log } \hat{\mu}\,.$$

The uniqueness of the Fourier transform (Theorem 2.1.4) implies the uniqueness of the n-th root μ_n of μ as asserted.

Finally, Corollary 2.3.4 together with Property 2.2.18.3 yields the last assertion of the theorem. ■

Theorem 2.3.6 *(Convergence of roots) For $\mu \in I(E)$ with sequence $(\mu_{\frac{1}{n}})_{n\geq 1}$ of n-th roots $\mu_{\frac{1}{n}}$ of μ,*

$$\mu_{\frac{1}{n}} \longrightarrow \varepsilon_0$$

as $n \to \infty$ in the weak topology τ_w.

Proof. Let $\delta > 0$. By Corollary 2.1.12

$$\sup\{|\mathrm{Log}\ \hat{\mu}(a)| : a \in V_\delta\} < \infty .$$

From Theorem 2.3.5 we infer that

$$\begin{aligned} \lim_{n\to\infty} (\mu_{\frac{1}{n}})^\wedge(a) &= \lim_{n\to\infty} \exp\left(\frac{1}{n}\ \mathrm{Log}\ \hat{\mu}(a)\right) \\ &= 1 \\ &= \hat{\varepsilon}_0(a) \end{aligned}$$

uniformly in $a \in V_\delta$. Since $\mu_{\frac{1}{n}} \prec \mu$ for all $n \in \mathbf{N}$, the sequence $(\mu_{\frac{1}{n}})_{n\geq 1}$ is shift tight by Property 2.2.16.5. But then Corollary 2.2.14 applies, and with the help of the uniqueness and continuity theorems 2.1.4 and 2.1.9 respectively the result follows.

Theorem 2.3.7 *(Closeness of $I(E)$ in $M^1(E)$). Let $(\mu_k)_{k\geq 1}$ be a sequence of measures in $I(E)$ such that*

$$\mu_k \longrightarrow \mu \in M^1(E)$$

as $k \to \infty$, with respect to τ_w. Then $\mu \in I(E)$, and

$$(\mu_k)_{\frac{1}{n}} \longrightarrow \mu_{\frac{1}{n}}$$

as $k \to \infty$, with respect to τ_w, for all $n \in \mathbf{N}$.

Proof. 1. Let $n \in \mathbf{N}$ be fixed. Then, by the continuity of the mappings $\mu \mapsto \mu^-$ and $(\mu, \nu) \mapsto \mu * \nu$ we obtain

$$|(\mu_k)_{\frac{1}{n}}|^2 \prec |\mu_k|^2$$

and

$$|\mu_k|^2 \longrightarrow |\mu|^2$$

as $k \to \infty$, with respect to τ_w. But then Property 2.2.16.5 and Theorem 2.2.19 apply and yield the τ_w-relative compactness of $(|(\mu_k)_{\frac{1}{n}}|^2)_{k\geq 1}$.

Moreover, Theorem 2.3.5 together with the continuity theorem 2.1.9 implies that for all $a \in E'$

$$\begin{aligned}\lim_{k\to\infty} (|(\mu_k)_{\frac{1}{n}}|^2)^\wedge(a) &= \lim_{k\to\infty} |((\mu_k)_{\frac{1}{n}})^\wedge(a)|^2 \\ &= \lim_{k\to\infty} |\hat{\mu}_k(a)|^{\frac{2}{n}} \\ &= |\hat{\mu}(a)|^{\frac{2}{n}}.\end{aligned}$$

Applying the continuity theorem 2.1.9 again there exists a measure $\nu_n \in M^1(E)$ satisfying

$$\hat{\nu}_n(a) = |\hat{\mu}(a)|^{\frac{2}{n}}$$

Since this equality holds for every $n \in \mathbf{N}$, $|\mu|^2 \in I(E)$, hence by Theorem 2.3.3

$$|\hat{\mu}(a)|^2 = (|\mu|^2)^\wedge(a) \neq 0$$

whenever $a \in E'$.

2. From 1. and from the existence of the function h (Theorem 2.1.10) we infer that Log $\hat{\mu}$ exists. Then Corollary 2.1.16 implies that

$$\lim_{k\to\infty} \text{Log } \hat{\mu}_k(a) = \text{Log } \hat{\mu}(a)$$

uniformly on bounded subsets of elements a of E'. It follows from Theorem 2.3.5 that

$$\begin{aligned}\lim_{k\to\infty} ((\mu_k)_{\frac{1}{n}})^\wedge(a) &= \lim_{k\to\infty} \exp\left(\frac{1}{n} \text{ Log } \hat{\mu}_k(a)\right) \\ &= \exp\left(\frac{1}{n} \text{ Log } \hat{\mu}(a)\right)\end{aligned}$$

uniformly on bounded subsets of elements a in E'. In addition, $((\mu_k)_{\frac{1}{n}})_{k\geq 1}$ is a shift tight sequence (by Property 2.2.16.5), consequently, by Corollary 2.2.14 there exists a measure $\mu_n \in M^1(E)$ such that

$$(\mu_k)_{\frac{1}{n}} \longrightarrow \mu_n$$

for all $n \in \mathbf{N}$ ($k \to \infty$, with respect to τ_w). But then the continuity of the convolution yields $\mu_n^n = \mu$ or $\mu \in I(E)$. By Theorem 2.3.5 we have that $\mu_n = \mu_{\frac{1}{n}}$ for all $n \in \mathbf{N}$. ■

Definition 2.3.8 *A family $(\mu_t)_{t \in \mathbf{R}_+}$ of measures in $M^1(E)$ is said to be a (continuous)* ***convolution semigroup*** *in $M^1(E)$ if it has the following properties:*

(a) $\mu_s * \mu_t = \mu_{s+t}$ for all $s, t \in \mathbf{R}_+$.

(b) $\mu_0 = \varepsilon_0$.

(c) The mapping $t \mapsto \mu_t$ from $\mathbf{R}_+$ into $M^1(E)$ is τ_w-continuous.

Evidently, the measures μ_t of a convolution semigroup $(\mu_t)_{t \in \mathbf{R}_+}$ in $M^1(E)$ are elements of $I(E)$, since

$$\mu_t = (\mu_{\frac{t}{n}})^n$$

for every $n \in \mathbf{N}$. For the converse we prove the subsequent

Theorem 2.3.9 *(Embedding) Let $\mu \in I(E)$. Then for every $t \in \mathbf{R}_+$ there exists exactly one measure $\mu_t \in M^1(E)$ such that*

$$\hat{\mu}_t(a) = \exp(t \ \mathrm{Log}\ \hat{\mu}(a))$$

whenever $a \in E'$, and $(\mu_t)_{t \in \mathbf{R}_+}$ is a convolution semigroup in $M^1(E)$.

In particular one has

$$\mu_1 = \mu\,.$$

Proof. 1. Let $t \in \mathbf{Q}_+^\times$ and $k, \ell, k', \ell' \in \mathbf{N}$ such that

$$\frac{k}{\ell} = t = \frac{k'}{\ell'}\,.$$

By Theorem 2.3.5 we obtain that for all $a \in E'$

$$\begin{aligned}\left((\mu_{\frac{1}{\ell}})^k\right)^\wedge(a) &= \left((\mu_{\frac{1}{\ell}})^\wedge(a)\right)^k = \left(\exp\frac{1}{\ell}\ \text{Log}\ \hat{\mu}(a)\right)^k \\ &= \exp\left(\frac{k}{\ell}\ \text{Log}\ \hat{\mu}(a)\right) = \exp(t\ \text{Log}\ \hat{\mu}(a)) \\ &= \left((\mu_{\frac{1}{\ell'}})^{k'}\right)^\wedge(a).\end{aligned}$$

The uniqueness theorem 2.1.4 then implies that

$$\mu_t = (\mu_{\frac{1}{\ell}})^k$$

is uniquely determined, and

$$\mu_s * \mu_t = \mu_{s+t}$$

whenever $s, t \in \mathbf{Q}_+^\times$.

2. Let $t \in \mathbf{R}_+$ and let $(t_n)_{n\geq 1}$ be a sequence in $\mathbf{Q}_+^\times$ such that $\lim_{n\to\infty} t_n = t$. Let, moreover, $m \in \mathbf{N}$ with $t_n < m$ for all $n \in \mathbf{N}$. Since by 1.

$$\mu_{t_n} * \mu_{m-t_n} = \mu_m$$

for all $n \in \mathbf{N}, (\mu_{t_n})_{n\geq 1}$ is shift tight. This follows from Theorem 2.2.7.

Now, let $\delta > 0$ be given. We know that

$$\sup\{|\text{Log}\ \hat{\mu}(a)| : a \in V_\delta\} < \infty .$$

But then 1. implies that

$$\begin{aligned}\lim_{n\to\infty} \hat{\mu}_{t_n}(a) &= \lim_{n\to\infty}\ \exp(t_n\ \text{Log}\ \hat{\mu}(a)) \\ &= \exp(t\ \text{Log}\ \hat{\mu}(a))\end{aligned}$$

uniformly for $a \in V_\delta$. It follows from Corollary 2.2.14 and from the uniqueness and continuity properties of the Fourier transform that there exists exactly one measure $\mu_t \in M^1(E)$ satisfying

$$\hat{\mu}_t(a) = \exp(t\ \text{Log}\ \hat{\mu}(a))$$

for all $a \in E'$. Moreover,

$$\lim_{n \to \infty} \mu_{t_n} = \mu_t$$

which shows the first assertion in the theorem.

3. As a consequence of the previous discussion we obtain that

$$\mu_s * \mu_t = \mu_{s+t}$$

for all $s, t \in \mathbf{R}_+$. The continuity of $t \mapsto \mu_t$ follows as in 2. by choosing for $t \in \mathbf{R}_+$ an arbitrary sequence $(t_n)_{n \geq 1}$ in $\mathbf{R}_+^\times$ such that $\lim_{n \to \infty} t_n = t$. ∎

2.4 Gauss and Poisson measures

In this section the prominent subsets of Gauss and Poisson measures within $I(E)$ will be studied.

Definition 2.4.1 *A measure $\rho \in M^1(E)$ is called a (symmetric)* ***Gauss measure*** *if there exists a symmetric linear mapping $R : E' \to E$ such that*

$$\hat{\rho}(a) = \exp\Big(-\frac{1}{2}\langle Ra, a\rangle\Big)$$

whenever $a \in E'$.

Here we apply the symmetry of a mapping $R : E' \to E$ in the sense that

$$\langle Ra, b\rangle = \langle a, Rb\rangle$$

for all $a, b \in E'$.

R is said to be the ***covariance operator*** *of ρ.*

By $G(E)$ we abbreviate the totality of all Gauss measures on E.

Examples 2.4.2

2.4.2.1 *Let $E := \mathbf{R}^p$. Then the p-**dimensional normal distribution** $N(0, C)$ with mean (vector) 0 and covariance (matrix) C is a Gauss measure on E.*

In fact, for all $a \in \mathbf{R}^p = (\mathbf{R}^p)'$ we have that

$$N(0, C)^\wedge(a) = \exp\Big(-\frac{1}{2}\langle Ca, a\rangle\Big),$$

and C is symmetric.

2.4.2.2 *Let $E := C(I)$ (equal to the space of continuous real-valued functions on the compact interval $I := [0, 1]$). Then the* ***Wiener measure*** *W is a Gauss measure on E.*

In fact, from the theory of Brownian motion follows that

$$\widehat{W}(\chi) = \exp\Big(-\frac{1}{2}\langle R\chi, \chi\rangle\Big)$$

with

$$(R\chi)(t) := \int (s \wedge t)\chi(ds)$$

for all $\chi \in C(I)' \cong M^b(I) - M^b(I),\ \ t \in I$. It turns out that R is a linear mapping $C(I)' \to C(I)$ which is shown to be symmetric (by Fubini's theorem).

2.4.3 Scholium on the Wiener measure. Let $(\Omega, \mathfrak{A}, \mathbf{P}, (B_t)_{t\in I})$ denote a Brownian motion in $\mathbf{R}$ with parameter set $I = [0,1]$. For every $t \in I$ there exists a real-valued random variable $\widetilde{B}_t$ on $(\Omega, \mathfrak{A}, \mathbf{P})$ with $\mathbf{P}([\widetilde{B}_t \neq B_t]) = 0$ such that for every $\omega \in \Omega$ the mapping $t \mapsto \widetilde{B}_t(\omega)$ from I into $\mathbf{R}$ is continuous. The process $(\Omega, \mathfrak{A}, \mathbf{P}, (\widetilde{B}_t)_{t\in I})$ is called a Brownian motion with continuous paths. As a result of this modification we may assume without loss of generality that the initial Brownian motion $(\Omega, \mathfrak{A}, \mathbf{P}, (B_t)_{t\in I})$ has continuous paths. It follows that the mapping

$$(\omega, t) \mapsto B_t(\omega)$$

from $\Omega \times I$ into $\mathbf{R}$ is $\mathfrak{A} \otimes \mathfrak{B}(I) - \mathfrak{B}(\mathbf{R})$-measurable, hence that the mapping $B : \Omega \to C(I)$ which sends ω onto the path $t \mapsto B_t(\omega)$ is $\mathfrak{A} - \mathfrak{B}(C(I))$-measurable. But then the Wiener measure $W := \mathbf{P}_B = B(\mathbf{P})$ is a measure in $M^1(C(I))$ with Fourier transform given by

$$\widehat{W}(\chi) = \exp\Big(-\frac{1}{2}\int (s \wedge t)\chi \otimes \chi(ds, dt)\Big)$$

for all $\chi \in C(I)'$.

Now let

$$(R\chi)(t) := \int (s \wedge t)\chi(ds)$$

for all $\chi \in C(I)',\ \ t \in I$. Since

$$|(R\chi)(t_1) - R\chi(t_2)| \leq |t_1 - t_1| \|\chi\|$$

for all $\chi \in C(I)',\ \ t_1, t_2 \in I,\ \ R\chi$ is Lipschitz continuous, in particular, $R\chi \in C(I)$. Clearly, R is a continuous linear mapping $C(I)' \to$

$C(I)$. Moreover,

$$\begin{aligned}\int (s\wedge t)\chi\otimes\chi(ds,dt) &= \int \Big(\int (s\wedge t)\chi(ds)\Big)\chi(dt)\\ &= \int (R\chi)(t)\chi(dt) = \langle R\chi,\chi\rangle\end{aligned}$$

and therefore

$$\widehat{W} = \exp\Big(-\frac{1}{2}\langle R\chi,\chi\rangle\Big)$$

whenever $\chi \in C(I)'$.

Properties 2.4.4 *of Gauss measures*

2.4.4.1 *R is uniquely determined and positive in the sense that*

$$\big\langle Ra, a\big\rangle \geq 0$$

for all $a \in E'$.

In fact, the uniqueness of R follows from the representation

$$\big\langle Ra, a\big\rangle = -\ \operatorname{Log}\ \hat{\rho}(a)$$

together with

$$\big\langle Ra, b\big\rangle = \frac{1}{4}\big\langle R(a+b), a+b\big\rangle - \frac{1}{4}\big\langle R(a-b), a-b\big\rangle,$$

both equalities being valid for all $a, b \in E'$.

The positivity of R is a consequence of

$$\hat{\rho}(a) \leq \hat{\rho}(0) = 1$$

valid for all $a \in E'$.

2.4.4.2 *Every Gauss measure is symmetric, since $\hat{\rho}$ is real-valued.*

2.4.4.3 *Let $\rho, \sigma \in G(E)$ with covariance operators R and S respectively. Then $\rho * \sigma \in G(E)$ with covariance operator $R + S$.*

In particular, $G(E)$ is a subsemigroup of $M^1(E)$.

As for the proof of this property we just compute

$$\begin{aligned}(\rho * \sigma)^\wedge(a) &= \hat{\rho}(a)\hat{\sigma}(a) \\ &= \exp\Big(-\frac{1}{2}\langle Ra, a\rangle\Big)\exp\Big(-\frac{1}{2}\langle Sa, a\rangle\Big) \\ &= \exp\Big(-\frac{1}{2}\Big(\langle Ra, a\rangle + \langle Sa, a\rangle\Big)\Big) \\ &= \exp\Big(-\frac{1}{2}\langle (R+S)a, a\rangle\Big)\end{aligned}$$

for all $a \in E'$ and observe that $R + S$ is a symmetric linear mapping $E' \to E$.

2.4.4.4 *Let $\rho \in G(E)$ with covariance operator R and let T be a continuous linear mapping from E into another Banach spaces F. Then $T(\rho) \in G(F)$ with covariance operator $T \circ R \circ T^t$.*

In fact, for all $b \in F'$ we have

$$\begin{aligned}T(\rho)^\wedge(b) = \hat{\rho} \circ T^t(b) = \hat{\rho}(T^t b) &= \exp\Big(-\frac{1}{2}\langle R(T^t b), T^t b\rangle\Big) \\ &= \exp\Big(-\frac{1}{2}\langle (T \circ R \circ T^t)b, b\rangle\Big),\end{aligned}$$

and $T \circ R \circ T^t$ is a symmetric mapping $F' \to F$.

2.4.4.5 $G(E) \subset I(E)$

Moreover, for every $\rho \in G(E)$ there exists a convolution semigroup $(\rho_t)_{t \in \mathbf{R}_+}$ given by

$$\rho_t := H_{\sqrt{t}}(\rho)$$

for all $t \in \mathbf{R}_+$ such that $\rho_1 = \rho$.

Here the symbol H_r denotes the homothetical mapping $x \mapsto rx$ (with $r \in \mathbf{R}_+$) on E.

For the proof of this property we first agree on R to be the covariance operator of ρ. Then by Property 2.4.4.4 $H_{\sqrt{t}}(\rho) \in G(E)$ with covariance operator tR. It follows that

$$\Big(H_{\sqrt{\frac{1}{n}}}(\rho)\Big)^n = \rho$$

for every $n \in \mathbf{N}$, hence that $\rho \in I(E)$. The remaining part of the statement follows from the embedding theorem 2.3.9.

2.4.4.6 *Let $s, t \in \mathbf{R}$ with $s^2 + t^2 = 1$, and let $T := T_{s,t}$ be defined by*

$$T(x, y) := (sx + ty, tx - sy)$$

for all $(x, y) \in E \times E$. T is a continuous linear mapping on $E \times E$.

Then for every $\rho \in G(E)$

$$T(\rho \otimes \rho) = \rho \otimes \rho\,.$$

In fact, let R denote the covariance operator of ρ, and let

$$q(a) := \langle Ra, a \rangle$$

for all $a \in E'$. Then

$$q(sa + tb) + q(ta - sb) = q(a) + q(b)$$

and

$$T^t(a, b) = (sa + tb, ta - sb)$$

whenever $a, b \in E'$. Here the identification $(E \times E)' = E' \times E'$ is applied.

It follows that for all $(a, b) \in E' \times E'$

$$\begin{aligned}(T(\rho \otimes \rho))^\wedge(a, b) &= (\rho \otimes \rho)^\wedge(T^t(a, b)) = \hat{\rho}(sa + tb)\hat{\rho}(ta - sb) \\ &= \exp\Big(-\frac{1}{2}[q(sa + tb) + q(ta - sb)]\Big) \\ &= \exp\Big(-\frac{1}{2}[q(a) + q(b)]\Big) \\ &= \hat{\rho}(a)\hat{\rho}(b) = (\rho \otimes \rho)^\wedge(a, b).\end{aligned}$$

Lemma 2.4.5 *(Fernique) For every $\rho \in G(E)$*

$$\int \|x\|^2 \rho(dx) < \infty\,.$$

Proof. 1. The mapping $T := T_{\frac{1}{\sqrt{2}},\frac{1}{\sqrt{2}}}$ defined as in Property 2.4.4.6 is invertible with $T^{-1} = T$. Now, let $x, y \in E$, $u, v \in \mathbf{R}_+$ with $\|x\| \leq u$ and $\|y\| > v$. Then

$$\left\|\frac{1}{\sqrt{2}}(x \pm y)\right\| \geq \frac{1}{\sqrt{2}}(\|y\| - \|x\|) > \frac{v-u}{\sqrt{2}},$$

and by Property 2.4.4.6 we obtain

$$\begin{aligned}
&\rho([\|\cdot\| \leq u])\rho([\|\cdot\| > v]) \\
&\qquad = \rho \otimes \rho([\|\cdot\| \leq u] \times [\|\cdot\| > v]) \\
&\qquad = T(\rho \otimes \rho)([\|\cdot\| \leq u] \times [\|\cdot\| > v]) \\
&\qquad = \rho \otimes \rho(T([\|\cdot\| \leq u] \times [\|\cdot\| > v])) \\
&\qquad \leq \rho \otimes \rho\Big(\Big[\|\cdot\| > \frac{v-u}{\sqrt{2}}\Big] \times \Big[\|\cdot\| > \frac{v-u}{\sqrt{2}}\Big]\Big) \\
&\qquad = \Big(\rho\Big(\Big[\|\cdot\| > \frac{v-u}{\sqrt{2}}\Big]\Big)\Big)^2.
\end{aligned}$$

2. Now we choose $v_0 := u$ sufficiently large such that

$$\alpha := \rho([\|\cdot\| \leq v_0]) > \frac{3}{4},$$

and we obtain

$$\rho([\|\cdot\| > v_0]) < \frac{1}{4}.$$

Moreover, let

$$v_n := \left(2^{\frac{n+1}{2}} - 1\right)\left(\sqrt{2} + 1\right)v_0$$

for all $n \in \mathbf{N}$. Then

$$v_{n+1} - v_0 = \sqrt{2}\ v_n$$

for all $n \in \mathbf{Z}_+$. The estimate achieved in 1. yields

$$\alpha\rho([\|\cdot\| > v_{n+1}]) \leq (\rho([\|\cdot\| > v_n]))^2$$

for all $n \in \mathbf{Z}_+$, hence by induction

$$\frac{1}{\alpha}\rho([\|\cdot\| > v_n]) \leq \left(\frac{1}{\alpha}\rho([\|\cdot\| > v_0])\right)^{2^{n-1}} < \left(\frac{1}{3}\right)^{2^{n-1}}.$$

3. It remains to be shown that

$$\int_{[\|\cdot\|>v_1]} \|x\|^2 \rho(dx) < \infty .$$

In fact, from $v_n \leq 4v_0 2^{\frac{n}{2}}$ and from the estimate in 2. we conclude that

$$\begin{aligned}\int_{[\|\cdot\|>v_1]} \|x\|^2 \rho(dx) &= \sum_{n\geq 1} \int_{[v_n<\|\cdot\|\leq v_{n+1}]} \|x\|^2 \rho(dx) \\ &\leq \sum_{n\geq 1} v_{n+1}^2 \rho([\|\cdot\| > v_n]) \\ &\leq 16v_0^2 \sum_{n\geq 1} 2^{n+1} \left(\frac{1}{3}\right)^{2^{n-1}} < \infty.\end{aligned}$$

■

With this preparation we can approach the more profound properties of Gauss measures on a (separable) Banach space.

Theorem 2.4.6 *(Integral representation of the covariance operator)* *Let $\rho \in G(E)$ with covariance operator R. Then*

(i) $Ra = \int \langle x, a \rangle x \rho(dx)$

for all $a \in E'$, where the integral is understood in the sense of Bochner.

(ii) R is a compact operator.

Proof. (i) At first we observe that

$$\|\langle x, a \rangle x\| \leq \|a\| \| x\|^2$$

for all $x \in E, a \in E'$. Therefore Lemma 2.4.5 implies that the mapping

$$x \mapsto \langle x, a \rangle x$$

is Bochner-integrable with respect to ρ.

We add a few remarks on *Bochner integrability*: Let $(\Omega, \mathfrak{A}, \mu)$ be a measure space, and let $f : \Omega \to E$ be a mapping from Ω into a separable Banach space B. f is said to be Bochner integrable if

(a) f is strongly measurable in the sense that there exist step functions $f_n : \Omega \to E$ such that

$$f_n \longrightarrow f$$

as $n \to \infty$ strongly μ-almost everywhere, and such that

(b)
$$\int \|f_n(\omega) - f(\omega)\| \mu(d\omega) \longrightarrow 0$$

as $n \to \infty$.

In this case

$$\int f(\omega)\mu(d\omega) := \lim_{n\to\infty} \int f_n(\omega)\mu(d\omega),$$

with the limit taken in the strong sense, is called the *Bochner integral of f with respect to μ.*

Moreover one has the useful criterion that a strongly measurable function $f : \Omega \to E$ is Bochner integrable with respect to μ if and only if $\|f\|$ is μ-integrable (in the usual sense).

We proceed with the proof of the theorem. Let

$$Sa := \int \langle x, a\rangle x \rho(dx)$$

for all $a \in E'$. S defines a linear mapping $E' \to E$. Moreover, for every $b \in E'$ we have

$$\big\langle Sa, b\big\rangle = \int \langle x, a\rangle\langle x, b\rangle \rho(dx)\,.$$

This representation shows that S is symmetric. Now, for all $a \in E'$

$$a(\rho)^\wedge(t) = \exp\Big(-\frac{t^2}{2}\big\langle Ra, a\big\rangle\Big) = N\big(0, \big\langle Ra, a\big\rangle\big)^\wedge(t)$$

whenever $t \in \mathbf{R}$, hence by Property 2.4.4.4 together with the uniqueness of the Fourier transform

$$a(\rho) = N\big(0, \langle Ra, a\rangle\big)\,.$$

But

$$\begin{aligned}\langle Ra, a\rangle &= \int t^2 N(0, \langle Ra, a\rangle)(dt)\\ &= \int t^2 a(\rho)(dt) = \int \langle x, a\rangle^2 \rho(dx) = \langle Sa, a\rangle\end{aligned}$$

for all $a \in E'$. Since R and S are symmetric, we obtain as in the proof of Property 2.4.4.1 that $R = S$.

(*ii*) Let $(a_n)_{n\geq 1}$ be a bounded sequence in E'. Then there exists a $\delta > 0$ such that $\|a_n\| < \delta$ for all $n \in \mathbf{N}$.

We know from Appendix B 14 that V_δ is compact and metrizable with respect to the topology $\sigma(E', E)$. Thus there exist a subsequence $(a_{n_k})_{k\geq 1}$ of $(a_n)_{n\geq 1}$ and an $a \in V_\delta$ such that

$$\lim_{k\to\infty} \langle x, a_{n_k}\rangle = \langle x, a\rangle$$

whenever $x \in E$. Applying (*i*) we now get

$$\|Ra_{n_k} - Ra\| \leq \int |\langle x, a_{nk}\rangle - \langle x, a\rangle| \|x\| \rho(dx)\,.$$

In addition we have that

$$|\langle x, a_{n_k}\rangle - \langle x, a\rangle| \|x\| \leq 2\delta \|x\|^2$$

for all $x \in E$. By Lebesgue's dominated convergence theorem follows

$$\lim_{k\to\infty} Ra_{n_k} = Ra,$$

and this finishes the proof that R is compact. ∎

Theorem 2.4.7 *(Characterization of Gauss measures). For every measure $\rho \in M^1(E)$ the following statements are equivalent:*

(i) $\rho \in G(E)$.

(ii) There exists a mapping $q : E' \to \mathbf{R}_+$ *with*

$$q(ta) = t^2 q(a)$$

for all $t \in \mathbf{R}$, $a \in E'$ *such that*

$$\hat{\rho}(a) = e^{-q(a)}$$

whenever $a \in E'$.

(iii) $a(\rho) \in G(\mathbf{R})$ *for all* $a \in E'$.

Proof. $(i) \Rightarrow (ii)$. Let $\rho \in G(E)$ with covariance operator R. Putting

$$q(a) = \frac{1}{2}\langle Ra, a \rangle$$

for all $a \in E'$ we immediately arrive at the assertion.

$(ii) \Rightarrow (iii)$. Let $a \in E'$. Then for all $t \in \mathbf{R}$

$$a(\rho)^\wedge(t) = \hat{\rho}(ta) = e^{-q(ta)} = e^{-t^2 q(a)} = N(0, 2q(a))^\wedge(t)\,,$$

hence

$$a(\rho) = N(0, 2q(a))$$

or $a(\rho) \in G(\mathbf{R})$.

$(iii) \Rightarrow (i)$. Let $a(\rho) = N(0, h(a))$ with a function $h : E' \to \mathbf{R}$. Then for all $a \in E'$

$$h(a) = \int t^2 N(0, h(a))(dt) = \int t^2 a(\rho)(dt) = \int \langle x, a \rangle^2 \rho(dx) \qquad (1)$$

and

$$\hat{\rho}(a) = a(\rho)^\wedge(1) = N(0, h(a))^\wedge(1) = \exp\Big(-\frac{1}{2}h(a)\Big) \qquad (2)$$

Now, for all $x \in E$, $a \in E'$ we define

$$(Ta)(x) := \langle x, a \rangle,$$

By (1) $Ta \in L^2(E, \rho)$ for all $a \in E'$, T is a linear mapping $E' \to L^2(E, \rho)$, and by (1) and (2)

$$\exp\Big(-\frac{1}{2}\|Ta\|^2\Big) = \hat{\rho}(a)$$

whenever $a \in E'$.

Now, by Property 2.1.6.4 of the Fourier transform T is a continuous mapping

$$(E', \tau(E', E)) \to (L^2(E, \rho), \|\cdot\|) .$$

Appendix B 13 (Arens, Mackey) provides us with the identification

$$(E', \tau(E', E))' \longleftrightarrow E .$$

Consequently, T^t can be regarded as a mapping $L^2(E, \rho) \to E$ and hence $R := T^t \circ T$ is a linear mapping $E' \to E$ with

$$\big\langle Ra, b \big\rangle = \big\langle T^t(Ta), b \big\rangle = \big\langle Ta, Tb \big\rangle$$

for all $a, b \in E'$. This shows that R is symmetric, and setting $a = b$ one obtains that

$$\hat{\rho}(a) = \exp\Big(-\frac{1}{2}\|Ta\|^2\Big) = \exp\Big(-\frac{1}{2}\big\langle Ra, a \big\rangle\Big)$$

for all $a \in E'$, and ρ has been shown to belong to $G(E)$. ■

Definition 2.4.8 *For any given measure $\lambda \in M^b(E)$ the measure*

$$e(\lambda) := e^{-\|\lambda\|} \sum_{n \geq 0} \frac{\lambda^n}{n!} \in M^1(E)$$

with $\lambda^0 := \varepsilon_0$ is called the ***Poisson measure with exponent*** *λ.*

By $P(E)$ we shall abbreviate the totality of all Poisson measures in E.

Discussion 2.4.9 *of the genesis of Poisson measure in classical probability theory. Here we have $E := \mathbf{R}^p$. Let $\lambda \in M^b(E) \setminus \{0\}$. Moreover, let*

(a) $(X_n)_{n\geq 1}$ *denote a sequence of independent E-valued random variables on a probability space $(\Omega, \mathfrak{A}, \mathbf{P})$ such that*

$$\mathbf{P}_{X_n} := \frac{1}{\|\lambda\|}\lambda$$

for all $n \geq 1$ and let

(b) N *denote a $\mathbf{Z}_+$-valued random variable on $(\Omega, \mathfrak{A}, \mathbf{P})$, independent of $(X)_{n\geq 1}$ and such that*

$$\mathbf{P}_N = \Pi(\|\lambda\|),$$

where $\Pi(\|\lambda\|)$ denotes the elementary Poisson distribution with parameter $\|\lambda\|$.

Now, for every $\omega \in \Omega$ let

$$S_N(\omega) := \sum_{k=1}^{N(\omega)} X_k(\omega)$$

(with $\sum_{k=1}^{0} X_k(\omega) := 0$). Then S_n is an E-valued random variable on $(\Omega, \mathfrak{A}, \mathbf{P})$ with

$$\mathbf{P}_{S_N} = e(\lambda)\,.$$

Properties 2.4.10 of *Poisson measures*

2.4.10.1 *For every $a \in E'$*

$$\begin{aligned} e(\lambda)^\wedge(a) &= \exp(\hat{\lambda}(a) - \hat{\lambda}(0)) \\ &= \exp\Big(\int \Big(e^{i\langle x,a\rangle} - 1\Big)\lambda(dx)\Big), \end{aligned}$$

hence

$$\text{Log } e(\lambda)^\wedge(a) = \hat{\lambda}(a) - \hat{\lambda}(0)\,.$$

The following computation serves as a proof of the first assertion:

$$\begin{aligned}\int e^{i\langle x,a\rangle} e(\lambda)(dx) &= e^{-\|\lambda\|} \sum_{n\geq 0} \frac{1}{n!} \int e^{i\langle x,a\rangle} \lambda^n(dx)\\ &= e^{-\|\lambda\|} \sum_{n\geq 0} \frac{1}{n!} (\lambda^n)^\wedge(a)\\ &= e^{-\|\lambda\|} \sum_{n\geq 0} \frac{1}{n!} (\hat{\lambda}(a))^n\\ &= e^{-\hat{\lambda}(0)} e^{\hat{\lambda}(a)} = e^{\hat{\lambda}(a)-\hat{\lambda}(0)} \quad (a \in E').\end{aligned}$$

The second statement follows from the existence of $h := \text{Log } \hat{\mu}$ (Theorem 2.1.10) together with the continuity of the Fourier transform.

2.4.10.2 *For* $\lambda_1, \lambda_2 \in M^b(E)$ *we have*

$$e(\lambda_1) * e(\lambda_2) = e(\lambda_1 + \lambda_2)\,.$$

In particular, $P(E)$ *is a subsemigroup of* $M^1(E)$.

For the proof one just notes that

$$\begin{aligned}(e(\lambda_1) * e(\lambda_2))^\wedge(a) &= e(\lambda_1)^\wedge(a) e(\lambda_2)^\wedge(a)\\ &= \exp(\hat{\lambda}_1(a) - \hat{\lambda}_1(0)) \exp(\hat{\lambda}_2(a) - \hat{\lambda}_2(0))\end{aligned}$$

(by Property 2.4.10.1)

$$\begin{aligned}&= \exp((\lambda_1 + \lambda_2)^\wedge(a) - (\lambda_1 + \lambda_2)^\wedge(0))\\ &= (e(\lambda_1 + \lambda_2))^\wedge(a)\end{aligned}$$

(again by Property 2.4.10.1), whenever $a \in E'$.

2.4.10.3 *$P(E) \subset I(E)$.*
Moreover, for any $e(\lambda) \in P(E)$ there exists a convolution semigroup $(e(\lambda)_t)_{t \in \mathbf{R}_+}$ given by

$$e(\lambda)_t := e(t\lambda)$$

for all $t \in \mathbf{R}_+$ such that $e(\lambda)_1 = e(\lambda)$.

This follows from Properties 2.4.10.1 and 2.4.10.2 together with the embedding theorem 2.3.9.

2.4.10.4 *Let $\lambda_1, \lambda_2 \in M^b(E)$ with $\lambda_1 \leq \lambda_2$. Then*

$$e(\lambda_1) \prec e(\lambda_2)\,.$$

In fact, looking at $\lambda_2 = \lambda_1 + (\lambda_2 - \lambda_1)$ with $\lambda_2 - \lambda_1 \in M^b(E)$ Property 2.4.10.2 yields

$$e(\lambda_2) = e(\lambda_1) * e(\lambda_2 - \lambda_2)$$

and hence the assertion.

2.4.10.5 *For every $\lambda \in M^b(E)$ we have $e(\lambda)^- = e(\lambda^-)$ since for all $a \in E'$*

$$\begin{aligned}(e(\lambda)^-)^\wedge(a) &= \overline{e(\lambda)^\wedge(a)} = \overline{\exp(\hat{\lambda}(a) - \hat{\lambda}(0))} \\ &= \exp\left(\overline{\hat{\lambda}(a)} - \overline{\hat{\lambda}(0)}\right) = \exp\left(\widehat{\lambda^-}(a) - \widehat{\lambda^-}(0)\right) \\ &= (e(\lambda^-))^\wedge(a).\end{aligned}$$

2.4.10.6 *Let $\lambda \in M^b(E)$. Then $e(\lambda + \lambda^-) = |e(\lambda)|^2$, since*

$$e(\lambda + \lambda^-) = e(\lambda) * e(\lambda^-) = e(\lambda) * e(\lambda)^-$$

by Property 2.4.10.5, and furthermore

$$= |e(\lambda)|^2.$$

Theorem 2.4.11 *(Denseness of $P(E)$ in $I(E)$) $P(E)$ is a τ_w-dense subsemigroup of $I(E)$.*

In particular, for any $\mu \in I(E)$ with sequence $(\mu_{\frac{1}{n}})_{n \geq 1}$ of n-th roots $\mu_{\frac{1}{n}}$ of μ we have that

$$e(n\mu_{\frac{1}{n}}) \longrightarrow \mu$$

as $n \to \infty$ with respect to the topology τ_w.

Proof. 1. Clearly,

$$e^{-n} \sum_{k=0}^{2n} \frac{n^k}{k!} \longrightarrow 1$$

as $n \to \infty$, since

$$\begin{aligned} e^{-n} \sum_{k>2n} \frac{n^k}{k!} &= \frac{e^{-n} n^{2n}}{(2n)!} \sum_{k \geq 1} \frac{n^k}{(2n+1) \cdot \ldots \cdot (2n+k)} \\ &\leq \frac{e^{-n} n^{2n}}{(2n)!} \sum_{k \geq 1} 2^{-k} = \frac{e^{-n} n^{2n}}{(2n)!} \\ &= \frac{e^{-2n} (2n)^{2n}}{(2n)!} \left(\frac{e}{4}\right)^n \leq \left(\frac{e}{4}\right)^n \leq \left(\frac{3}{4}\right)^n \end{aligned}$$

for all $n \geq 1$.

2. The sequence $(e(n\mu_{\frac{1}{n}}))_{n \geq 1}$ is τ_w-relatively compact.

In fact, for $\varepsilon > 0$ there exists by 1. an $n_0 \in \mathbf{N}$ such that

$$e^{-n} \sum_{k=0}^{2n} \frac{n^k}{k!} \geq 1 - \varepsilon$$

for all $n \geq n_0$. Since the mapping $t \mapsto \mu_t$ from $\mathbf{R}_+$ into $M^1(E)$ is continuous, the set $\{\mu_t : t \in [0, 2]\}$ is relatively compact. But then Prohorov's theorem 1.3.7 applies, and there exists a set $K \in \mathcal{K}(E)$ with

$$\mu_t(K) \geq 1 - \varepsilon$$

for all $t \in [0, 2]$. This implies that for all $n \geq n_0$

$$e(n\mu_{\frac{1}{n}})(K) = e^{-n} \sum_{k \geq 0} \frac{n^k}{k!} \mu_{\frac{k}{n}}(K) \geq e^{-n} \sum_{k=0}^{2n} \frac{n^k}{k!} \mu_{\frac{k}{n}}(K)$$
$$\geq e^{-n} \left(\sum_{k=0}^{2n} \frac{n^k}{k!} \right) (1 - \varepsilon) \geq (1 - \varepsilon)^2.$$

3. For all $z \in \mathbf{C}$ one has

$$n\left(e^{\frac{z}{n}} - 1\right) \longrightarrow z$$

as $n \to \infty$. It follows that

$$\begin{aligned} \lim_{n \to \infty} e(n\mu_{\frac{1}{n}})^\wedge(a) &= \lim_{n \to \infty} \exp((n\mu_{\frac{1}{n}})^\wedge(a) - n) \\ &= \lim_{n \to \infty} \exp\Big(n\Big[\exp\Big(\frac{1}{n}\text{Log } \hat{\mu}(a)\Big) - 1\Big]\Big) \\ &= \exp(\text{Log } \hat{\mu}(a)) = \hat{\mu}(a) \end{aligned}$$

whenever $a \in E'$. We now apply 2. in order to use the continuity theorem 2.1.9 yielding

$$\lim_{n \to \infty} e(n\mu_{\frac{1}{n}}) = \mu\,. \qquad \blacksquare$$

Theorem 2.4.12 *(Continuity of the Poisson mapping). Let $(\lambda_n)_{n \geq 1}$ be a sequence of measures in $M^b(E)$.*

(i) If $(\lambda_n)_{n \geq 1}$ is τ_w-relatively compact, then so is $(e(\lambda_n))$.

(ii) The mapping $e : M^b(E) \to M^1(E)$ is continuous in the sense that the τ_w-convergencc $\lambda_n \to \lambda \in M^b(E)$ implies the τ_w-convergence

$$e(\lambda_n) \longrightarrow e(\lambda)$$

as $n \to \infty$.

Proof. (i) By Prohorov's theorem 1.3.7

$$d := \sup_{n\geq 1} \|\lambda_n\| < \infty .$$

For given $\varepsilon > 0$ there exists a $k_0 \in \mathbf{N}$ satisfying

$$\sum_{k>k_0} \frac{d^k}{k!} \leq \frac{\varepsilon}{2} .$$

But since $\{\lambda_n^k : 0 \leq k \leq k_0, n \in \mathbf{N}\}$ is τ_w-relatively compact (by the continuity of the convolution) Prohorov's theorem 1.3.7 provides us with the existence of a set $K \in \mathcal{K}(E)$ such that

$$\lambda_n^k(K^c) \leq \frac{\varepsilon}{2e}$$

whenever $0 \leq k \leq k_0, \;\; n \in \mathbf{N}$. Therefore

$$\begin{aligned} e(\lambda_n)(K^c) &= e^{-\|\lambda_n\|} \sum_{k\geq 0} \frac{1}{k!} \lambda_n^k(K^c) \\ &\leq \sum_{k=0}^{k_0} \frac{1}{k!} \lambda_n^k(K^c) + \sum_{k>k_0} \frac{1}{k!} \lambda_n^k(E) \\ &\leq \frac{\varepsilon}{2e} \sum_{k=0}^{k_0} \frac{1}{k!} + \sum_{k>k_0} \frac{d^k}{k!} \leq \frac{\varepsilon}{2} + \frac{\varepsilon}{2} = \varepsilon. \end{aligned}$$

Another application of Prohorov's theorem 1.3.7 yields (i).

(ii) We apply Property 2.4.10.1 in order to obtain

$$\lim_{n\to\infty} e(\lambda_n)^\wedge(a) = e(\lambda)^\wedge(a)$$

whenever $a \in E'$. But then (i) together with the continuity theorem 2.1.9 implies the assertion. ∎

Remark 2.4.13 *The converse of statement (i) of Theorem 2.4.12 does not hold in general. In fact, by Theorem 2.4.11 we have for any* $\mu \in I(E)$ *that*

$$e(n\mu_{\frac{1}{n}}) \longrightarrow \mu$$

as $n \to \infty$, *but*

$$\sup_{n \geq 1} \|n\mu_{\frac{1}{n}}\| = \sup_{n \geq 1}(n\mu_{\frac{1}{n}})(E)$$
$$= \sup_{n \geq 1} \; n = \infty.$$

Résumé 2.4.14 *A measure* $\mu \in M^1(E)$ *is called (continuously)* ***embeddable*** *if there exists a (continuous) convolution semigroup* $(\mu_t)_{t \in \mathbf{R}_+}$ *in* $M^1(E)$ *such that* $\mu_1 = \mu$.

By $EM(E)$ we denote the totality of all embeddable measures on E. Then $EM(E)$ is a subsemigroup of $M^1(E)$, and $G(E)$ as well as $P(E)$ are subsemigroups of the semigroup $I(E)$ which by the embedding Theorem 2.3.9 is contained in $EM(E)$. While $I(E)$ is τ_w-closed in $M^1(E)$ by Theorem 2.3.7, $P(E)$ is τ_w-dense in $I(E)$ by Theorem 2.4.11.

3

The Structure of Infinitely Divisible Probability Measures

3.1 The Ito-Nishio theorem

This section is devoted to studying the convergence behavior of sums of independent random variables taking their values in a separable Banach space E. The discussion will provide a complete version of Lévy's continuity theorem which extends Theorem 2.1.9.

Definition 3.1.1 *A subset Z of E is said to be a (measurable)* ***cylinder set*** *if there exist a finite set $\{a_1, ..., a_n\}$ in E' and a set $B \in \mathfrak{B}(\mathbf{R}^n)$ such that*

$$Z = Z(a_1, ..., a_n; B) := \{x \in E : (\langle x, a_1 \rangle, ..., \langle x, a_n \rangle) \in B\} .$$

Let $\mathfrak{Z}(E)$ denote the system of cylinder sets of E.

Properties 3.1.2 *of cylinder sets*

3.1.2.1 $\mathfrak{Z}(E) \subset \mathfrak{B}(E)$.

3.1.2.2 $\mathfrak{Z}(E) = \mathfrak{B}(E)$ *if and only if dim* $E < \infty$.

3.1.2.3 $\mathfrak{Z}(E)$ *is an algebra.*

In fact, we note that $E = Z(a; \mathbf{R})$ for arbitrary $a \in E'$ and hence belongs to $\mathfrak{Z}(E)$. Moreover, for $Z := Z(a_1, ..., a_n; B) \in \mathfrak{Z}(E)$ also $Z^c = Z(a_1, ..., a_n; B^c) \in \mathfrak{Z}(E)$ and with $Z' := Z(a'_1, ..., a'_m; B')$ for $\{a'_1, ..., a'_m\}$, $B' \in \mathfrak{B}(\mathbf{R})^m$ we have that

$$Z \cap Z' = Z(a_1, ..., a_n, a'_1, ..., a'_m; B \times B') \in \mathfrak{Z}(E).$$

3.1.2.4 $\mathfrak{Z}(E)$ *is translation invariant.*

This property follows directly from

$$Z + y = Z(a_1, ..., a_n; B + (\langle y, a_1 \rangle, ..., \langle y, a_n \rangle))$$

for $Z := Z(a_1, ..., a_n; B)$ with $\{a_1, ..., a_n\} \subset E'$, $B \in \mathfrak{B}(\mathbf{R})^n$ and $y \in E$.

3.1.2.5 $\sigma(\mathfrak{Z}(E)) = \mathfrak{B}(E)$.

By Property 3.1.2.1 it suffices to show that $\mathcal{O}(E) \subset \sigma(\mathfrak{Z}(E))$. Moreover, applying the assumption that E is separable the proof is reduced to showing that $B(x, r)^- \in \sigma(\mathfrak{Z}(E))$ for all $x \in E$, $r \in \mathbf{R}_+^\times$. Finally, by Property 3.1.2.4 it remains to verify that $B(0, r)^- \in \sigma(\mathfrak{Z}(E))$ for every $r \in \mathbf{R}_+^\times$.

Let $(x_n)_{n \geq 1}$ be a sequence dense in E. From Appendix B 5 (Banach, Hahn) we obtain a sequence $(a_n)_{n \geq 1}$ in E' with $\|a_n\| = 1$ such that $\langle x_n, a_n \rangle = \|x_n\|$ for all $n \geq 1$. Now, let

$$B_r := \bigcap_{n \geq 1} \{x \in E : \langle x, a_n \rangle \leq r\}.$$

Clearly, $B_r \in \sigma(\mathfrak{Z}(E))$ and hence the proof is finished once $B(0, r)^- = B_r$ has been established. At first we note that $B(0, r) \subset B_r$. For the reversed inclusion we pick $x \in B_r$ to which there exists a subsequence $(x_{n_k})_{k\geq 1}$ of $(x_n)_{n\geq 1}$ such that $\lim_{k\to\infty} x_{n_k} = x$. But the inequalities

$$\|x_{n_k}\| = \langle x_{n_k}, a_{n_k}\rangle = \langle x_{n_k} - x, a_{n_k}\rangle + \langle x, a_{n_k}\rangle \leq \|x_{n_k} - x\| + r$$

valid for all $k \geq 1$ imply that $\|x\| \leq r$, hence that $x \in B(0,r)^-$.

Discussion 3.1.3 *of types of convergence for E-valued random variables.*

Let $(Y_n)_{n\geq 1}$ be a sequence of E-valued random variables on a probability space $(\Omega, \mathfrak{A}, \mathbf{P})$.

3.1.3.1 *$(Y_n)_{n\geq 1}$ converges $\mathbf{P}$-a.s. (in symbols $Y_n \overset{\mathbf{P}-a.s.}{\longrightarrow}$) if and only if for all $\varepsilon > 0$*

$$\lim_{m\to\infty} \mathbf{P}([\sup_{n\geq m} \|Y_n - Y_m\| > \varepsilon]) = 0\,.$$

3.1.3.2 *$(Y_n)_{n\geq 1}$ converges $\mathbf{P}$-stochastically (in symbols $Y_n \overset{\mathbf{P}-stoch}{\longrightarrow}$) if and only if for all $\varepsilon > 0$*

$$\lim_{m\to\infty} \sup_{n\geq m} \mathbf{P}([\|Y_n - Y_m\| > \varepsilon]) = 0\,.$$

3.1.3.3 *If Y denotes the $\mathbf{P}$-a.s. or $\mathbf{P}$-stoch limit of $(Y_n)_{n\geq 1}$ then the limiting relations hold for Y instead of Y_m in 3.1.3.1 and 3.1.3.2 respectively.*

We note that **P**-a.s. convergence implies **P**-stoch convergence, and **P**-stoch convergence implies convergence of $(Y_n)_{n\geq 1}$ in distribution which is defined as weak convergence of the sequence $(\mathbf{P}_{Y_n})_{n\geq 1}$ of distributions $\mathbf{P}_{Y_n}$ of Y_n as measures in $M^1(E)$ (in symbols $Y_n \overset{d}{\longrightarrow}$).

Theorem 3.1.4 *(Ottaviani's inequality). Let $(Y_k)_{1\leq k\leq n}$ be a sequence of independent E-valued random variables on $(\Omega, \mathfrak{A}; \mathbf{P})$. Then for all $\varepsilon > 0$ we have the inequality*

$$\begin{aligned}1- &\max_{1\leq k<n} \mathbf{P}([\|Y_{k+1} + \ldots + Y_n\| \geq \varepsilon])\mathbf{P}([\max_{1\leq k<n} \|Y_1 + \ldots + Y_k\| > 2\varepsilon]) \\ &\leq \mathbf{P}([\|Y_1 + \ldots + Y_n\| > \varepsilon]).\end{aligned}$$

Proof. For every $k = 1, ..., n-1$ let

$$B_k := [\|Y_1\| < 2\varepsilon, ..., \|Y_1 + ... + Y_{k-1}\| \le 2\varepsilon, \|Y_1 + ... + Y_k\| > 2\varepsilon],$$

$$C_k := [\|Y_{k+1} + ... + Y_n\| < \varepsilon], \text{ and}$$

$$D \ := [\|Y_1 + ... + Y_n\| > \varepsilon].$$

Then $B_1, ..., B_{n-1}$ are pairwise disjoint, and B_k, C_k are independent for $1 \le k \le n-1$. Since

$$\|Y_1 + ... + Y_n\| \ge \|Y_1 + ... + Y_k\| - \|Y_{k+1} + ... + Y_n\|,$$

we obtain

$$\bigcup_{k=1}^{n-1} (B_k \cap C_k) \subset D,$$

and hence

$$\begin{aligned}
\mathbf{P}(D) &\ge \sum_{k=1}^{n-1} \mathbf{P}(B_k \cap C_k) = \sum_{k=1}^{n} \mathbf{P}(B_k)\mathbf{P}(C_k) \\
&\ge (\min_{1 \le k < n} \mathbf{P}(C_k)) \left(\sum_{k=1}^{n-1} \mathbf{P}(B_k) \right) \\
&= \min_{1 \le k < n} (1 - \mathbf{P}(C_k^c))\mathbf{P}\left(\bigcup_{k=1}^{n-1} B_k \right) \\
&= (1 - \max_{1 \le k < n} \mathbf{P}(C_k^c)) \cdot \mathbf{P}([\max_{1 \le k < n} \|Y_1 + ... + Y_k\| > 2\varepsilon]).
\end{aligned}$$

■

Corollary 3.1.5 *For each $\ell \ge 1$ let $S_\ell := \sum_{k=1}^{\ell} Y_k$. Then*

$$\begin{aligned}
1 - \sup_{n \ge 1} \big(\max_{1 \le k < n} \mathbf{P}([\|S_n - S_k\| \ge \varepsilon]) \big) \mathbf{P}([\sup_{k \ge 1} \|S_k\| > 2\varepsilon]) \\
\le \ \sup_{n \ge 1} \mathbf{P}([\|S_n\| > \varepsilon])
\end{aligned}$$

for all $\varepsilon > 0$.

Proof. Applying the theorem to each of the sequences $(Y_k)_{1\leq k\leq n}$ $(n \geq 1)$ we obtain for all $\varepsilon > 0$

$$1 - \sup_{n\geq 1}(\max_{1\leq k<n} \mathbf{P}([\|S_n - S_k\| \geq \varepsilon]))\mathbf{P}([\max_{1\leq k<m} \|S_k\| > 2\varepsilon])$$
$$\leq \sup_{n\geq 1} \mathbf{P}([\|S_n\| > \varepsilon])$$

whenever $m \geq 1$. But we have

$$[\max_{1\leq k<m} \|S_k\| > 2\varepsilon] \uparrow [\sup_{k\geq 1} \|S_k\| > 2\varepsilon],$$

hence by the continuity from below of the measure $\mathbf{P}$ the assertion follows. ■

Theorem 3.1.6 *(Equivalence of types of convergence). Let $(X_k)_{k\geq 1}$ be a sequence of independent E-valued random variables on $(\Omega, \mathfrak{A}, \mathbf{P})$ with distributions $\lambda_k = \mathbf{P}_{X_k} (k \geq 1)$, and let $(S_n)_{n\geq 1}$ be the corresponding sequence of n-the partial sums $S_n = \sum_{k=1}^{n} X_k (n \geq 1)$. For every $n \geq 1$ we abbreviate*

$$\mu_n := \mathbf{P}_{S_n} = \lambda_1 * \ldots * \lambda_n \, .$$

Then the following statements are equivalent:

(i) $(S_n)_{n\geq 1}$ *converges* $\mathbf{P}$*-a.s..*

(ii) $(S_n)_{n\geq 1}$ *converges* $\mathbf{P}$*-stoch.*

(iii) $(S_n)_{n\geq 1}$ *converges in distribution.*

Proof. $(i) \Rightarrow (ii) \Rightarrow (iii)$ are well-known implications valid for arbitrary sequences of E-valued random variables on $(\Omega, \mathfrak{A}, \mathbf{P})$. (For the second implication see Application 1.2.15.)
$(ii) \Rightarrow (i)$ Let $\varepsilon > 0$. For $\delta \in]0, \frac{1}{2}]$ there exists an $m_0 \geq 1$ such that

$$\mathbf{P}([\|X_{m+1} + \ldots + X_n\| \geq \varepsilon]) = \mathbf{P}([\|S_n - S_m\| \geq \varepsilon]) < \delta$$

whenever $m_0 \leq m < n$. But Corollary 3.1.5 applied to the sequence of E-valued random variables $Y_n := X_{m+n}$ $(n \geq 1)$ yields

$$\begin{aligned}
&\frac{1}{2}\mathbf{P}([\sup_{k\geq 1} \|S_{m+k} - S_m\| > 2\varepsilon]) \\
&\quad \leq (1-\delta)\mathbf{P}([\sup_{k\geq 1} \|S_{m+k} - S_m\| > 2\varepsilon]) \\
&\quad \leq 1 - \delta\mathbf{P}([\sup_{k\geq 1} \|S_{m+k} - S_m\| > 2\varepsilon]) \\
&\quad \leq 1 - \sup_{n\geq 1}(\max_{1\leq k<n} \mathbf{P}([\|S_{m+n} - S_{m+k}\| > \varepsilon])) \\
&\qquad \cdot \mathbf{P}([\sup_{k\geq 1} \|S_{m+k} - S_m\| > 2\varepsilon]) \\
&\quad \leq \sup_{n\geq 1} \mathbf{P}([\|S_{m+n} - S_m\| > \varepsilon]) \leq \delta,
\end{aligned}$$

hence

$$\frac{1}{2}\mathbf{P}([\sup_{n\geq m} \|S_n - S_m\| > 2\varepsilon]) \leq \delta$$

for all $m \geq m_0$, and this implies (i).
$(iii) \Rightarrow (ii)$. For $1 \leq m < n$ we have

$$\mu_{mn} := \lambda_{m+1} * \ldots * \lambda_n = \mathbf{P}_{S_n - S_m} = \mathbf{P}_{X_{m+1}+\ldots+X_n}$$

and consequently

$$\mathbf{P}([\|S_n - S_m\| \geq \varepsilon]) = \mu_{mn}(B(0,\varepsilon)^c)\,.$$

Now we assume that $S_n \overset{\mathbf{P}-stoch}{\longrightarrow}$ is not fulfilled. Then there exist subsequences $(m_k)_{k\geq 1}$, $(n_k)_{k\geq 1}$ in $\mathbf{N}$ with $m_k < n_k$ and $\mu_{m_k,n_k}(B(0,\varepsilon)^c) \geq \varepsilon$ for all $k \geq 1$. But since $\mu_{m_k} * \mu_{m_k,n_k} = \mu_{n_k}$ for all $k \geq 1$ and $\tau_w - \lim_{n\to\infty} \mu_n = \mu \in M^1(E)$ by assumption, the sequence $(\mu_{m_k,n_k})_{k\geq 1}$ is τ_w-relatively compact. Without loss of generality we may therefore assume that $\tau_w - \lim_{k\to\infty} \mu_{m_k,n_k} = \nu \in M^1(E)$. But then $\mu * \nu = \mu$ by (iii) of Theorem 1.4.9 and hence $\nu = \varepsilon_0$ by Lemma 2.2.2. Finally, the sequence of inequalities

$$0 = \varepsilon_0(B(0,\varepsilon)^c) \geq \limsup_{k\to\infty} \mu_{m_k,n_k}(B(0,\varepsilon)^c) \geq \varepsilon > 0$$

yields the desired contradiction, and (*ii*) of the theorem has been established. ■

Corollary 3.1.7 *Let the sequence* $(\mu_n)_{n\geq 1}$ *of distributions* $\mathbf{P}_{S_n}$ *be shift tight in* $M^1(E)$. *Then there exists a sequence* $(x_n)_{n\geq 1}$ *in* E *such that*

$$S_n - x_n \mathbf{1}_\Omega \xrightarrow{\mathbf{P}-a.s.}$$

Proof. By Theorem 2.2.25 the sequence $(\nu_n)_{n\geq 1}$ with $\nu_n := |\mu_n|^2$ for all $n \geq 1$ is τ_w-relatively compact. Since $|\hat{\nu}_n| = |\hat{\mu}_n|^2$ by Property 2.2.24.3 and $\nu_n \prec \nu_{n+1}$, Property 2.2.16.7 yields $0 \leq \hat{\nu}_{n+1} \leq \hat{\nu}_n$ for all $n \geq 1$. By the continuity theorem 2.1.9 this implies that $(\nu_n)_{n\geq 1}$ is τ_w-convergent.

In the following we aim at representing the sequence $(\nu_n)_{n\geq 1}$ as a sequence of distributions of E-valued random variables on an appropriate probability space. Let

$$Y_n(\omega_1, \omega_2) := X_n(\omega_1)$$

and

$$Z_n(\omega_1, \omega_2) := X_n(\omega_2)$$

whenever $(\omega_1, \omega_2) \in \Omega \times \Omega$ $(n \geq 1)$. Clearly, $\{Y_1, Z_1, Y_2, Z_2, ...\}$ is a sequence of independent E-valued random variables on the probability space $(\Omega \times \Omega, \mathfrak{A} \otimes \mathfrak{A}, \mathbf{P} \otimes \mathbf{P})$ such that

$$(\mathbf{P} \otimes \mathbf{P})_{Y_n} = (\mathbf{P} \otimes \mathbf{P})_{Z_n} = \mathbf{P}_{X_n} = \lambda_n$$

for all $n \geq 1$. As a consequence we obtain that the E-valued random variable

$$\sum_{k=1}^{n}(Y_k - Z_k) = \left(\sum_{k=1}^{n} Y_k\right) - \left(\sum_{k=1}^{n} Z_k\right)$$

has ν_n as its distribution $(n \geq 1)$ (See Application 1.4.5 and Property 1.4.4.4). It follows that

$$\sum_{k=1}^{n}(Y_k - Z_k) \xrightarrow{d} \quad \text{as } n \to \infty .$$

Moreover, since $(Y_n - Z_n)_{n\geq 1}$ is a sequence of independent E-valued random variables, the theorem applies and

$$\sum_{k=1}^{n}(Y_k - Z_k) \overset{\mathbf{P}\otimes\mathbf{P}-a.s.}{\longrightarrow} W$$

where W denotes an E-valued random variable on $(\Omega\times\Omega, \mathfrak{A}\otimes\mathfrak{A}, \mathbf{P}\otimes\mathbf{P})$. But then there exists $Q \in \mathfrak{A}\otimes\mathfrak{A}$ with $(\mathbf{P}\otimes\mathbf{P})(Q) = 1$ such that

$$\lim_{n\to\infty}(S_n(\omega_1) - S_n(\omega_2)) = W(\omega_1, \omega_2)$$

for all $(\omega_1, \omega_2) \in Q$. An application of Fubini's theorem yields

$$1 = (\mathbf{P}\otimes\mathbf{P})(Q) = \int \mathbf{P}(Q_{\omega_2})\mathbf{P}(d\omega_2)$$

and hence provides us with an $\omega_2 \in \Omega$ such that $\mathbf{P}(Q_{\omega_2}) = 1$. Choosing $x_n := S_n(\omega_2)$ for all $n \geq 1$ we obtain the sequence $(x_n)_{n\geq 1}$ in E required in the assertion. ■

The following two Lemmata are designed to prepare the main result of the section.

Lemma 3.1.8 *Let $X, Y, Y_1, Y_2, ..$ be E-valued random variables on $(\Omega, \mathfrak{A}, \mathbf{P})$. For every $n \geq 1$ the random variables X and Y_n are assumed to be independent. Moreover, by hypothesis*

$$\langle Y_n, a\rangle \overset{\mathbf{P}-stoch}{\longrightarrow} \langle Y, a\rangle$$

for every $a \in E'$. Then X and Y are independent.

Proof. Let $a_1, ..., a_k \in E'$. For each $n \geq 1$ we introduce the $\mathbf{R}^k$-valued random variable

$$\varphi_n := (\langle Y_n, a_1\rangle, ..., \langle Y_n, a_k\rangle)\,.$$

Then $\varphi_n \overset{\mathbf{P}-stoch}{\longrightarrow} \varphi := (\langle Y, a_1\rangle, ..., \langle Y, a_k\rangle)$, since

$$\mathbf{P}([\|\varphi_n - \varphi\|_\infty > \varepsilon]) \leq \sum_{j=1}^{k}\mathbf{P}([|\langle Y_n, a_j\rangle - \langle Y, a_j\rangle| > \varepsilon]),$$

where by hypothesis the right hand side tends to 0 for all $\varepsilon > 0$.

Next we note that by assumption the random variables

$$\psi := (\langle X, a_1\rangle, ..., \langle X, a_k\rangle)$$

and φ_n are independent, and

$$\varphi_n \otimes \psi \overset{\mathbf{P}-stoch}{\longrightarrow} \varphi \otimes \psi\,.$$

The equivalence theorem 3.1.6 together with Theorem 1.4.8 yields

$$\mathbf{P}_{\varphi\otimes\psi} = \tau_w - \lim_{n\to\infty} \mathbf{P}_{\varphi_n\otimes\psi} = \tau_w - \lim_{n\to\infty} (\mathbf{P}_{\varphi_n} \otimes \mathbf{P}_{\psi}) = \mathbf{P}_{\varphi} \otimes \mathbf{P}_{\psi},$$

thus φ and ψ are independent.

Now, let $Z_1, Z_2 \in \mathfrak{Z}(E)$. There exist $a_1, ..., a_k \in E'$ and $B_1, B_2 \in \mathfrak{B}(\mathbf{R}^k)$ such that

$$Z_j = Z_j(a_1, ..., a_k; B_j)$$

for $j = 1, 2$. From the above chain of equalities we deduce

$$\begin{aligned} \mathbf{P}([Y \in Z_1, X \in Z_2]) &= \mathbf{P}([\varphi \in B_1, \psi \in B_2]) \\ &= \mathbf{P}([\varphi \in B_1])\mathbf{P}([\psi \in B_2]) \\ &= \mathbf{P}([Y \in Z_1,])\mathbf{P}([X \in Z_2]). \end{aligned}$$

Since $\mathfrak{Z}(E)$ is a $\cap$-stable generator of $\mathfrak{B}(E)$ by Properties 3.1.2.3 and 3.1.2.5 we have shown that X and Y are independent. ■

Lemma 3.1.9 *Let* $(\Omega, \mathfrak{A}, \mathbf{P}, (Y_a)_{a\in E'})$ *and* $(E, \mathfrak{B}(E), \mu, (\langle \cdot, a\rangle)_{a\in E'})$ *denote two equivalent stochastic processes (with parameter set* E' *and state space* $\mathbf{R}$*). Then there exists an* E*-valued random variable* Y *on* $(\Omega, \mathfrak{A}, \mathbf{P})$ *such that*

$$\mathbf{P}([Y_a = \langle Y, a\rangle]) = 1$$

whenever $a \in E'$.

In particular, $\mathbf{P}_Y = \mu$.

Proof. With $\mathbf{R}^{\mathbf{N}}$ carrying the product topology we have that $\mathfrak{B}(\mathbf{R}^{\mathbf{N}}) = \mathfrak{B}(\mathbf{R})^{\otimes \mathbf{N}}$. Let $(a_n)_{n\geq 1}$ be a sequence in E' with

$$\bigcap_{n\geq 1} \{x \in E : \langle x, a_n \rangle \leq r\} = B(0,r)^-$$

for every $r \in \mathbf{R}_+^\times$. This follows from Appendix B 5 (Banach, Hahn) which indeed yields a sequence $(a_n)_{n\geq 1}$ in E' satisfying $\|a_n\| = 1$ and $\langle x, a_n \rangle = \|x\|$ for all $n \geq 1$.

By

$$\varphi(x) := (\langle x, a_n \rangle)_{n\geq 1}$$

for all $x \in E$ we introduce an injective continuous linear mapping from E into $\mathbf{R}^{\mathbf{N}}$. Now, $\mu \in M^1(E)$ is tight by Theorem 1.1.6. Hence there exists an increasing sequence $(K_n)_{n\geq 1}$ in $\mathcal{K}(E)$ such that for $C := \bigcup_{n\geq 1} K_n$ we obtain that $\mu(C) = 1$. Clearly, $C \in \mathfrak{B}(E)$ and $\varphi(C) = \bigcup_{n\geq 1} \varphi(K_n) \in \mathfrak{B}(\mathbf{R}^{\mathbf{N}})$. But then $\mathrm{Res}_{K_n}\varphi$ is a homeomorphism from K_n onto $\varphi(K_n)$ and hence $\mathrm{Res}_C\varphi$ turns out to be a bimeasurable bijection from $(C, C \cap \mathfrak{B}(E))$ onto $(\varphi(C), \varphi(C) \cap \mathfrak{B}(\mathbf{R}^{\mathbf{N}}))$. Consequently, there is a measurable mapping ψ from $(\mathbf{R}^{\mathbf{N}}, \mathfrak{B}(\mathbf{R}^{\mathbf{N}}))$ into $(E, \mathfrak{B}(E))$ given by $\psi(\varphi(x)) = x$ for all $x \in C$. Now we put

$$\Lambda(\omega) := (Y_{a_k}(\omega))_{n\geq 1}$$

for all $\omega \in \Omega$. Λ is a measurable map from $(\Omega, \mathfrak{A})$ into $(\mathbf{R}^{\mathbf{N}}, \mathfrak{B}(\mathbf{R}^{\mathbf{N}}))$. For every finite sequence $\{B_1, ..., B_k\}$ in $\mathfrak{B}(\mathbf{R})$ we have by assumption that

$$\begin{aligned}\mathbf{P}([\Lambda \in B_1 \times ... \times B_k {\times} \mathbf{R} \times \mathbf{R} \times ...] \\ &= \mathbf{P}([Y_{a_j} \in B_j \ \text{ for } \ j = 1, ..., k]) \\ &= \mu([\langle ., a_j \rangle \in B_j \ \text{ for } \ j = 1, ..., k]) \\ &= \mu([\varphi \in B_1 \times ... \times B_k \times \mathbf{R} \times \mathbf{R} \times ...]).\end{aligned}$$

But the system of sets of the form $B_1 \times ... \times B_k \times \mathbf{R} \times \mathbf{R} \times ...$ is a $\cap$-stable generator of $\mathfrak{B}(\mathbf{R}^{\mathbf{N}}) = (\mathfrak{B}(\mathbf{R}))^{\otimes \mathbf{N}}$, hence $\mathbf{P}_\Lambda = \mu_\varphi$.

Analogously one shows that

$$\mathbf{P}_{Y_a \otimes \Lambda} = \mu_{\langle ., a \rangle \otimes \varphi}$$

for all $a \in E'$.

Now, for $a \in E'$ we form the set

$$B_a := \{(t_n)_{n\geq 0} \in \mathbf{R} \times \mathbf{R}^{\mathbf{N}} : t_0 = \langle \psi((t_n)_{n\geq 1}), a\rangle\} \in \mathfrak{B}(\mathbf{R} \times \mathbf{R}^{\mathbf{N}})$$

and introduce the E-valued random variable $Y := \psi \circ \Lambda$. Then

$$\begin{aligned}\mathbf{P}([Y_a = \langle Y, a\rangle]) &= \mathbf{P}([Y_a = \langle \psi \circ \Lambda, a\rangle]) \\ &= \mathbf{P}([Y_a \otimes \Lambda \in B_a]) \\ &= \mu([\langle ., a\rangle \otimes \varphi \in B_a]) \\ &= \mu([\langle ., a\rangle = \langle \psi \circ \varphi, a\rangle]) \\ &= \mu([\langle ., a\rangle = \langle \psi \circ \varphi, a\rangle] \cap C) \\ &= \mu(C) = 1,\end{aligned}$$

and the first statement of the Lemma has been established.

For the remaining statement we take

$$Z := Z(a_1, ..., a_n; B) \in \mathfrak{Z}(E)$$

(for $a_1, ..., a_n \in E'$, $B \in \mathfrak{B}(\mathbf{R}^n)$). Then

$$\begin{aligned}\mathbf{P}_Y(Z) &= \mathbf{P}([(\langle Y, a_1\rangle, ..., \langle Y, a_n\rangle) \in B]) \\ &= \mathbf{P}([Y_{a_1} \otimes ... \otimes Y_{a_n} \in B]) \\ &= \mu(Z).\end{aligned}$$

Applying Properties 3.1.2.3 and 3.1.2.5 this implies that $\mathbf{P}_Y = \mu$. ■

Theorem 3.1.10 *(Ito, Nishio) Let $(X_k)_{k\geq 1}$ be a sequence of independent E-valued random variables X_k on $(\Omega, \mathfrak{A}, \mathbf{P})$ with the property that $\mathbf{P}_{X_k}$ is symmetric for all $k \geq 1$, and let $(S_n)_{n\geq 1}$ be the corresponding sequence of n-th partial sums $\sum_{k=1}^{n} X_k$.*

The following statements are equivalent:

(i) $(S_n)_{n\geq 1}$ converges $\mathbf{P}$-almost surely.

(ii) $(S_n)_{n\geq 1}$ *converges* $\mathbf{P}$*-stochastically.*

(iii) $(S_n)_{n\geq 1}$ *converges in distribution.*

(iv) $(\mathbf{P}_{S_n})_{n\geq 1}$ *is uniformly tight.*

(v) There exists an E*-valued random variable* S *on* $(\Omega, \mathfrak{A}, \mathbf{P})$ *such that*

$$\langle S_n, a\rangle \overset{\mathbf{P}-stoch}{\longrightarrow} \langle S, a\rangle$$

for all $a \in E'$.

(vi) There exists a measure $\mu \in M^1(E)$ *such that*

$$\hat{\mathbf{P}}_{S_n}(a) \longrightarrow \hat{\mu}(a)$$

for all $a \in E'$.

Proof. From the equivalence theorem 3.1.6 we deduce the equivalences
$(i) \Leftrightarrow (ii) \Leftrightarrow (iii)$
and from the continuity theorem 2.1.9 the implication
$(iii) \Rightarrow (iv)$.

We proceed to the proof of implication
$(iv) \Rightarrow (i)$. By assumption the sequence $(\mathbf{P}_{S_n})_{n\geq 1}$ is uniformly tight, consequently Corollary 3.1.7 becomes applicable and hence there exists a sequence $(x_n)_{n\geq 1}$ in E such that

$$S_n - x_n \mathbf{1}_\Omega \overset{\mathbf{P}-a.s.}{\longrightarrow}$$

as $n \to \infty$. Since the random variables X_k are independent and their distributions $\mathbf{P}_{X_n}$ are symmetric for all $k \geq 1$, we obtain

$$\mathbf{P}_{\bigotimes_{n\geq 1} X_n} = \bigotimes_{n\geq 1} \mathbf{P}_{X_n} = \bigotimes_{n\geq 1} \mathbf{P}^-_{X_n} = \bigotimes_{n\geq 1} \mathbf{P}_{-X_n} = \mathbf{P}_{\bigotimes_{n\geq 1}(-X_n)},$$

and therefore for every $\varepsilon > 0$ and all $k \geq 1$

$$\begin{aligned}
&\mathbf{P}([\sup_{m\geq k} \|(S_m - x_m) - (S_k - x_k)\| > \varepsilon]) \\
&\quad = \mathbf{P}([\sup_{m\geq k} \|X_{k+1} + ... + X_m - (x_m - x_k)\| > \varepsilon]) \\
&\quad = \mathbf{P}_{\bigotimes_{n\geq 1} X_n}(\{(y_n)_{n\geq 1} \in E^{\mathbf{N}} : \\
&\qquad \sup_{m\geq k} \|y_{k+1} + ... + y_m - (x_m - x_k)\| > \varepsilon\}) \\
&\quad = \mathbf{P}_{\bigotimes_{n\geq 1}(-X_n)}(\{(y_n)_{n\geq 1} \in E^{\mathbf{N}} : \\
&\qquad \sup_{m\geq k} \|y_{k+1} + ... + y_m - (x_m - x_k)\| > \varepsilon\}) \\
&\quad = \mathbf{P}([\sup_{m\geq k} \| - X_{k+1} - ... - X_m - (x_m - x_k)\| > \varepsilon]) \\
&\quad = \mathbf{P}([\sup_{m\geq k} \|(-S_m - x_m) - (-S_k - x_k)\| > \varepsilon]).
\end{aligned}$$

But this implies the limiting relation

$$-S_n - x_n \mathbf{1}_\Omega \overset{\mathbf{P}-a.s.}{\longrightarrow}$$

as $n \to \infty$. Observing that for every $n \geq 1$

$$S_n = \frac{1}{2}(S_n - x_n \mathbf{1}_\Omega) - \frac{1}{2}(-S_n - x_n \mathbf{1}_\Omega)$$

we obtain the desired statement (i).

$(v) \Rightarrow (iv)$. Let $\mu := \mathbf{P}_S$. By the inner regularity of μ (see Theorem 1.1.2(ii)) for every $\varepsilon > 0$ there exists a $K \in \mathcal{K}(E)$ such that $\mu(K^c) < \varepsilon$. The set $K_1 := \{\frac{1}{2}(x - y) : x, y \in K\}$ is compact. Now, given $k \geq 1$ we note that for every $n \geq 1$ the E-valued random variables $X := S_k$ and $Y_n := S_{k+n} - S_k$ are independent, and by hypothesis

$$\langle Y_n, a\rangle = \langle S_{k+n}, a\rangle - \langle S_k, a\rangle \overset{\mathbf{P}-stoch}{\longrightarrow} \langle S, a\rangle - \langle S_k, a\rangle = \langle S - S_k, a\rangle$$

whenever $a \in E'$. Applying Lemma 3.1.8 this implies that the random variables X and $Y := S - S_k$ are independent. But with $X + Y = S$ and $\mu_k := \mathbf{P}_{S_k} = \mathop{\ast}_{j=1}^{k} \mathbf{P}_{X_k}$ we obtain

$$\begin{aligned}
1 - \varepsilon < \mu(K) &= \mu_k * \mathbf{P}_{S-S_k}(K) \\
&= \int \mu_k(K - x)\mathbf{P}_{S-S_k}(dx),
\end{aligned}$$

hence there exists an $x_k \in E$ such that $\mu_k(K - x_k) > 1 - \varepsilon$. This, however, implies that

$$\begin{aligned}\mu_k(K_1^c) &= \mathbf{P}([S_k \notin K_1]) \\ &\leq \mathbf{P}([S_k + x_k \notin K] \cup [-S_k + x_k \notin K]) \\ &\leq \mathbf{P}([S_k \notin K - x_k]) + \mathbf{P}([-S_k \notin K - x_k]) \\ &= 2\mu_k((K - x_k)^c) < 2\varepsilon.\end{aligned}$$

Since the inequalities are valid for each $k \geq 1, (iv)$ has been established.

$(vi) \Rightarrow (v)$ For each $a \in E'$ $((\langle X_n, a\rangle)_{n\geq 1}$ is a sequence of independent real-valued random variables on $(\Omega, \mathfrak{A}, \mathbf{P})$, and for all $n \geq 1$

$$\langle X_1, a\rangle + ... + \langle X_n, a\rangle = \langle S_n, a\rangle .$$

But then

$$\hat{\mathbf{P}}_{\langle S_n, a\rangle}(t) = \widehat{a(\mathbf{P}_{S_n})}(t) = \hat{\mathbf{P}}_{S_n}(a^t(t)) = \hat{\mathbf{P}}_{S_n}(ta)$$

and

$$\widehat{a(\mu)}(t) = \hat{\mu}(a^t(t)) = \hat{\mu}(ta)$$

whenever $a \in E',\ \ t \in \mathbf{R}$. Now, by hypothesis

$$\lim_{n\to\infty} \hat{\mathbf{P}}_{\langle S_n, a\rangle}(t) = \widehat{a(\mu)}(t)$$

for all $t \in \mathbf{R}$, and by the classical continuity theorem which follows from Theorem 4.3.8 we have

$$\mathbf{P}_{\langle S_n, a\rangle} \xrightarrow{\tau_w} a(\mu)$$

or

$$\langle S_n, a\rangle \xrightarrow{d} \quad .$$

The equivalence theorem 3.1.6 now yields

$$\langle S_n, a\rangle \xrightarrow{\mathbf{P}-a.s.} Y_a,$$

where Y_a is a real-valued random variable on $(\Omega, \mathfrak{A}, \mathbf{P})$.

On the other hand the two stochastic processes

$$(\Omega, \mathfrak{A}, \mathbf{P}, (Y_a)_{a\in E'}) \text{ and } (E, \mathfrak{B}(E), \mu, (\langle ., a\rangle)_{a\in E'})$$

are equivalent.

In fact, let $a_1, ..., a_k \in E'$. Then for all $(t_1, ..., t_k) \in \mathbf{R}^k$ assumption (vi) together with an application of Lebesgue's dominated convergence theorem implies that

$$\hat{\mathbf{P}}_{(Y_{a_1},...,Y_{a_k})}(t_1, ..., t_k) = \int \exp\left(i\sum_{j=1}^{k} t_j Y_{a_j}\right) d\mathbf{P}$$

$$= \lim_{n\to\infty} \int \exp\left(i\sum_{j=1}^{k} t_j \langle S_n, a_j\rangle\right) d\mathbf{P}$$

$$= \lim_{n\to\infty} \int \exp\left(i\langle S_n, \sum_{j=1}^{k} t_j a_j\rangle\right) d\mathbf{P}$$

$$= \lim_{n\to\infty} \hat{\mathbf{P}}_{S_n}\left(\sum_{j=1}^{k} t_j a_j\right) = \int \exp\left(i\sum_{j=1}^{k} t_j \langle ., a_j\rangle\right) d\mu$$

$$= \hat{\mu}_{(\langle .,a_1\rangle,...,\langle .,a_k\rangle)}(t_1, ..., t_k)$$

hence that

$$\mathbf{P}_{(Y_{a_1},...,Y_{a_k})} = \mu_{(\langle .,a_1\rangle,...,\langle .,a_k\rangle)} \, .$$

Finally, we apply Lemma 3.1.9 and obtain an E-valued random variable $S := Y$ on $(\Omega, \mathfrak{A}, \mathbf{P})$ such that

$$\mathbf{P}([Y_a = \langle S, a\rangle]) = 1$$

for all $a \in E'$. Consequently,

$$\langle S_n, a\rangle \overset{\mathbf{P}-a.s.}{\longrightarrow} \langle S, a\rangle,$$

thus

$$\langle S_n, a\rangle \overset{\mathbf{P}-stoch}{\longrightarrow} \langle S, a\rangle$$

whenever $a \in E'$. But this is (v). ■

Corollary 3.1.11 *Let $(\mu_n)_{n\geq 1}$ be a sequence of symmetric measures in $M^1(E)$ such that $\mu_n \prec \mu_{n+1}$ for all $n \geq 1$, and let $\mu \in M^1(E)$ such that*

$$\lim_{n\to\infty} \hat{\mu}_n(a) = \hat{\mu}(a)$$

for all $a \in E'$. Then

$$\mu_n \xrightarrow{\tau_w} \mu \,.$$

Proof. By assumption, for every $n \geq 2$ there exists a $\lambda_n \in M^1(E)$ with $\mu_n = \lambda_n * \mu_{n-1}$. Moreover we put $\lambda_1 := \mu_1$. Taking Fourier transforms and applying Properties 2.1.6.7 and 2.2.18.3 we see that λ_n is symmetric for all $n \geq 1$.

Now, let $(X_n)_{n\geq 1}$ be a sequence of independent E-valued random variables on $(\Omega, \mathfrak{A}, \mathbf{P})$ such that $\mathbf{P}_{X_n} = \lambda_n$ for all $n \geq 1$. Then the distribution of $S_n := \sum_{k=1}^n X_k$ has the form

$$\mathbf{P}_{S_n} = \mathop{\Huge *}_{k=1}^{n} \mathbf{P}_{X_k} = \mathop{\Huge *}_{k=1}^{n} \lambda_k = \mu_n \,.$$

Therefore (vi) of the theorem becomes applicable, and we obtain that

$$\mu_n \xrightarrow{\tau_w} \nu \in M^1(E) \,.$$

It remains to apply the uniqueness and continuity properties of the Fourier transform (Theorems 2.1.4 and 2.1.9 respectively) in order to see that $\nu = \mu$. ■

Remark 3.1.12 *Without the assumption of symmetry for the distributions $\mathbf{P}_{X_k}$ of the random variables X_k $(k \geq 1)$ the implication $(v)\Rightarrow(i)$ of the Ito-Nishio theorem 3.1.10 does not remain valid.*

In fact, let E denote an infinite-dimensional separable Banach space with orthonormal basis $\{e_n : n \geq 1\}$. Let $(\Omega, \mathfrak{A}, \mathbf{P})$ be a probability space, and let

$$X_n(\omega) := e_n - e_{n-1}$$

for all $\omega \in \Omega$, $n \geq 1$ (where $e_0 := 0$). Then $(X_n)_{n\geq 1}$ is a sequence of independent E-valued random variables on $(\Omega, \mathfrak{A}, \mathbf{P})$. Moreover

$$S_n(\omega) = \sum_{k=1}^{n} X_k(\omega) = e_n$$

for all $\omega \in \Omega$, $n \geq 1$. Finally, we put $S(\omega) := 0$ for all $\omega \in \Omega$. Then

$$\lim_{n\to\infty} \langle S_n, a\rangle = \lim_{n\to\infty} \langle e_n, a\rangle = 0 = \langle S, a\rangle$$

whenever $a \in E'$, and this implies (v).

But

$$\lim_{n\to\infty} S_n(\omega) \neq S(\omega),$$

since $\|S_n(\omega)\| = 1$ for all $n \geq 1$ $(\omega \in \Omega)$, hence (i) does not hold.

As an immediate application of the Ito-Nishio theorem 3.1.10 we discuss the representation of E-valued Gaussian random variables as a.s.-convergent random series (in E).

Theorem 3.1.13 *(Random series representation of Gaussian random variables).*

Let ρ be a Gauss measure on E. Then there exist a sequence $(x_n)_{n\geq 1}$ in E and a sequence $(\xi_n)_{n\geq 1}$ of independent real-valued random variables on a probability space $(\Omega, \mathfrak{A}, \mathbf{P})$ with $\mathbf{P}_{\xi_n} = N(0,1)$ for all $n \in \mathbf{N}$ such that the series $\sum_{n\geq 1} \xi_n x_n$ is $\mathbf{P}$-a.s. convergent, and

$$\mathbf{P}_{\sum_{n\geq 1} \xi_n x_n} = \rho\,.$$

Proof. We recall the proof of Theorem 2.4.7 and realize that given $\rho \in G(E)$ there is a continuous linear mapping $T : (E', \tau(E', E)) \to (L^2(E,\rho), \|\,.\,\|)$ such that

$$\hat{\rho}(a) = \exp(-\frac{1}{2}\|Ta\|^2)$$

for all $a \in E'$. Moreover, T^t can be regarded as a mapping from $L^2(E,\rho)$ into E. For a given orthonormal basis $\{f_n : n \in \mathbf{N}\}$ of

$L^2(E, \rho)$ we define the sequence $(x_n)_{n\geq 1}$ in E by $x_n := T^t f_n$ for all $n \in \mathbf{N}$.

Now let $(\xi_n)_{n\geq 1}$ be a sequence of independent, identically distributed real-valued random variables on $(\Omega, \mathfrak{A}; \mathbf{P})$ with $\mathbf{P}_{\xi_n} = N(0,1)$ for all $n \in \mathbf{N}$. Employing the notation $\lambda_n := \mathbf{P}_{\xi_n x_n}$ we obtain for all $a \in E'$ that

$$\begin{aligned}\hat{\lambda}_n(a) &= \int e^{i\langle \xi_n x_n, a\rangle} d\mathbf{P} = \int e^{i\langle \xi_n f_n, Ta\rangle} d\mathbf{P} \\ &= \int e^{ir\langle f_n, Ta\rangle} N(0,1)(dr) \\ &= N(0,1)^\wedge(\langle f_n, Ta\rangle) = e^{-\frac{1}{2}\langle f_n, Ta\rangle^2}\end{aligned}$$

for all $a \in E'$ and that λ_n is symmetric ($n \in \mathbf{N}$). For $\mu_n := \lambda_1 * ... * \lambda_n$ we have

$$\begin{aligned}\hat{\mu}_n(a) &= \hat{\lambda}_1(a) \cdot \ldots \cdot \hat{\lambda}_n(a) \\ &= \exp\left\{ -\frac{1}{2}\sum_{j=1}^{n} \langle f_j, Ta\rangle^2 \right\}\end{aligned}$$

whenever $a \in E'$ ($n \in \mathbf{N}$). But then Parseval's identity yields

$$\begin{aligned}\lim_{n\to\infty} \hat{\mu}_n(a) &= \exp\left\{ -\frac{1}{2}\sum_{j\geq 1} \langle f_j, Ta\rangle^2 \right\} \\ &= \exp\left\{ -\frac{1}{2}\|Ta\|^2 \right\} = \hat{\rho}(a)\end{aligned}$$

for all $a \in E'$, and from the Ito-Nishio theorem 3.1.10 together with the continuity theorem 2.1.9 the assertion follows. ∎

3.2 Fourier expansion and construction of Brownian motion

As an application of the results of the previous section we shall return to the discussion of the Wiener measure introduced as a Gauss measure in Example 2.4.2.2 and its rise from Brownian motion in $\mathbf{R}$. In particular we shall establish the Fourier expansion of Brownian motion in $\mathbf{R}$ and construct Brownian motions in $\mathbf{R}$ with continuous paths.

In preparing the tools we recall the notion of ***p-dimensional standard normal distribution*** as the measure

$$N(0, I_p) := N(0,1)^{\otimes p}$$

in $M^1(\mathbf{R}^p)$, where I_p denotes the p-dimensional unit matrix.

It is well-known that for any linear mapping T from $\mathbf{R}^q$ into $\mathbf{R}^p$ the Fourier representation

$$\begin{aligned}(T(N(0, I_q)))^\wedge(a) &= \exp\left(-\frac{1}{2}\|T^t a\|^2\right)\\ &= \exp\left(-\frac{1}{2}\langle (TT^t)a, a\rangle\right)\end{aligned}$$

holds whenever $a \in \mathbf{R}^p$.

For any $p \times p$-matrix C over $\mathbf{R}$, $q \geq 1$ and any linear mapping T from $\mathbf{R}^q$ into $\mathbf{R}^p$ satisfying $TT^t = C$ the measure

$$N(0, C) := T(N(0, I_q))$$

in $M^1(\mathbf{R}^p)$ turns out to be the (p-dimensional) ***normal distribution with mean*** (vector) 0 ***and covariance*** (matrix) C.

Remarks 3.2.1

3.2.1.1 *By the uniqueness of the Fourier transform (Theorem 2.1.4) the measure $N(0, C)$ is uniquely determined by its covariance C.*

3.2.1.2 *The covariance $C = TT^t$ is obviously symmetric and positive semidefinite. Moreover, given a symmetric and positive semidefinite $p \times p$-matrix C over $\mathbf{R}$ there exists a $p \times p$-matrix R over $\mathbf{R}$ such that $C = RR^t$, and consequently*

$$N(0, C) = R(N(0, I_p))$$

or

$$N(0, C)^\wedge(a) = \exp\left(-\frac{1}{2}\langle Ca, a\rangle \right)$$

for all $a \in \mathbf{R}^p$.

3.2.1.3 *Let $X = (X_1, ..., X_p)^t$ be an $\mathbf{R}^p$-valued random variable on a probability space $(\Omega, \mathfrak{A}, \mathbf{P})$ with $\mathbf{P}_X = N(0, C)$. Then $\|X\|^2$ is $\mathbf{P}$-integrable, and for the expectation vector and covariance (matrix) of X we obtain*

$$E(X) := (E(X_1), ..., E(X_p))^t = 0$$

and

$$C(X) := (\text{Cov}\,(X_i, X_j))_{1 \le i,j \le p} = C$$

respectively.

The following

Properties 3.2.2 *of the normal distribution will be used in the discussion below.*

3.2.2.1 *With $D(r_1, ..., r_p)$ denoting the diagonal matrix containing $r_1, ..., r_p \in \mathbf{R}$ as diagonal elements we have*

$$N(0, D(\sigma_1^2, ..., \sigma_p^2)) = \bigotimes_{i=1}^{p} N(0, \sigma_i^2)$$

3.2.2.2 *Let $X_1, ..., X_p$ denote independent $\mathbf{R}$-valued random variables on $(\Omega, \mathfrak{A}, \mathbf{P})$ with $\mathbf{P}_{X_i} = N(0, \sigma_i^2)$ for all $i = 1, ..., p$. Then for $X = (X_1, ..., X_p)^t$ one obtains*

$$\mathbf{P}_X = N(0, D(\sigma_1^2, ..., \sigma_p^2))$$

3.2.2.3 *Let* $X = (X_1, ..., X_p)^t$ *be an* $\mathbf{R}^p$*-valued random variable on* $(\Omega, \mathfrak{A}, \mathbf{P})$ *with* $\mathbf{P}_X = N(0, C)$. *Then the real-valued random variables* $X_1, ..., X_p$ *are independent if and only if* $X_1, ..., X_p$ *are uncorrelated.*

3.2.2.4 *For any linear mapping* S *from* $\mathbf{R}^p$ *into* $\mathbf{R}^q$

$$S(N(0, C)) = N(0, SCS^t)$$

3.2.2.5 *Let* $(C_n)_{n \geq 0}$ *be a sequence of symmetric and positive semidefinite* $p \times p$*-matrices over* $\mathbf{R}$. *Then the following statements are equivalent:*

(i) $N(0, C_n) \xrightarrow{\tau_w} N(0, C_0)$.

(ii) $C_n \longrightarrow C_0$, *and the sequence* $(N(0, C_n))_{n \geq 1}$ *is* τ_w*-relatively compact (in* $M^1(\mathbf{R}^p)$*).*

(iii) $C_n \longrightarrow C_0$.

While the proof of the implication $(iii) \Rightarrow (i)$ is an immediate consequence of the classical continuity theorem for measures on $\mathbf{R}^p$ the implication $(i) \Rightarrow (ii)$ makes use of the continuity theorem 2.1.9.

Now let $(\Omega, \mathfrak{A}, \mathbf{P}, (B_t)_{t \in I})$ denote a Brownian motion in $\mathbf{R}$ with parameter set $I = [0, 1]$. By $\mathcal{H}(B)$ we abbreviate the closed linear subspace of $L^2(\Omega, \mathfrak{A}, \mathbf{P})$ generated by the family $\{B_t : t \in I\}$ (of Brownian variables). It can be shown that $\mathcal{H}(B)$ is a separable Hilbert space with scalar product defined by

$$\langle \xi, \eta \rangle := \int \xi \eta \, d\mathbf{P}$$

for all $\xi, \eta \in \mathcal{H}(B)$.

Theorem 3.2.3 *There is a unique linear isometry* S *form* $L^2(I) := = L^2(I, \mathfrak{B}(I), \lambda_I)$ *onto* $\mathcal{H}(B)$ *satisfying*

$$S(\mathbf{1}_{[0,t]}) = B_t$$

for all $t \in I$.

Proof. At first let f be an elementary (or step) function on I of the form

$$f = \sum_{j=1}^{n} f_j \mathbf{1}_{[t_{j-1}, t_j[}$$

for a subdivision $0 = t_0 < t_1 < ... < t_n = 1$ of I and coefficients $f_1, ..., f_n \in \mathbf{R}$.

Moreover, let

$$S(f) := \sum_{j=1}^{n} f_j (B_{t_j} - B_{t_{j-1}}) .$$

Then $f \mapsto S(f)$ is a linear mapping from the space $\mathcal{T}(I)$ of all elementary functions on I into the Hilbert space $\mathcal{H}(B)$ such that

$$S(\mathbf{1}_{[0,t]}) = B_t .$$

Therefore $S(\mathcal{T}(I))$ is dense in $\mathcal{H}(B)$. On the other hand $\mathcal{T}(I)$ is dense in $L^2(I)$.

Now, for every $f \in \mathcal{T}(I)$ we have that

$$\|S(f)\|_2 = \|f\|_2 .$$

Picking $f \in \mathcal{T}(I)$ of the form $f = \sum_{j=1}^{n} f_j \mathbf{1}_{[t_{j-1}, t_j[}$ we see that $f^2 = \sum_{j=1}^{n} f_j^2 \mathbf{1}_{[t_{j-1}, t_j[}$, and consequently

$$\begin{aligned}\int_\Omega S(f)^2 d\mathbf{P} &= \sum_{i=1}^{n} \sum_{j=1}^{n} f_i f_j \int (B_{t_i} - B_{t_{i-1}})(B_{t_j} - B_{t_{j-1}}) d\mathbf{P} \\ &= \sum_{j=1}^{n} f_j^2 (t_j - t_{j-1}) = \int_I f(t)^2 dt,\end{aligned}$$

where for the latter equality independence and normal distribution of the increments of the given Brownian motion have been applied. Thus S is a linear isometry between the dense subspaces $\mathcal{T}(I)$ and

$S(\mathcal{T}(I))$ of $L^2(I)$ and $\mathcal{H}(B)$ respectively, and hence extendable to a linear isometry S from $L^2(I)$ onto $\mathcal{H}(B)$.

Finally we note that the set $\{\mathbf{1}_{[0,t]} : t \in I\}$ is total in $L^2(I)$. As a consequence S is unique. ∎

Definition 3.2.4 *For every $f \in L^2(I)$ the unique element $S(f)$ of $L^2(\Omega, \mathfrak{A}, \mathbf{P})$ (constructed in Theorem 3.2.3) is called the* ***stochastic integral*** *of f with respect to the Brownian motion $(\Omega, \mathfrak{A}, \mathbf{P}, (B_t)_{t\in I})$ and is abbreviated by*

$$\int_I f(t)dB_t$$

or

$$\int fdB .$$

Properties 3.2.5 *of the stochastic integral*

3.2.5.1 *For $f, g \in L^2(I)$ we have*

$$E\left(\int fdB\right) = 0$$

and

$$\text{Cov}\left(\int fdB, \int gdB\right) = \langle f, g\rangle .$$

In particular

$$\text{Var}\left(\int fdB\right) = \|f\|_2^2 .$$

For the proof we note that the mapping $\xi \mapsto \int \xi\, d\mathbf{P}$ from $L^2(\Omega, \mathfrak{A}, \mathbf{P})$ into $\mathbf{R}$ is continuous and identically zero on the dense subspace $S(\mathcal{T}(I))$ of $\mathcal{H}(B)$. Therefore it vanishes on $\mathcal{H}(B)$. Now let $f, g \in L^2(I)$. Since $S(f)$ and $S(g)$ are centered and since S is an isometry we obtain that

$$\text{Cov}\,(S(f), S(g)) = \langle S(f), S(g)\rangle = \langle f, g\rangle .$$

The remaining identity follows by choosing $g := f$.

3.2.5.2 *For any finite family $\{f^{(i)} : i = 1, ..., k\}$ of functions in $L^2(I)$ the $\mathbf{R}^k$-valued random variable*

$$\left(\int f^{(1)} dB, ..., \int f^{(k)} dB\right)^t$$

has a (k-dimensional) normal distribution with mean 0 and covariance $(\langle f^{(i)}, f^{(j)} \rangle)_{1 \le i,j \le k}$.

Moreover, the real-valued random variables $\int f^{(1)} dB, ..., \int f^{(k)} dB$ are independent if and only if the functions $f^{(1)}, ..., f^{(k)} \in L^2(I)$ are pairwise orthogonal.

The second statement follows from the first one with the help of Property 3.2.2.3. For the proof of the first statement we first take functions $f^{(1)}, ..., f^{(k)} \in \mathcal{T}(I)$ of the form

$$f^{(i)} := \sum_{j=1}^{m} f_j^{(i)} 1_{[t_{j-1}, t_j[}$$

with $f_j^{(i)} \in \mathbf{R}$ and $0 = t_0 < t_1 < ... < t_n = 1$ $(1 \le i \le k, 1 \le j \le m)$. By

$$R((x_j)_{1 \le j \le m}) := \left(\sum_{j=1}^{m} f_j^{(i)} x_j\right)_{1 \le i \le k}$$

for all $(x_j)_{1 \le j \le m} \in \mathbf{R}^m$ a linear mapping R from $\mathbf{R}^m$ into $\mathbf{R}^k$ is defined. From the properties of Brownian motion the distribution of the $\mathbf{R}^m$-valued random variable $X = (B_{t_j} - B_{t_{j-1}})_{1 \le j \le m}$ on $(\Omega, \mathfrak{A}, \mathbf{P})$ is $N(0, C)$ with $C := D(t_1 - t_0, ..., t_m - t_{m-1})$ (See Property 3.2.2.2). But now

$$(S(f^{(i)}))_{1 \le i \le k} = R \circ X,$$

hence by Property 3.2.2.4 $(S(f^{(i)}))_{1 \le i \le k}$ is normally distributed.

Next we pick $f^{(i)}, ..., f^{(k)} \in L^2(I)$ arbitrarily. There are sequences $(f_n^{(i)})_{n \ge 1}$ in $\mathcal{T}(I)$ such that $\lim_{n \to \infty} \|f_n^{(i)} - f^{(i)}\|_2 = 0$ whenever $i = 1, ..., k$. An application of Theorem 3.2.3 yields

$$\lim_{n \to \infty} \|S(f_n^{(i)}) - S(f^{(i)})\|_2 = 0\,.$$

Therefore the sequence $(S(f_n^{(i)}))_{n\geq 1}$ converges stochastically towards $S(f^{(i)})$ for all $i = 1, ..., k$, and consequently the sequence $((S(f_n^{(1)}), ..., S(f_n^{(k)})))_{n\geq 1}$ converges stochastically towards $(S(f^{(1)}), ..., S(f^{(k)}))$. Employing Application 1.2.15 we arrive at

$$\mathbf{P}_{(S(f_n^{(i)}))_{1\leq i\leq k}} \xrightarrow{\tau_w} \mathbf{P}_{(S(f^{(i)}))_{1\leq i\leq k}}\,.$$

From the above discussion together with Property 3.2.5.1

$$\mathbf{P}_{(S(f_n^{(i)}))_{1\leq i\leq k}} = N(0, C_n),$$

where $C_n := (\langle f_n^{(i)}, f_n^{(j)}\rangle)_{1\leq i,j\leq k}$ (for all $n \geq 1$).

Let $C := (\langle f^{(i)}, f^{(j)}\rangle)_{1\leq i,j\leq k}$. Then $\lim_{n\to\infty} \|f_n^{(i)} - f^{(i)}\|_2 = 0$ $(1 \leq i \leq k)$ implies $\lim_{n\to\infty} C_n = C$. But now Property 3.2.2.5 yields

$$\mathbf{P}_{(S(f_n^{(i)}))_{1\leq i\leq k}} \xrightarrow{\tau_w} N(0, C),$$

hence

$$\mathbf{P}_{(S(f^{(i)}))_{1\leq i\leq k}} = N(0, C)$$

which completes the proof.

Theorem 3.2.6 *(Fourier expansion of Brownian motion). Let $(f_n)_{n\geq 1}$ be an orthonormal basis of $L^2(I)$. For every $n \in \mathbf{N}$ let*

$$\xi_n := \int f_n \, dB\,.$$

Then $(\xi_n)_{n\geq 1}$ is a sequence of independent, identically distributed real-valued random variables on $(\Omega, \mathfrak{A}, \mathbf{P})$ with $\mathbf{P}_{\xi_n} = N(0, 1)$ for every $n \in \mathbf{N}$, and

$$B_t = \sum_{n\geq 1} \left(\int_0^t f_n(u) du \right) \xi_n$$

uniformly in $t \in I$ $\mathbf{P}$*-a.s..*

Proof. By Theorem 3.2.3 $(\xi_n)_{n\geq 1}$ is an orthonormal basis of $\mathcal{H}(B)$. Let

$$b_n(t) := \langle \mathbf{1}_{[0,t]}, f_n \rangle = \int_0^t f_n(u)du$$

for all $t \in I$ $(n \in \mathbf{N})$. Then

$$\mathbf{1}_{[0,t]} = \sum_{n\geq 1} b_n(t) f_n$$

in $L^2(I)$. Again employing Theorem 3.2.3 we arrive at

$$\begin{aligned} B_t &= \int \mathbf{1}_{[0,t]} dB \\ &= \sum_{n\geq 1} b_n(t) \int f_n dB \\ &= \sum_{n\geq 1} b_n(t) \xi_n \end{aligned}$$

in $L^2(\Omega, \mathfrak{A}, \mathbf{P})$. Now by Property 3.2.5.2 $\mathbf{P}_{\xi_n} = N(0,1)$ for all $n \in \mathbf{N}$, and the sequence $(\xi_n)_{n\geq 1}$ is independent. Since L^2-convergence implies stochastic convergence the equivalence theorem 3.1.6 yields that

$$B_t = \sum_{n\geq 1} \left(\int_0^t f_n(u)du \right) \xi_n \quad \mathbf{P} - a.s. \ (t \in I)$$

In order to show the required *uniform* convergence $\mathbf{P}$-a.s. we look at the sequence $(X_n)_{n\geq 1}$ of $C(I)$-valued random variables

$$\omega \mapsto \left(t \mapsto \left(\int_0^t f_n(u)du \right) \xi_n(\omega) \right)$$

on $(\Omega, \mathfrak{A}, \mathbf{P})$. Since $\mathbf{P}_{\xi_n} = N(0,1)$, $\mathbf{P}_{X_n}$ is symmetric for all $n \in \mathbf{N}$, and the sequence $(X_n)_{n\geq 1}$ is clearly independent.

Let $S_n := \sum_{i=1}^n X_i$ for every $\mathbf{N}$. Then our assertion reads as $S_n \overset{\mathbf{P}-a.s.}{\longrightarrow} B$, where B is the $(\mathfrak{A} - \mathfrak{B}(C(I))$-measurable) $C(I)$-valued random variable $\omega \mapsto (t \mapsto B_t(\omega))$ on $(\Omega, \mathfrak{A}, \mathbf{P})$. By the Ito-Nishio theorem 3.1.10 (and since $C(I)' = M^b(I) - M^b(I)$) it suffices to show that

$$\langle S_n, \mu \rangle \overset{\mathbf{P}-stoch}{\longrightarrow} \langle B, \mu \rangle$$

for all $\mu \in M^b(I)$, and for this limit relation it suffices in turn to show that

$$\lim_{n \to \infty} \int |\langle S_n, \mu \rangle - \langle B, \mu \rangle| d\mathbf{P} = 0$$

whenever $\mu \in M^b(I)$. Now we apply Fubini's theorem in order to obtain

$$\begin{aligned}
\int & |\langle S_n \mu \rangle - \langle B, \mu \rangle| d\mathbf{P} \\
&= \int_\Omega \left| \int_0^1 (S_n(\omega)(t) - B_t(\omega)) \mu(dt) \right| \mathbf{P}(d\omega) \\
&\leq \int_0^1 \left(\int_\Omega |S_n(\omega)(t) - B_t(\omega)| \mathbf{P}(d\omega) \right) \mu(dt) \\
&\leq \int_0^1 \left(\int_\Omega |S_n(\omega)(t) - B_t(\omega)|^2 \mathbf{P}(d\omega) \right)^{\frac{1}{2}} \mu(dt)
\end{aligned}$$

for all $\mu \in M^b(I)$ $(n \in \mathbf{N})$. But

$$\int_\Omega [S_n(\omega)(t) - B_t(\omega)]^2 \mathbf{P}(d\omega) = \sum_{k > n} b_k(t)^2$$

and

$$\sum_{k \geq 1} b_k(t)^2 = \sum_{k \geq 1} \langle \mathbf{1}_{[0,t]}, f_n \rangle^2 = \|\mathbf{1}_{[0,t]}\|_2^2 = t$$

for all $n \in \mathbf{N}$ $(t \in I)$. By the dominated convergence theorem this implies

$$\lim_{n \to \infty} \int_0^1 \left(\int_\Omega [S_n(\omega)(t) - B_t(\omega)]^2 \mathbf{P}(d\omega) \right) \mu(dt) = 0$$

for all $\mu \in M^b(I)$, hence the assertion. ■

Remark 3.2.7 *Employing the orthonormal basis $(f_n)_{n\geq 1}$ of $L^2(I)$ defined by*

$$f_n(t) := \sqrt{2}\ \sin(n\pi t)$$

for which obviously

$$\int_0^t f_n(u)du = \frac{\sqrt{2}}{n\pi}(1 - \cos(n\pi t))$$

holds (for all $t \in I,\ \ n \geq 1$). Theorem 3.2.6 yields that

$$B_t = \frac{\sqrt{2}}{n\pi}\sum_{n\geq 1}(1 - \cos(n\pi t))\xi_n$$

holds uniformly in $t \in I$ $\mathbf{P}$-a.s..

In the remaining part of this section we shall establish the existence of Brownian motion in $\mathbf{R}$ with continuous paths. In fact, the process $(B_t)_{t\in I}$ constructed in Remark 3.2.7 has continuous paths due to the uniform convergence of the series and the continuity of the functions $f_n,\ \ n \geq 1$.

Lemma 3.2.8 *For every $n \in \mathbf{Z}_+$ and $k = 1, ..., 2^n$ let*

$$\begin{aligned} h_n^{(k)}\ :=&\ \sqrt{2^n}1_{[(2k-2)2^{-n-1},(2k-1)2^{-n-1}[} \\ &-\sqrt{2^n}1_{[(2k-1)2^{-n-1},\ (2k)2^{-n-1}[}. \end{aligned}$$

Moreover, let $f_1 := \mathbf{1}_{[0,1]}$ and

$$f_{2^n+k} := h_n^{(k)}$$

for all $n \in \mathbf{Z}_+,\ \ k = 1, ..., 2^n$.

*Then the so defined **Haar system** $(f_n)_{n\geq 1}$ is an orthonormal basis of $L^2(I)$.*

Proof. From

$$(h_n^{(k)})^2 = 2^n 1_{[(k-1)2^{-n},\ k2^{-n}[}$$

follows $\|f_m\|_2 = 1$ for all $m \in \mathbf{N}$, and $f_m f_n = 0$ a.e. or $f_m f_n = c f_{m \wedge n}$ a.e. for all $m, n \in \mathbf{N}$ with $m \neq n$, where $c = c(m,n)$ denotes an appropriate constant, implies $\langle f_m, f_n \rangle = 0$ for $m \neq n$. Thus $(f_n)_{n \geq 1}$ is an orthonormal system in $L^2(I)$. In order to see that $(f_n)_{n \geq 1}$ is also a basis of $L^2(I)$ we pick $f \in L^2(I)$ with $\langle f, f_n \rangle = 0$ for all $n \in \mathbf{N}$ and set $F(t) := \int_0^t f(s)ds$ for all $t \in I$. Clearly $F \in C(I)$. Moreover, by induction one shows that $F(k2^{-n}) = 0$ for all $n \in \mathbf{Z}_+$ and all $k = 0, 1, ..., 2^n$. Since F is continuous, we obtain

$$0 = F(t) = \int_0^t f(s)ds$$

for all $t \in I$, hence that $f = 0$ a.e.. ■

Lemma 3.2.9 *For every $t \in I$ and $m \geq 1$ let*

$$b_m(t) := \langle \mathbf{1}_{[0,t]}, f_m \rangle = \int_0^t f_m(s)ds\,.$$

Then $b_m(t) \geq 0$ for all $t \in I$, $m \in \mathbf{N}$ and

$$\sum_{2^n < k \leq 2^{n+1}} b_k(t) \leq \frac{1}{2} 2^{-\frac{n}{2}}$$

whenever $t \in I$, $n \in \mathbf{Z}_+$.

Proof. Evidently $b_m(t) \geq 0$ for all $t \in I, m \in \mathbf{N}$, and for $t \in [(j-1)2^{-n},\ j2^{-n}[$ we have that

$$\begin{aligned}\sum_{2^n < k \leq 2^{n+1}} b_k(t) &= b_j(t) \\ &\leq b_j((2j-1)2^{-n-1}) \\ &= 2^{-(n+1)}\sqrt{2^n} \\ &= \frac{1}{2} 2^{-\frac{n}{2}}\end{aligned}$$

whenever $1 \leq j \leq 2^n$. ■

Lemma 3.2.10 *Let $(\xi_n)_{n\geq 1}$ be a sequence of independent, identically distributed real-valued random variables on a probability space $(\Omega, \mathfrak{A}, \mathbf{P})$ with $\mathbf{P}_{\xi_n} = N(0,1)$ for all $n \in \mathbf{N}$. Moreover, let $(X_n)_{n\geq 1}$ be the sequence of $C(I)$-valued random variables on $(\Omega, \mathfrak{A}, \mathbf{P})$ defined by*

$$X_n(\omega) := b_n(.)\xi_n(\omega)$$

for all $\omega \in \Omega$, where the sequence $(b_n)_{n\geq 1}$ is given as in Lemma 3.2.9. Finally, let $S_n := \sum_{i=1}^n X_i$ for all $n \in \mathbf{N}$.

Then there exists a $C(I)$-valued random variable Z on $(\Omega, \mathfrak{A}, \mathbf{P})$ such that

$$S_n \overset{\mathbf{P}-a.s.}{\longrightarrow} Z\,.$$

Proof. At first we establish the inequalities

$$\begin{aligned}\mathbf{P}([|\xi_n| > u]) &= \frac{2}{\sqrt{2\pi}}\int_u^\infty e^{-\frac{v^2}{2}}\,dv = \sqrt{\frac{2}{\pi}}\int_{\frac{u^2}{2}}^\infty (2r)^{-\frac{1}{2}}e^{-r}\,dr \\ &= \sqrt{\frac{2}{\pi}}\Big[-e^{-r}(2r)^{-\frac{1}{2}}\Big]\Big|_{\frac{u^2}{2}}^\infty - \sqrt{\frac{2}{\pi}}\int_{\frac{u^2}{2}}^\infty (2r)^{-\frac{3}{2}}e^{-r}\,dr \\ &\leq \sqrt{\frac{2}{\pi}}\frac{1}{u}e^{-\frac{u^2}{2}}\end{aligned}$$

valid for all $u \in \mathbf{R}_+^\times$ $(n \in \mathbf{N})$. But then

$$\sum_{n\geq 2}\mathbf{P}\Big(\Big[|\xi_n| > \sqrt{3\ln n}\Big]\Big) \leq \sqrt{\frac{2}{3\pi}}\sum_{n\geq 2} n^{-\frac{3}{2}}(\ln n)^{-\frac{1}{2}} < \infty\,.$$

An application of the Borel-Cantelli Lemma implies that

$$\mathbf{P}\Big(\varliminf_{n\geq 2}\Big[|\xi_n| \leq \sqrt{3\ln n}\Big]\Big) = 1\,.$$

Therefore for $\mathbf{P}$-almost every $\omega \in \Omega$ there exists an $n(\omega) \in \mathbf{N}$ such that $|\xi_m(\omega)| \leq \sqrt{3\ln m}$ for all $m > 2^{n(\omega)}$, and hence we obtain

$$\alpha_n(\omega) := \max_{2^n < k \leq 2^{n+1}} |\xi_k(\omega)| \leq \sqrt{3\ln 2^{n+1}} = (n+1)^{\frac{1}{2}}\sqrt{3\ln 2}$$

whenever $n \geq n(\omega)$. But now Lemma 3.2.8 yields

$$\begin{aligned}\sum_{m>2^{n(\omega)}} |b_m(t)\xi_m(\omega)| &= \sum_{n\geq n(\omega)} \sum_{2^n<k\leq 2^{n+1}} b_k(t)|\xi_k(\omega)| \\ &\leq \sum_{n\geq n(\omega)} \left(\sum_{2^n<k\leq 2^{n+1}} b_k(t) \right) \alpha_n(\omega) \\ &\leq \sqrt{\frac{3\ln 2}{2}} \sum_{n\geq n(\omega)} (n+1)^{\frac{1}{2}} 2^{-\frac{n}{2}} < \infty\end{aligned}$$

for all $t \in I$ and $\mathbf{P}$-almost all $\omega \in \Omega$. Consequently the series $\sum_{m\geq 1} b_m(t)\xi_m(\omega)$ converges uniformly in $t \in I$ for $\mathbf{P}$-almost all $\omega \in \Omega$, and the proof is complete. ■

Theorem 3.2.11 *(Existence of Brownian motion with continuous paths)*

Let Z denote the $C(I)$-valued random variable on $(\Omega, \mathfrak{A}, \mathbf{P})$ established as the $\mathbf{P}$-a.s. limit of the sequence $(S_n)_{n\geq 1}$ introduced in Lemma 3.2.10. For every $t \in I$ we consider the real-valued random variable Z_t defined on $(\Omega, \mathfrak{A}, \mathbf{P})$ by

$$Z_t(\omega) := \sum_{n\geq 1} S_n(t)(\omega) := \sum_{n\geq 1} b_n(t)\xi_n(\omega)$$

whenever $\omega \in \Omega$.

Then $(\Omega, \mathfrak{A}, \mathbf{P}, (Z_t)_{t\in I})$ is a Brownian motion on $\mathbf{R}$ having continuous paths.

Proof. Since $Z(\omega) \in C(I)$ for all $\omega \in \Omega$, the process $(\Omega, \mathfrak{A}, \mathbf{P}, (Z_t)_{t\in I})$ has continuous paths.

Clearly, $Z_0(\omega) = 0$ for $\mathbf{P}$-almost all $\omega \in \Omega$. It remains to show that the process $(\Omega, \mathfrak{A}, \mathbf{P}, (Z_t)_{t\in I})$ admits stationary independent increments $Z_t - Z_s$ with $N(0, t-s)$ as their distributions $(s, t \in I,\ s < t)$. Let $t_1, ..., t_k \in I$ with $t_1 < t_2 < ... < t_k$ be fixed for the sequel. For every $n \in \mathbf{N}$ let $T_n := (b_j(t_i))_{1\leq i\leq k, 1\leq j\leq n}$ and $\eta_n := (\xi_1, ..., \xi_n)^t$.

By Property 3.2.2.2 $\mathbf{P}_{\eta_n} = N(0, I_n)$, and by Property 3.2.2.4 we deduce from $(S_n(t_1), ..., S_n(t_k))^t = T_n\eta_n$ that

$$\mathbf{P}_{(S_n(t_1),...,S_n(t_k))} = N(0, T_nT_n^t)\,.$$

We now apply Lemma 3.2.10 in order to obtain that

$$\mathbf{P}_{(S_n(t_1),...,S_n(t_k))} \xrightarrow{\tau_w} \mathbf{P}_{(Z_{t_1},...,Z_{t_k})}\,.$$

On the other hand

$$\lim_{n\to\infty} T_nT_n^t = (t_i \wedge t_j)_{1\le i,j\le k} =: C$$

as follows easily from the equalities

$$\begin{aligned}\lim_{n\to\infty}(T_nT_n^t)_{ij} &= \lim_{n\to\infty}\sum_{m=1}^{n} b_m(t_i)b_m(t_j) \\ &= \sum_{m\ge 1}\langle \mathbf{1}_{[0,t_i]}, f_m\rangle\langle \mathbf{1}_{[0,t_j]}, f_m\rangle \\ &= \langle \mathbf{1}_{[0,t_i]}, \mathbf{1}_{[0,t_j]}\rangle = t_i \wedge t_j\end{aligned}$$

valid for all $1 \le i \le k,\ \ 1 \le j \le n$.

But then Property 3.2.2.5 yields $\mathbf{P}_{(Z_{t_1},...,Z_{t_k})} = N(0, C)$. Introducing the linear mapping $R : \mathbf{R}^k \to \mathbf{R}^k$ given by

$$R(x_1, ..., x_k) := (x_1, x_2 - x_1, ..., x_k - x_{k-1})$$

for all $(x_1, ..., x_k) \in \mathbf{R}^k$ we see that $RCR^t = D(t_1, t_2 - t_1, ..., t_k - t_{k-1})$, and once again applying Property 3.2.2.4 we obtain that

$$\begin{aligned}\mathbf{P}_{(Z_{t_1}, Z_{t_2}-Z_{t_1},...,Z_{t_k}-Z_{t_{k-1}})} &= R(N(0, C)) \\ &= N(0, D(t_1, t_2 - t_1, ..., t_k - t_{k-1})).\end{aligned}$$

Now, Properties 3.2.2.3 and 3.2.2.1 provide the final arguments. ∎

3.3 Symmetric Lévy measures and generalized Poisson measures

In this section we shall extend the notion of Poisson measure by admitting not necessarily bounded though at least σ-finite exponents. Let $M^\sigma(E)$ denote the set of all σ-finite measures on $\mathfrak{B}(E)$.

Definition 3.3.1 *A measure $\lambda \in M^\sigma(E)$ is said to be a* ***symmetric Lévy measure*** *if it has the following properties:*

(a) λ is symmetric in the sense that $\lambda(-B) = \lambda(B)$ for all $B \in \mathfrak{B}(E)$.

(b) $\lambda(\{0\}) = 0$.

(c) There exists a measure $\tilde{e}(\lambda) \in M^b(E)$ such that

$$\widehat{\tilde{e}(\lambda)}(a) = \exp\left\{\int(\cos\langle x, a\rangle - 1)\lambda(dx)\right\}$$

for all $a \in E'$.

By $L_s(E)$ we abbreviate the totality of symmetric Lévy measures on E.

Remarks 3.3.2

3.3.2.1 *From the uniqueness of the Fourier transform (Theorem 2.1.4) follows that $\tilde{e}(\lambda)$ is uniquely determined by part (c) of its definition.*

3.3.2.2 *$\tilde{e}(\lambda)$ is a symmetric measure in $M^1(E)$ as one concludes from the facts that $\widehat{\tilde{e}(\lambda)}$ is real-valued and $\tilde{e}(\lambda)(E) = \widehat{\tilde{e}(\lambda)}(0) = 1$.*

3.3.2.3 *The correspondence $\lambda \mapsto \tilde{e}(\lambda)$ between the sets $L_s(E)$ and $\{\tilde{e}(\lambda) : \lambda \in L_s(E)\}$ is one-to-one.*

This assertion follows with some more effort as Theorem 3.3.11.

We will also see later that at this stage

3.3.2.4 *there is no need to assume that*

$$\int (1 - \cos\langle x, y\rangle)\lambda(dx) < \infty$$

or that $\widehat{\tilde{e}(\lambda)}(a) \neq 0$ *for all* $a \in E'$.

3.3.2.5 *Let* λ *be a symmetric measure in* $M^b(E)$ *satisfying part (b) of Definition 3.3.1. Then* $\lambda \in L_s(E)$, *and* $\tilde{e}(\lambda)$ *is the Poisson measure* $e(\lambda)$ *with exponent* λ *(as defined in 2.4.8).*

This observation motivates the extension of Poisson measures to generalized Poisson measures to be introduced later.

3.3.2.6 *For* $\lambda_1, \lambda_2 \in L(E)$ *the sum* $\lambda_1 + \lambda_2 \in L_s(E)$, *and*

$$\tilde{e}(\lambda_1 + \lambda_2) = \tilde{e}(\lambda_1) * \tilde{e}(\lambda_2)\,.$$

In fact, for every $a \in E'$

$$(\tilde{e}(\lambda_1) * \tilde{e}(\lambda_2))^\wedge(a) = \exp\left\{\int (\cos < x, y > -1)(\lambda_1 + \lambda_2)(dx)\right\},$$

hence by the uniqueness of the Fourier transform (Theorem 2.1.4) the assertion follows.

We now turn our attention to the study of sequences in the set $P(E)$ of Poisson measures on E.

For $\lambda \in M^\sigma(E)$ and $\delta \in \mathbf{R}_+^\times$ we introduce the abbreviations

$$\lambda|_\delta := \lambda_{U_\delta}$$

and

$$\lambda|^\delta := \lambda_{U_\delta^c},$$

where $U_\delta := \{x \in E : \|x\| \leq \delta\}$. Clearly, $\lambda|_\delta + \lambda|^\delta = \lambda$. Moreover, let

$$C(\lambda) := \{\delta \in \mathbf{R}_+^\times : \lambda(\partial U_\delta) = 0\}.$$

Evidently $C(\lambda)^c$ is a countable set.

Theorem 3.3.3 *Let $(\lambda_n)_{n\geq 1}$ be a sequence in $M^b(E)$ such that $(e(\lambda_n))_{n\geq 1}$ is relatively shift compact. Then for every $\delta \in \mathbf{R}_+^\times$ the sequence $(\lambda_n|^\delta)_{n\geq 1}$ is τ_w-relatively compact.*

Proof. 1. For $\delta > 0$ we have that $\lambda_n|_\delta + \lambda_n|^\delta = \lambda_n$, hence that $e(\lambda_n|^\delta) \prec e(\lambda_n)$. But then Property 2.2.16.5 yields that $(e(\lambda_n|^\delta))_{n\geq 1}$ is relatively shift compact. As a consequence we may assume without loss of generality that $\lambda_n(U_\delta) = 0$, hence that $\lambda_n|^\delta = \lambda_n$ for all $n \in \mathbf{N}$. Moreover, by Theorem 2.2.25 $(|e(\lambda_n)|^2)_{n\geq 1}$ is τ_w-relatively compact, and from Property 2.4.10.6 we obtain $|e(\lambda_n)|^2 = e(\lambda_n + \lambda_n^-)$ whenever $n \in \mathbf{N}$. Therefore, in view of Theorem 2.2.22 we may assume without loss of generality that λ_n is symmetric for all $n \in \mathbf{N}$.
2. We show that $d := \sup_{n\geq 1} \|\lambda_n\| < \infty$.

Suppose that $d = \infty$. By an eventual choice of an appropriate subsequence we may assume without loss of generality that $\|\lambda_n\| \geq n$ for all $n \in \mathbf{N}$ and that

$$e(\lambda_n) \xrightarrow{\tau_w} \mu \in M^1(E)$$

(See part 1. of this proof). For every $n \in \mathbf{N}$ we now consider the measure $\sigma_n := \frac{1}{\|\lambda_n\|}\lambda_n$ which is symmetric and satisfies $n\sigma_n \leq \lambda_n$ as well as $\sigma_n(U_\delta) = 0$. Moreover, we have that $e(\sigma_n)^n \prec e(\lambda_n)$ for all $n \in \mathbf{N}$. Applying Property 2.2.16.5 and Theorem 2.2.19 yields the τ_w-relative compactness of $(e(\sigma_n))_{n\geq 1}$. Let ν be an accumulation point of $(e(\sigma_n))_{n\geq 1}$. From

$$e(\sigma_n)^m \prec e(\lambda_n)$$

for all $n \geq m$ follows that $\nu^m \prec \mu$ for all $m \in \mathbf{N}$ and hence that $(\nu^m)_{m\geq 1}$ is τ_w-relatively compact. We now infer from Lemma 2.2.1 that $\nu = \varepsilon_0$ and thus that

$$e(\sigma_n) \xrightarrow{\tau_w} \varepsilon_0 .$$

Since $0 \notin \partial(U_\delta^c)$, U_δ^c is an ε_0-continuity set, and the desired contradiction follows from the inequalities

$$\begin{aligned} 0 = \varepsilon_0(U_\delta^c) &= \lim_{n\to\infty} e(\sigma_n)(U_\delta^c) \\ &\geq \frac{1}{e} \limsup_{n\geq 1} \ \sigma_n(U_\delta^c) = \frac{1}{e}. \end{aligned}$$

3. Finally, let $\varepsilon > 0$. From part 1. of this proof we conclude that there exists a set $K \in \mathcal{K}(E)$ such that $e(\lambda_n)(K^c) \leq \varepsilon$ for all $n \in \mathbf{N}$. Now we apply part 2. of this proof in order to arrive at the estimate

$$\lambda_n(K^c) \leq e^{\|\lambda_n\|} e(\lambda_n)(K^c) \leq e^d \varepsilon$$

valid for all $n \in \mathbf{N}$. But then $(\lambda_n)_{n\geq 1}$ is a τ_w-relatively compact sequence by Prohorov's theorem 1.3.7. ■

Theorem 3.3.4 *Let $(\lambda_n)_{n\geq 1}$ be a sequence in $M^b(E)$ such that $(e(\lambda_n))_{n\geq 1}$ is relatively shift compact. For every $n \in \mathbf{N}$ let*

$$f_n(a) := \int_{U_1} \langle x, a\rangle^2 \lambda_n(dx)$$

whenever $a \in V_1$. Then $(f_n)_{n\geq 1}$ is relatively compact in $C(V_1)$.

Proof. Replacing λ_n by $\lambda_n + \lambda_n^-$ changes f_n by the factor 2. Therefore we may assume without loss of generality that λ_n is symmetric and that $(e(\lambda_n))_{n\geq 1}$ is τ_w-relatively compact. This can be justified by referring to Property 2.4.10.6 and Theorem 2.2.25 respectively. Thus, with the help of Theorem 2.1.8, we obtain that $(e(\lambda_n)^\wedge)_{n\geq 1}$ is relatively compact in $C(V_1)$. But now we observe that $1 - \cos\ t \geq \frac{t^2}{3}$ for all $t \in [-1, 1]$. It follows that

$$f_n(a) \leq -3 \int (\cos\langle x, a\rangle - 1)\lambda_n(dx) = -3 \operatorname{Log}\ e(\lambda_n)^\wedge(a)$$

for all $a \in E'$. Moreover, $e(\lambda_n)^\wedge(a) > 0$ for all $n \in \mathbf{N}$. Thus the relative compactness of $(e(\lambda_n)^\wedge)_{n\geq 1}$ in $C(V_1)$ implies that

$$\inf_{n\geq 1} e(\lambda_n)^\wedge(a) > 0$$

for all $a \in V_1$.

In fact, for some subsequence $(\lambda_{n_k})_{k\geq 1}$ of $(\lambda_n)_{n\geq 1}$ we have that

$$e(\lambda_{n_k}) \xrightarrow{\tau_w} \mu \in I(E)$$

where $\hat{\mu}(a) \neq 0$ for all $a \in E'$.

But then $\sup_{n\geq 1} |f_n(a)| < \infty$ for all $a \in V_1$. On the other hand we know that the set $\{f_n : n \in \mathbf{N}\}$ is equicontinuous in 0 (with respect to $\tau(E', E)$).

This is easily deduced from Property 2.1.6.5 together with the fact that $e(\lambda_n)^\wedge(0) = 1$ for all $n \in \mathbf{N}$.

Finally,

$$|f_n(a)^{\frac{1}{2}} - f_n(b)^{\frac{1}{2}}| \leq f_n(a-b)$$

for all $a, b \in V_1$ satisfying $a - b \in V_1$ (and all $n \in \mathbf{N}$).

In fact, for such $a, b \in V_1$

$$\int_{U_1} \langle x, a\rangle \langle x, b\rangle \lambda_n(dx) \leq f_n(a)^{\frac{1}{2}} f_n(b)^{\frac{1}{2}},$$

hence

$$\begin{aligned}
&|f_n(a)^{\frac{1}{2}} - f_n(b)^{\frac{1}{2}}|^2 \\
&\quad \leq f_n(a) + f_n(b) - 2 f_n(a)^{\frac{1}{2}} f_n(b)^{\frac{1}{2}} \\
&\quad \leq \int_{U_1} \langle x, a\rangle^2 \lambda_n(dx) + \int_{U_1} \langle x, b\rangle^2 \lambda_n(dx) - 2 \int_{U_1} \langle x, a\rangle \langle x, b\rangle \lambda_n(dx) \\
&\quad = \int_{U_1} (\langle x, y\rangle - \langle x, b\rangle)^2 \lambda_n(dx) \\
&\quad = \int_{U_1} \langle x, a - b\rangle^2 \lambda_n(dx) = f_n(a-b).
\end{aligned}$$

From the inequality just established we infer that $\{f_n : n \in \mathbf{N}\}$ is equicontinuous everywhere on V_1 (with respect to $\tau(E', E)$). The Arzelà-Ascoli theorem yields the assertion. ■

Theorem 3.3.5 *Let $(\lambda_n)_{n\geq 1}$ be a sequence of measures in $L_s(E) \cap M^b(E)$ such that $\lambda_n \uparrow \lambda \in L_s(E)$. Then*

$$\tilde{e}(\lambda_n) = e(\lambda_n) \xrightarrow{\tau_w} \tilde{e}(\lambda).$$

Proof. By Property 2.4.10.4 we have that $\tilde{e}(\lambda_n) \prec \tilde{e}(\lambda_{n+1})$ for all $n \in \mathbf{N}$. Since $\lambda_n \leq \lambda$, the Radon-Nikodym theorem provides a measurable function $f_n : E \to I$ such that $\lambda_n = f_n \cdot \lambda$ $(n \in \mathbf{N})$. From $\lambda_n \uparrow \lambda$ we infer that $f_n \uparrow \mathbf{1}_E$ λ-a.e.. Now the monotone convergence theorem applies and we obtain that

$$\begin{aligned}
&\lim_{n\to\infty} \int (1 - \cos\langle x, a\rangle)\lambda_n(dx) \\
&\quad = \lim_{n\to\infty} \int (1 - \cos\langle x, a\rangle) f_n(x)\lambda(dx) \\
&\quad = \int (1 - \cos\langle x, a\rangle)\lambda(dx),
\end{aligned}$$

hence that

$$\lim_{n\to\infty} \tilde{e}(\lambda_n)^\wedge(a) = \tilde{e}(\lambda)^\wedge(a)$$

for all $a \in E'$. But every measure $\tilde{e}(\lambda_n)$ is symmetric $(n \in \mathbf{N})$. Thus, by the Ito-Nishio theorem 3.1.10 the assertion follows. ∎

Properties 3.3.6 *of symmetric Lévy measures*

Let $\lambda \in L_s(E)$.

3.3.6.1 *For every $\delta \in \mathbf{R}_+^\times$ we have $\lambda(U_\delta^c) < \infty$.*

In fact, λ being σ-finite there exists a sequence $(\lambda_n)_{n\geq 1}$ of symmetric measures in $M^b(E)$ with $\lambda_n \uparrow \lambda$. But then Theorem 3.3.5 implies

$$\tilde{e}(\lambda_n) \xrightarrow{\tau_w} \tilde{e}(\lambda),$$

and hence Theorem 3.3.3 that

$$\lambda(U_\delta^c) = \sup_{n\geq 1} \lambda_n(U_\delta^c) = \sup_{n\geq 1}(\lambda_n|^\delta)(E) < \infty.$$

3.3.6.2 *For every sequence $(\delta_n)_{n\geq 1}$ in $\mathbf{R}_+^\times$ with $\delta_n \downarrow 0$ we have that*

$$\tilde{e}(\lambda|^{\delta_n}) \xrightarrow{\tau_w} \tilde{e}(\lambda)$$

For a proof we note that from Property 3.3.6.1 together with Remark 3.3.2.5 follows that $\lambda_n := \lambda|^{\delta_n} \in L_s(E) \cap M^b(E)$ for every $n \in \mathbf{N}$. Evidently $\lambda_n \uparrow \lambda$, so that Theorem 3.3.5 yields the assertion.

3.3.6.3

$$\sup\left\{\int_{U_1} \langle x, a\rangle^2 \lambda(dx) : a \in V_1\right\} < \infty$$

In order to see this let $(\delta_n)_{n\geq 1}$ be a sequence in $\mathbf{R}_+^\times$ with $\delta_n \downarrow 0$ and let $\lambda_n := \lambda|^{\delta_n}$ for all $n \in \mathbf{N}$. Then $\lambda_n \in M^b(E)$ for all $n \in \mathbf{N}$ by Property 3.3.6.1 and

$$\tilde{e}(\lambda_n) \xrightarrow{\tau_w} \tilde{e}(\lambda)$$

by Property 3.3.6.2. But then by Theorem 3.3.4 there exists an $\alpha \in \mathbf{R}_+^\times$ satisfying

$$\int_{U_1} \langle x, a\rangle^2 \lambda_n(dx) \leq \alpha$$

for all $a \in V_1,\ n \in \mathbf{N}$, and since $\lambda_n \uparrow \lambda$ we obtain that

$$\int_{U_1} \langle x, a\rangle^2 \lambda(dx) \leq \alpha$$

(appealing to the proof of Theorem 3.3.5).

3.3.6.4 *(See Remark 3.3.2.4)*

$$\int (1 - \cos\langle x, a\rangle)\lambda(dx) < \infty$$

and

$$\lim_{t\to\infty} \int \frac{1 - \cos\ t\langle x, a\rangle}{t^2} \lambda(dx) = 0$$

for all $a \in E'$.

For the proof let, given $a \in E'$,

$$f_a(x) := \langle x, a \rangle^2 \mathbf{1}_{U_1}(x) + 2 \cdot \mathbf{1}_{U_1^c}(x)$$

whenever $x \in E$. From Properties 3.3.6.1 and 3.3.6.3 we infer that f_a is λ-integrable. Since $1 - \cos s \le \frac{s^2}{2}$ for all $s \in \mathbf{R}$ we have that

$$\frac{1 - \cos\ t\langle x, a \rangle}{t^2} \le f_a(x)$$

for all $x \in E$, $t \ge 1$. The special choice $t = 1$ yields the first assertion. The remaining one is implied by the limit relation

$$\lim_{t \to \infty} \frac{1 - \cos\ t\langle x, a \rangle}{t^2} = 0$$

valid for all $x \in E$ with the help of Lebesgue's dominated convergence theorem.

Theorem 3.3.7 *(Characterization of symmetric Lévy measures).*

Let λ be a symmetric measure in $M^\sigma(E)$ with $\lambda(\{0\}) = 0$. Then the following statements are equivalent:

(i) $\lambda \in L_s(E)$.

(ii) For each $\delta \in \mathbf{R}_+^\times$ we have that $\lambda(U_\delta^c) < \infty$, and for some (each) sequence $(\delta_n)_{n \ge 1}$ in $\mathbf{R}_+^\times$ with $\delta_n \downarrow 0$ the sequence $(\tilde{e}(\lambda|^{\delta_n}))_{n \ge 1}$ is τ_w-relatively compact.

(iii) There exists a sequence $(\lambda_n)_{n \ge 1}$ of symmetric measures in $M^b(E)$ with $\lambda_n \uparrow \lambda$ such that the sequence $(\tilde{e}(\lambda_n))_{n \ge 1}$ is τ_w-relatively compact.

Proof. *(i)* $\Rightarrow$ *(ii)* follows directly from Properties 3.3.6.1 and 3.3.6.2.

(ii) $\Rightarrow$ *(iii)*. The choice $\lambda_n := \lambda|^{\delta_n}$ for all $n \in \mathbf{N}$ yields the implication.

(iii) $\Rightarrow$ *(i)*. The defining properties *(a)* and *(b)* of Definition 3.3.1 of a symmetric Lévy measure are part of the assumptions. From $\lambda_n \uparrow \lambda$ we obtain that for all $a \in E'$

$$\lim_{n\to\infty} \int (1 - \cos\langle x, a\rangle)\lambda_n(dx) = \int (1 - \cos\langle x, a\rangle)\lambda(dx)$$

(See the proof of Theorem 3.3.5). Then Properties 3.3.6.1 and 3.3.6.4 imply that

$$\lim_{n\to\infty} \tilde{e}(\lambda_n)^\wedge(a) = \exp\left\{\int (\cos\langle x, a\rangle - 1)\lambda(dx)\right\}$$

for all $a \in E'$. From the continuity theorem follows property *(c)* of the definition 3.3.1 and hence *(i)*. ■

Corollary 3.3.8 *Suppose that* $\lambda \in L_s(E)$.

(i) Let σ *be a symmetric measure on* E *with* $\sigma \leq \lambda$. *Then* $\sigma \in L_s(E)$, *and*

$$\tilde{e}(\sigma) \prec \tilde{e}(\lambda).$$

(ii) $\tilde{e}(\lambda) \in I(E)$.

Moreover, $\tilde{e}(\lambda)$ *is (continuously) embeddable* ($\in EM(E)$) *with (continuous) embedding convolution semigroup* $(\tilde{e}(\lambda)_t)_{t\geq 0}$, *where*

$$\tilde{e}(\lambda)_t := \tilde{e}(t\lambda)$$

for all $t \in \mathbf{R}$.

Proof. *(i)*. Obviously the measure σ belongs to $M^\sigma(E)$ and fulfills properties *(a)* and *(b)* of the Lévy measure. For every $\delta \in \mathbf{R}_+^\times$ we deduce from Property 3.3.6.1 that $\sigma(U_\delta^c) \leq \lambda(U_\delta^c) < \infty$. Now let $(\delta_n)_{n\geq 1}$ be a sequence in $\mathbf{R}_+^\times$ with $\delta_n \downarrow 0$. Then for every $n \in \mathbf{N}$ we obtain that

$$\tilde{e}(\sigma|^{\delta_n}) * \tilde{e}((\lambda - \sigma)|^{\delta_n}) = \tilde{e}(\lambda|^{\delta_n})$$

holds and that $\tilde{e}(\sigma|^{\delta_n})$ is symmetric. (See Properties 2.4.10.2 and 2.4.10.5 and observe that $0 \leq \lambda - \sigma \leq \lambda$). From Property 3.3.6.2 and Theorem 2.2.21 we infer that the sequence $(e(\sigma|^{\delta_n}))_{n\geq 1}$ is τ_w-relatively compact. Thus Theorem 3.3.7 implies the assertion.

(*ii*). In view of Remark 3.3.2.6 and (*i*) of this corollary we have $t\lambda \in L_s(E)$ and

$$\tilde{e}(s\lambda) * \tilde{e}(t\lambda) = \tilde{e}((s+t)\lambda)$$

for all $s, t \in \mathbf{R}_+$. This shows that $\tilde{e}(\lambda) \in I(E)$. The remaining part of the statement follows from

$$\text{Log } \tilde{e}(\lambda)^\wedge(a) = \int (\cos\langle x, a\rangle - 1)\lambda(dx)$$

valid for all $a \in E'$ and from the embedding theorem 2.3.9 (Here we rely on Properties 2.1.6.4, 3.3.6.4 and on Theorem 2.1.10 for detailed arguments.).

Definition 3.3.9 *For $\lambda \in L_s(E)$ the measure $\tilde{e}(\lambda) \in M^1(E)$ introduced in Definition 3.3.1 is called the* ***generalized Poisson measure with exponent*** *λ.*

Clearly,

$$\text{Log } \tilde{e}(\lambda)^\wedge(a) = \int (\cos\langle x, a\rangle - 1)\lambda(dx)$$

for every $a \in E'$.

Lemma 3.3.10 *Let $\lambda_1, \lambda_2 \in L_s(E)$ satisfying*

$$\int_B (1 - \cos\langle x, a\rangle)\lambda_1(dx) = \int_B (1 - \cos\langle x, a\rangle)\lambda_2(dx)$$

for all $B \in \mathfrak{B}(E)$ and $a \in E'$. Then $\lambda_1 = \lambda_2$.

Proof. Remark 3.3.2.1 and property (*b*) of the Lévy measure together with the fact that $U^c_{\frac{1}{n}} \uparrow E \setminus \{0\}$ holds enable us to assume without loss of generality that λ_1 and λ_2 belong to $M^b(E)$.

Now we define for every $a \in E'$

$$g_a(x) := \sum_{m\geq 1} \frac{1}{2^{m+1}}\Big(1 - \cos\Big(\frac{\langle x, a\rangle}{m}\Big)\Big)$$

whenever $x \in E$. Clearly $g_a \in C(E)$, $0 \leq g_a \leq 1$, and $g_a(x) = 0$ if and only if $\langle x, a\rangle = 0$ $(x \in E)$. From Appendix B 5 (Banach, Hahn) we deduce the existence of a sequence $(a_j)_{j\geq 1}$ in E' with the property that $\langle x, a_j\rangle = 0$ for all $j \in \mathbf{N}$ implies that $x = 0$. In fact, for every sequence $(x_n)_{n\geq 1}$ dense in E there exists a sequence $(a_n)_{n\geq 1}$ in E' satisfying $\|a_n\| = 1$ and $\langle x, a_n\rangle = \|x_n\|$ for all $n \in \mathbf{N}$.

Now let

$$h(x) := \sum_{j\geq 1} \frac{1}{2^j} g_{a_j}(x)$$

for all $x \in E$. Then $h \in C(E)$, $0 \leq h \leq 1$, and $h(x) = 0$ if and only if $x = 0$ $(x \in E)$. By assumption

$$\int_B h d\lambda_1 = \int_B h d\lambda_2$$

for all $B \in \mathfrak{B}(E)$, hence $\lambda_1(B) = \lambda_2(B)$ for all $B \in \mathfrak{B}(E)$ with $0 \notin B$. Property (b) of a Lévy measure leads to the assertion. ■

Theorem 3.3.11 *(Injectivity of the generalized Poisson mapping; see Remark 3.3.2.3).*

Let $\lambda_1, \lambda_2 \in L_s(E)$ *with* $\tilde{e}(\lambda_1) = \tilde{e}(\lambda_2)$. *Then* $\lambda_1 = \lambda_2$.

Proof. For each $a \in E'$ let

$$f_a(x) := 1 - \cos\langle x, a\rangle$$

whenever $x \in E$. By assumption we have that

$$\begin{aligned}\exp\Big(-\int f_a d\lambda_1\Big) &= \tilde{e}(\lambda_1)^\wedge(a)\\ &= \tilde{e}(\lambda_2)^\wedge(a)\\ &= \exp\Big(-\int f_a d\lambda_2\Big),\end{aligned}$$

hence after an application of Property 3.3.6.4 (of Lévy measures) that

$$\int f_a d\lambda_1 = \int f_a d\lambda_2$$

for all $a \in E'$. We note that

$$e^{i(\alpha+\beta)} + e^{i(\alpha-\beta)} = e^{i\alpha}(e^{i\beta} + e^{-i\beta}) = 2e^{i\alpha}\cos\beta$$

(for $\alpha, \beta \in \mathbf{R}$), hence that

$$f_{a+b} + f_{a-b} - 2f_a = 2\cos\langle .,a\rangle f_b$$

for all $a, b \in E'$. But then

$$\begin{aligned}(f_b \cdot \lambda_1)^\wedge(a) &= \int \cos\langle .,a\rangle f_b \ d\lambda_1 \\ &= \int \cos\langle .,a\rangle f_b \ d\lambda_2 \\ &= (f_b \cdot \lambda_2)^\wedge(a)\end{aligned}$$

for all $a, b \in E'$. The uniqueness theorem 2.1.4 yields $f_b \cdot \lambda^1 = f_b \cdot \lambda^2$ for all $b \in E'$, and Lemma 3.3.10 completes the proof. ■

Theorem 3.3.12 *Let $(\lambda_n)_{n\geq 1}$ be a sequence in $L_s(E)$ such that the corresponding sequence $(\tilde{e}(\lambda_n))_{n\geq 1}$ of generalized Poisson measures is τ_w-relatively compact in $M^1(E)$. Then*

(i) for every $\delta \in \mathbf{R}_+^\times$ the sequence $(\lambda_n|^\delta)_{n\geq 1}$ is τ_w-relatively compact in $M^b(E)$,

(ii) there exist a subsequence $(\lambda_{n_k})_{k\geq 1}$ of $(\lambda_n)_{n\geq 1}$ and a $\lambda \in L_s(E)$ such that

$$\lambda_{n_k}|^\delta \xrightarrow{\tau_w} \lambda|^\delta$$

for all $\delta \in C(\lambda)$, and

(iii) the sequence $(f_n)_{n\geq 1}$ defined by

$$f_n(a) := \int_{U_1} \langle x, a\rangle^2 \lambda_n(dx)$$

for all $a \in V_1$, $n \in \mathbf{N}$, *is relatively compact in* $C(V_1)$.

Proof. (i). From Corollary 3.3.8(i) we infer that $\tilde{e}(\lambda_n|^\delta) \prec \tilde{e}(\lambda_n)$ for all $n \in \mathbf{N}$. Consequently, by Theorem 2.2.21 the sequence $(\tilde{e}(\lambda_n|^\delta))_{n\geq 1}$ is τ_w-relatively compact. Property 3.3.6.1 yields that $\lambda_n|^\delta \in M^b(E)$ for all $n \in \mathbf{N}$. The desired assertion now follows from Theorem 3.3.3.

(ii). The assumption together with part(i) of this proof supply us with a subsequence $(\lambda_{n_k})_{k\geq 1}$ of $(\lambda_n)_{n\geq 1}$ satisfying

$$\tilde{e}(\lambda_{n_k}) \xrightarrow{\tau_w} \mu \in M^1(E)$$

and

$$\lambda_{n_k}|^{\frac{1}{j}} \xrightarrow{\tau_w} \lambda^{(j)} \in M^b(E)$$

for all $j \in \mathbf{N}$ (Here the usual diagonal procedure has been applied). The set $C := \bigcap_{j\geq 1} C(\lambda^{(j)})$ has a countable complement in $\mathbf{R}_+^\times$. Let $j, m \in \mathbf{N}$ and $\delta \in C$ with $\delta > \frac{1}{j}$ and $\delta > \frac{1}{m}$. By Corollary 1.2.10 we obtain that

$$\lambda^{(j)}|^\delta = \lambda^{(m)}|^\delta .$$

Now let $(\delta_j)_{j\geq 1}$ be a sequence in C with $\delta_j \downarrow 0$ and $\delta_j > \frac{1}{j}$ for all $j \in \mathbf{N}$. We put

$$\lambda(B) := \sup_{j\geq 1}(\lambda^{(j)}|^{\delta_j})(B)$$

for all $B \in \mathfrak{B}(E)$. Observing that $\lambda^{(j)}|^{\delta_j} = \lambda^{(j+1)}|^{\delta_j} \leq \lambda^{(j+1)}|^{\delta_{j+1}}$ for all $j \in \mathbf{N}$ we see that λ is a symmetric measure in $M^\sigma(E)$ with $\lambda(\{0\}) = 0$. Moreover, for given $\delta \in C(\lambda)$ we choose $j \in \mathbf{N}$ with $\delta_j < \delta$. Then $\delta \in C(\lambda^{(j)})$, and from Corollary 1.2.10 we deduce that

$$\lambda_{n_k}|^\delta = (\lambda_{n_k}|^{\frac{1}{j}})^\delta \xrightarrow{\tau_w} \lambda^{(j)}|^\delta = (\lambda^{(j)}|^{\delta_j})^\delta = (\lambda|^{\delta_j})^\delta = \lambda|^\delta,$$

where $\lambda|^\delta \in M^b(E)$. Thus, it remains to show that $\lambda \in L_s(E)$. For this we first note that by Theorem 2.4.12(ii) we have that

$$\tilde{e}(\lambda_{n_k}|^{\delta_j}) \xrightarrow{\tau_w} \tilde{e}(\lambda|^{\delta_j})$$

for all $j \in \mathbf{N}$. Next we employ Corollary 3.3.8(i) and Remark 3.3.2.6 in order to obtain

$$\tilde{e}(\lambda_{n_k}) = \tilde{e}(\lambda_{n_k}|_{\delta_j}) * \tilde{e}(\lambda_{n_k}|^{\delta_j})$$

valid for all $i, j \in \mathbf{N}$. Therefore Corollary 2.2.4 together with the fact that $\tilde{e}(\lambda|^{\delta_j})^\wedge(a) \neq 0$ for all $a \in E'$ (Property 2.4.10.1) yields the existence of a measure $\nu_j \in M^1(E)$ such that

$$\mu = \nu_j * \tilde{e}(\lambda|^{\delta_j})$$

for all $j \in \mathbf{N}$. Now Theorem 2.2.21 implies that the sequence $(e(\lambda|^{\delta_j}))_{j \geq 1}$ is τ_w-relatively compact, since the measures $e(\lambda|^{\delta_j})$ are symmetric by Property 2.4.10.5. Finally, Theorem 3.3.7 yields the desired statement.

(*iii*) is shown in analogy to the proof of Theorem 3.3.4. ■

Theorem 3.3.13 *(Construction of symmetric Lévy measures).*

Let λ be a symmetric measure in $M^\sigma(E)$ satisfying the following conditions:

(a) $\lambda(\{0\}) = 0$.

(b) $\int (1 \wedge \|x\|)\lambda(dx) < \infty$.

Then $\lambda \in L_s(E)$.

Proof. 1. Let $\lambda \in M^b(E)$ such that $\int \|x\|\lambda(dx) < \infty$. Then

$$\int \|x\| e(\lambda)(dx) \leq \int \|x\|\lambda(dx).$$

In order to see this we first establish the inequality

$$\int \|x\|\lambda^k(dx) \leq k\lambda(E)^{k-1} \int \|x\|\lambda(dx)$$

valid for all $k \in \mathbf{N}$. The proof runs by induction. While the case $k = 1$ is clear, we need only treat the step from k to $k+1$. But this is easily done:

$$\begin{aligned}\int \|x\| \lambda^{k+1}(dx) &= \int\int \|x+y\| \lambda^k(dx)\lambda(dy) \\ &\le \int\int \|x\|\lambda^k(dx)\lambda(dy) + \int\int \|y\|\lambda^k(dx)\lambda(dy) \\ &= \lambda(E)\int \|x\|\lambda^k(dx) + \lambda^k(E)\int \|y\|\lambda(dy) \\ &\le \lambda(E)k\lambda(E)^{k-1}\int \|x\|\lambda(dx) + \lambda(E)^k \int \|x\|\lambda(dx) \\ &= (k+1)\lambda(E)^k \int \|x\|\lambda(dx).\end{aligned}$$

Finally,

$$\begin{aligned}\int \|x\| e(\lambda)(dx) &= e^{-\lambda(E)} \sum_{k\ge 0} \frac{1}{k!}\int \|x\|\lambda^k(dx) \\ &\le e^{-\lambda(E)} \sum_{k\ge 1}\frac{1}{k!}(k\lambda(E)^{k-1}\int \|x\|\lambda(dx)) \\ &= e^{-\lambda(E)}\left(\sum_{k\ge 1}\frac{1}{(k-1)!}\lambda(E)^{k-1}\right)\int \|x\|\lambda(dx) \\ &= \int \|x\|\lambda(dx).\end{aligned}$$

Here we note that

$$\int \|x\|\lambda^0(dx) = \int \|x\|\varepsilon_0(dx) = 0\,.$$

2. From the assumptions on the symmetric measure λ we now deduce that $\lambda \in L_s(E)$. We have $\lambda(U_1^c) \le \int (1 \wedge \|x\|)\lambda(dx) < \infty$ and $\lambda = \lambda|_1 + \lambda|^1$. Without loss of generality we assume that $\lambda(U_1^c) = 0$. With $\lambda_n := \lambda|^{\frac{1}{n}}$ for all $n \in \mathbf{N}$ we obtain that $\lambda_1 = 0$ and $\lambda_n \uparrow \lambda$. Moreover,

$$\lambda_n(E) = \lambda(U_{\frac{1}{n}}^c) \le n \int (1 \wedge \|x\|)\lambda(dx) < \infty,$$

hence $\lambda \in M^\sigma(E)$ and $\lambda_n - \lambda_{n-1} \in M^b(E)$ for all $n \geq 2$.

Let $(X_n)_{n\geq 2}$ be a sequence of independent E-valued random variables on a probability space $(\Omega, \mathfrak{A}, \mathbf{P})$ such that

$$\mathbf{P}_{X_n} := e(\lambda_n - \lambda_{n-1}) \ (n \geq 2).$$

Then

$$\begin{aligned}\mathbf{P}_{\sum_{j=m}^n X_j} &= \mathop{\ast}_{j=m}^{n} \mathbf{P}_{X_j} \\ &= \mathop{\ast}_{j=m}^{n} e(\lambda_j - \lambda_{j-1}) \\ &= e(\lambda_n - \lambda_{m-1}) \ (2 \leq m \leq n).\end{aligned}$$

In particular,

$$\mathbf{P}_{\sum_{j=2}^n X_j} = e(\lambda_n) \ (n \geq 2).$$

But

$$\begin{aligned}&\int \Big\| \sum_{j=m}^{n} X_j \Big\| d\mathbf{P} \\ &\qquad = \int \|x\| e(\lambda_n - \lambda_{m-1})(dx) \\ &\qquad \leq \int \|x\| (\lambda_n - \lambda_{m-1})(dx) \ \text{(by part 1. of this proof)} \\ &\qquad = \int_{U^c_{\frac{1}{n}}} \|x\| \lambda(dx) - \int_{U^c_{\frac{1}{m-1}}} \|x\| \lambda(dx) \\ &\qquad \leq \int_{U_{\frac{1}{m-1}}} \|x\| \lambda(dx).\end{aligned}$$

Since $\int_{U_1} \|x\| \lambda(dx) < \infty$, Lebesgue's dominated convergence theorem yields

$$\lim_{m\to\infty} \int_{U_{\frac{1}{m-1}}} \|x\| \lambda(dx) = 0.$$

Now, the inequality

$$\mathbf{P}\left(\left[\left\|\sum_{j=m}^{n} X_j\right\| > \varepsilon\right]\right) \leq \frac{1}{\varepsilon}\int\left\|\sum_{j=m}^{n} X_j\right\| d\mathbf{P}$$

valid for all $2 \leq m \leq n$ implies that

$$\lim_{m\to\infty}\sup_{n\geq m}\mathbf{P}\left(\left[\left\|\sum_{j=m}^{n} X_j\right\| > \varepsilon\right]\right) = 0$$

whatever $\varepsilon > 0$. Therefore

$$\sum_{j=2}^{m} X_j \overset{\mathbf{P}-stoch}{\longrightarrow} X,$$

where X is an E-valued random variable on $(\Omega, \mathfrak{A}, \mathbf{P})$, hence

$$\sum_{j=2}^{n} X_j \overset{d}{\longrightarrow} \quad (\text{as } n \to \infty)\,.$$

It follows that

$$e(\lambda_n) \overset{\tau_w}{\longrightarrow} \quad ,$$

and from Theorem 3.3.7 that $\lambda \in L_s(E)$ with $\mathbf{P}_X = e(\lambda)$. ■

Discussion 3.3.14 *of assumption (b) of Theorem 3.3.13.*

There exist measures $\lambda \in L_s(E)$ with

$$\int (1 \wedge \|x\|^2)\lambda(dx) = \infty\,.$$

In fact, let $I := [0,1]$, $E := C(I)$ and $(x_n)_{n\geq 1}$ a sequence in $C(I)$ defined by $x_n(0) = x_n(2^{-n}) = x_n(2^{-(n-1)}) = x_n(1) =$

0, $x_n(\frac{3}{2}2^{-n}) = n^{-\frac{1}{8}}$ and extended linearly in between these arguments. Then $\|x_n\| = n^{-\frac{1}{8}}$ for all $n \in \mathbf{N}$. The measure

$$\lambda := \sum_{n\geq 1} n^{-\frac{3}{4}}(\varepsilon_{x_n} + \varepsilon_{-x_n}) \in M^\sigma(E)$$

is symmetric and satisfies $\lambda(\{0\}) = 0$. Moreover,

$$\int_{U_1} \|x\|^2 \lambda(dx) = \sum_{n\geq 1} n^{-\frac{3}{4}} 2n^{-\frac{1}{4}} = 2\sum_{n\geq 1} n^{-1} = \infty$$

and $\lambda(U_\delta^c) < \infty$ for all $\delta \in \mathbf{R}_+^\times$.

We now consider sequences $(\xi_n)_{n\geq 1}$ and $(\xi'_n)_{n\geq 1}$ of independent real-valued random variables on $(\Omega, \mathfrak{A}, \mathbf{P})$ with

$$\mathbf{P}_{\xi_n} = \mathbf{P}_{\xi'_n} = \Pi\left(n^{-\frac{3}{4}}\right)$$

for all $n \in \mathbf{N}$. Then the E-valued random variables $\xi_n x_n$ and $\xi'_n x_n$ are independent and have distributions

$$\mathbf{P}_{\xi_n x_n} = \mathbf{P}_{\xi'_n x_n} = e\left(n^{-\frac{3}{4}}\varepsilon_{x_n}\right)(n \in \mathbf{N}).$$

In fact, if ξ is a real-valued random variable on $(\Omega, \mathfrak{A}, \mathbf{P})$ with $\mathbf{P}_\xi = \Pi(\alpha)(= e(\alpha\varepsilon_1))$ and $y \in E$ then

$$\mathbf{P}_{\xi y} = e(\alpha\varepsilon_y),$$

since for all $a \in E'$ we have

$$\begin{aligned}
(\mathbf{P}_{\xi y})^\wedge(a) &= \int e^{i\langle x,a\rangle}\mathbf{P}_{\xi y}(dx) = \int e^{i\xi(\omega)\langle y,a\rangle}\mathbf{P}(d\omega) \\
&= \int e^{it\langle y,a\rangle}\mathbf{P}_\xi(dt) = \int e^{it\langle y,a\rangle}\Pi(\alpha)(dt) \\
&= \Pi(\alpha)^\wedge(\langle y,a\rangle) = \exp(\alpha(e^{i\langle y,a\rangle} - 1)) = e(\alpha\varepsilon_y)^\wedge(a).
\end{aligned}$$

Next we obtain that

$$\begin{aligned}\mathbf{P}_{Y_n} &:= \mathbf{P}_{\xi_n x_n - \xi'_n x_n} \\ &= e\Big(n^{-\frac{3}{4}}\varepsilon_{x_n}\Big) * e\Big(n^{-\frac{3}{4}}\varepsilon_{x_n}\Big)^{-} \\ &= e\Big(n^{-\frac{3}{4}}(\varepsilon_{x_n} + \varepsilon_{-x_n})\Big)(n \in \mathbf{N}),\end{aligned}$$

where the $C(I)$-valued random variables $Y_n := \xi_n x_n - \xi'_n x_n$ are independent (on $(\Omega, \mathfrak{A}, \mathbf{P})$). It follows that

$$\begin{aligned}\mathbf{P}_{\sum_{j=1}^n Y_j} &= \mathop{\ast}_{j=1}^{n} \mathbf{P}_{Y_j} = \mathop{\ast}_{j=1}^{n} e\Big(j^{-\frac{3}{4}}(\varepsilon_{x_j} + \varepsilon_{-x_j})\Big) \\ &= e\left(\sum_{j=1}^{n} j^{-\frac{3}{4}}(\varepsilon_{x_j} + \varepsilon_{-x_j})\right) = e\Big(\lambda|^{(n+1)^{-\frac{1}{8}}}\Big) = e\Big(\lambda|^{\delta_n}\Big),\end{aligned}$$

where $\delta_n := (n+1)^{-\frac{1}{8}}$ for all $n \in \mathbf{N}$. But now

$$[|\xi_n - \xi'_n| > 1] \subset [\xi_n > 1] \cup [\xi'_n > 1],$$

hence

$$\begin{aligned}\sum_{n\geq 1} \mathbf{P}([|\xi_n - \xi'_n| > 1]) &\leq 2\sum_{n\geq 1} \mathbf{P}([\xi_n > 1]) \\ &= 2\sum_{n\geq 1} \Pi(n^{-\frac{3}{4}}(\{2,3,...\}) \leq 2\sum_{n\geq 1} n^{-\frac{3}{2}} < \infty.\end{aligned}$$

Here we use the estimate $\Pi(\alpha)(\{2,3,...\}) \leq \alpha^2$ which is obviously obtained from

$$\Pi(\alpha)(\{0,1\}) = e^{-\alpha}(1+\alpha) \geq (1-\alpha)(1+\alpha) = 1-\alpha^2$$

for $\alpha \in \mathbf{R}_+^{\times}$. Now the Borel-Cantelli Lemma applies and yields

$$\mathbf{P}(\limsup_{n\to\infty}[|\xi_n - \xi'_n| > 1]) = 0$$

or

$$\mathbf{P}(\liminf_{n\to\infty}[|\xi_n - \xi'_n| \le 1]) = 1\,.$$

Therefore, for $\mathbf{P}$-almost all $\omega \in \Omega$ there exists an $n(\omega) \in \mathbf{N}$ such that $|\xi_n(\omega) - \xi'_n(\omega)| \le 1$ whenever $n \ge n(\omega)$. This implies that for $n \ge m \ge n(\omega)$

$$\left\| \sum_{j=m}^{n} Y_j(\omega) \right\| = \left\| \sum_{j=m}^{n} (\xi_j(\omega) - \xi'_j(\omega))x_j \right\| \le m^{-\frac{1}{8}},$$

since $[x_j \ne 0] \cap [x_k \ne 0] = \emptyset$ for $j \ne k$. But then

$$\sum_{j\ge 1} Y_j \overset{\mathbf{P}-a.s.}{\longrightarrow} Y,$$

where Y is a $C(I)$-valued random variable on $(\Omega, \mathfrak{A}, \mathbf{P})$, hence

$$e(\lambda|^{\delta_n}) = \mathbf{P}_{\sum_{j=1}^{n} Y_j} \overset{\tau_w}{\longrightarrow} \mathbf{P}_Y\,.$$

Theorem 3.3.7 yields that $\lambda \in L_s(E)$ with $e(\lambda) = \mathbf{P}_Y$.

3.4 The Lévy-Khintchine decomposition

The aim of the subsequent discussion is to establish the canonical decomposition of infinitely divisible probability measures on a separable Banach space as convolutions of measures of Poisson type, Gauss measures and of Dirac measures. The first named measures will be defined via not necessarily symmetric Lévy measures. For symmetric Lévy measures the canonical decomposition can be derived from the material of the previous section. In the nonsymmetric case the proof depends on an additional technique: the centralization of generalized Poisson measures.

For any measure $\lambda \in M^b(E)$ we consider the Bochner integral $x(\lambda) \in E$ given by

$$x(\lambda) := -\int_{U_1} x\,\lambda(dx)\,.$$

Theorem 3.4.1 *Let $(\lambda_n)_{n\geq 1}$ be a sequence in $M^b(E)$ such that the sequence $(e(\lambda_n))_{n\geq 1}$ is relatively shift compact. Then the sequence $(e(\lambda_n) * \varepsilon_{x(\lambda_n)})_{n\geq 1}$ is τ_w-relatively compact.*

Proof. Applying Theorem 3.3.3 it suffices to show that the sequence $(e(\lambda_n|^1) * \varepsilon_{x(\lambda_n)})_{n\geq 1}$ is τ_w-relatively compact. We therefore may assume without loss of generality that $\lambda_n(U_1^c) = 0$ for all $n \geq 1$. From the inequalities

$$\begin{aligned} &|1 - (e(\lambda_n)*\varepsilon_{x(\lambda_n)})^\wedge(a)| \\ &\qquad = \left|1 - \exp\left(\int_{U_1} (e^{i\langle x,a\rangle} - 1 - i\langle x,a\rangle)\lambda_n(dx)\right)\right| \\ &\qquad \leq \exp\left(\frac{1}{2}\int_{U_1} \langle x,a\rangle^2 \lambda_n(dx)\right) - 1 \end{aligned}$$

valid for all $a \in V_1$ we deduce that the sequence $((e(\lambda_n)*\varepsilon_{x(\lambda_n)})^\wedge)_{n\geq 1}$ is equicontinuous in $C(V_1)$. Theorem 3.3.4 together with Theorem 2.2.12 yield the assertion. ■

Definition 3.4.2 *For any (bounded) measure $\lambda \in M^b(E)$ the measure*

$$e_s(\lambda) := e(\lambda) * \varepsilon_{x(\lambda)}$$

is called the ***exponential*** of λ.

Introducing the kernel K by

$$K(x, a) := e^{i\langle x,a\rangle} - 1 - i\langle x, a\rangle 1_{U_1}(x)$$

for all $x \in E, a \in E'$ we observe that

$$e_s(\lambda)\hat{}\,(a) = \exp\Big(\int_E K(x, a)\lambda(dx)\Big)$$

whenever $a \in E'$.

Clearly, $e_s(\lambda) \in I(E)$ and $e_s(\lambda) = e(\lambda)$ provided λ is symmetric. In this case $x(\lambda) = 0$.

Comparing the Fourier transform of the exponential $e_s(\lambda)$ of λ with that of the Poisson measure $e(\lambda)$ with exponent λ the desired generalization to nonsymmetric measures λ relies on a systematic replacement of the kernel $(x, a) \mapsto \cos\langle x, a\rangle - 1$ by the kernel K.

Theorem 3.4.3 *Let $(\lambda_n)_{n\geq 1}$ be a τ_w-relatively compact sequence in $M^b(E)$. Then for any $\delta > 0$ the sequence $(x_n)_{n\geq 1}$ defined by*

$$x_n = x_n^\delta := \int_{U_\delta} x\, \lambda_n(dx)$$

for all $n \in \mathbf{N}$ is relatively compact in E.

Proof. From Prohorov's theorem 1.3.7 we infer that for $\varepsilon > 0$ there exists a compact, convex and balanced set $K \subset E$ such that

$$\lambda_n(K^c) < \frac{\varepsilon}{\delta}$$

for all $n \in \mathbf{N}$. But this implies that

$$\Big\| \int_{U^\delta \cap K^c} x \, \lambda_n(dx) \Big\| < \varepsilon$$

for all $n \in \mathbf{N}$. Thus it remains to show that the sequence $(y_n)_{n \geq 1}$ defined by

$$y_n := \int_{U^\delta \cap K} x \, \lambda_n(dx)$$

for all $n \in \mathbf{N}$ admits a finite ε-net. In fact, if this statement has been established, the sequence $(x_n)_{n \geq 1}$ admits a finite 2ε-net, and this being true for every $\varepsilon > 0$ yields the assertion.

In order to finish the proof we just consider an $a \in E'$ such that $|\langle x, a \rangle| \leq 1$ for all $x \in K$. We obtain that

$$|\langle y_n, a \rangle| \leq \lambda_n(K) \leq \lambda_n(E) \leq d := \sup_{n \geq 1} \lambda_n(E),$$

hence by an application of the bipolar theorem that $y_n \in dK$ for all $n \in \mathbf{N}$. But this implies that $(y_n)_{y \geq 1}$ is relatively compact. ■

Corollary 3.4.4 *Let $(\lambda_n)_{n \geq 1}$ be a sequence in $M^b(E)$ such that*

$$\lambda_n \xrightarrow{\tau_w} \lambda .$$

Then for every $\delta \in C(\lambda)$ we have

$$x_n := \int_{U_\delta} x \, \lambda_n(dx) \to \int_{U_\delta} x \, \lambda(dx)$$

as $n \to \infty$.

Proof. From the theorem we know that $(x_n)_{n \geq 1}$ being relatively compact admits a subsequence $(x_{n'})$ such that

$$\langle x_{n'}, a \rangle \longrightarrow \langle y, a \rangle$$

for $y \in E$ whenever $a \in E'$. On the other hand Theorem 1.2.9 implies that

$$\langle x_{n'}, a\rangle \longrightarrow \int_{U_\delta} \langle x, a\rangle\, \lambda(dx)$$

for all $a \in E'$. Since the limits of $(\langle x_{n'}, a\rangle)$ coincide for all subsequences of $(x_n)_{n\geq 1}$, the assertion has been proved. ■

Facts 3.4.5 *of exponentials of bounded measures*

3.4.5.1 *For $(\lambda_n)_{n\geq 1}$ in $M^b(E)$ the sequence $(e_s(\lambda_n))_{n\geq 1}$ is relatively shift compact if and only if it is τ_w-relatively compact, and both properties are equivalent to the relative shift compactness of the sequence $(e(\lambda_n))_{n\geq 1}$.*

This fact follows directly from Theorem 3.4.1.

3.4.5.2 *For $(\lambda_n)_{n\geq 1}$ in $M^b(E)$ with $\lambda_n \xrightarrow{\tau_w} \lambda$ and $1 \in C(\lambda)$, we have*

$$e_s(\lambda_n) \xrightarrow{\tau_w} e_s(\lambda)\,.$$

This statement is a restricted version of the τ_w-continuity of the mapping $\lambda \mapsto e_s(\lambda)$ from $M^b(E)$ into $M^1(G)$ the restriction being crucial.

As for the proof of Fact 3.4.5.2 we just observe that

$$e_s(\lambda_n) = e(\lambda_n) * \varepsilon_{x(\lambda_n)},$$

where

$$x(\lambda_n) := -\int_{U_1} x\, \lambda_n(dx)$$

for all $n \geq 1$, and apply Theorem 2.4.12*(ii)* together with Corollary 3.4.4.

Theorem 3.4.6 *For any sequence $(\lambda_n)_{n\geq 1}$ in $M^b(E)$ with $\lambda_n \xrightarrow{\tau_w} \lambda$ the following statements are equivalent:*

(i) $e_s(\lambda_n) \xrightarrow{\tau_w}$ *(in $M^1(E)$).*

(ii) $x(\lambda_n) \longrightarrow$ *(in E) (as* $n \to \infty$*).*

Moreover, if either of these statements is available, we have that

$$e_s(\lambda_n) \xrightarrow{\tau_w} e_s(\lambda) * \varepsilon_z,$$

where

$$\begin{aligned} z &= \lim_{n\to\infty} x(\lambda_n) - x(\lambda) \\ &= \int_{U_1} x\,\lambda(dx) - \lim_{n\to\infty} \int_{U_1} x\,\lambda_n(dx) \\ &= \lim_{\substack{\delta\downarrow 1 \\ \delta\in C(\lambda)}} \lim_{n\to\infty} \int_{U_\delta\cap U_1^c} x\,\lambda_n(dx). \end{aligned}$$

Clearly, if $1 \in C(\lambda)$ *then* $z = 0$.

Proof. For every $n \geq 1$ we have that

$$e_s(\lambda_n) = e(\lambda_n) * \varepsilon_{x(\lambda_n)}\,.$$

Now we apply Theorem 2.4.12*(ii)* together with Corollary 2.2.4 in order to obtain the equivalence $(i) \iff (ii)$. As to the remaining statements of the theorem we suppose that

$$e_s(\lambda_n) \xrightarrow{\tau_w} \mu$$

for some $\mu \in M^1(E)$. But then $\lim_{n\to\infty} x(\lambda_n)$ exists, and

$$\begin{aligned} \mu &= \lim_{n\to\infty} (e(\lambda_n) * \varepsilon_{x(\lambda_n)}) \\ &= e(\lambda) * \varepsilon_{\lim_{n\to\infty} x(\lambda_n)} \\ &= e_s(\lambda) * \varepsilon_{\lim_{n\to\infty} x(\lambda_n) - x(\lambda)}\,. \end{aligned}$$

The double limit representation of $z = \lim_{n\to\infty} x(\lambda_n) - x(\lambda)$ follows from Corollary 3.4.4. ■

Definition 3.4.7 *A measure $\lambda \in M^\sigma(E)$ is called a* **Lévy measure** *if $\lambda + \lambda^-$ is a symmetric Lévy measure (in the sense of Definition 3.3.1).*

The totality of Lévy measures will be abbreviated by $L(E)$.

The following

Properties 3.4.8 *of a Lévy measure λ are proved similarly to those for a symmetric one (See 3.3.6)*

3.4.8.1 $\lambda(\{0\}) = 0$.

3.4.8.2 *For each $\delta > 0$ we have that $\lambda(U_\delta^c) < \infty$.*

3.4.8.3 *For $(\delta_n)_{n\geq 1}$* in $\mathbf{R}_+^\times$ *the sequence $(e(\lambda_n|^{\delta_n}))_{n\geq 1}$ is relatively shift compact.*

3.4.8.4 $\sup\left\{ \int_{U_1} \langle x, a\rangle^2 \lambda(dx) : a \in V_1 \right\} < \infty$.

3.4.8.5 *For $(\delta_n)_{n\geq 1}$* in $\mathbf{R}_+^\times$ *with $\delta_n \downarrow 0$ the sequence $(e(\lambda|^{\delta_n}))_{n\geq 1}$* is *$\tau_w$-relatively compact.*

3.4.8.6 *If $\sigma \in M^\sigma(E)$ satisfies $\sigma \leq \lambda$, then σ is also a Lévy measure.*

The only properties deserving additional arguments are 3.4.8.3 and 3.4.8.5. For 3.4.8.3 we note that by Theorem 3.3.7 the sequence $(e(\lambda + \lambda^-)|^{\delta_n}))_{n\geq 1}$ is τ_w-relatively compact. But since

$$e(\lambda|^{\delta_n}) \prec e((\lambda + \lambda^-)|^{\delta_n})$$

for all $n \geq 1$, the sequence $(e(\lambda|^{\delta_n}))_{n\geq 1}$ is relatively shift compact. Property 3.4.8.5 is a direct consequence of Fact 3.4.5.1.

Theorem 3.3.7 can be extended to nonsymmetric Lévy measures.

Theorem 3.4.9 *(Characterization of Lévy measures)*

For any $\lambda \in M^\sigma(E)$ satisfying $\lambda(\{0\}) = 0$ the following statements are equivalent:

(*i*) $\lambda \in L(E)$.

(*ii*) (*a*) *For each* $a \in E'$

$$\int |K(x,a)|\lambda(dx) < \infty,$$

and

(*b*) *there exists a measure* $\tilde{e}_s(\lambda) \in M^1(E)$ *such that*

$$\widehat{\tilde{e}_s(\lambda)}(a) = \exp\Big(\int K(x,a)\lambda(dx)\Big)$$

for all $a \in E'$.

(*iii*) *There exists a sequence* $(\lambda_n)_{n\geq 1}$ *in* $M^b(E)$ *with* $\lambda_n \uparrow \lambda$ *such that* $(e_s(\lambda_n))_{n\geq 1}$ is τ_w*-relatively compact.*

(*iv*) *For each* $\delta > 0$

(*a*) $\lambda(U_\delta^c) < \infty$, *and*

(*b*) *for some (each) sequence* $(\delta_n)_{n\geq 1}$ *with* $\delta_n \downarrow 0$ *the sequence* $(e_s(\lambda|^{\delta_n}))_{n\geq 1}$ *is* τ_w*-relatively compact.*

Proof. As the implication *(i)* ⇒ *(iv)* follows with the help of Theorem 3.3.3, *(iv)* ⇒ *(iii)* is trivial, and *(iii)* ⇒ *(ii)* is deduced from Properties 3.3.6 (as in the proof of Theorem 3.3.7) by adapting the arguments to the measures $e_s(\lambda)$ and the kernel K; it remains to show implication *(ii)* ⇒ *(i)*. For completeness we prove the full equivalence *(i)* ⇔ *(ii)*.

(i) ⇒ *(ii)*. Let $\lambda \in L(E)$. From Properties 3.4.8.2 and 3.4.8.4 we conclude that

$$\int |K(x,a)|\lambda(dx) < \infty$$

for all $a \in E'$, since $|K(x,a)| \leq \frac{1}{2}\langle x,a\rangle^2$ for all $x \in E$ satisfying $\|x\| \leq 1$. Moreover, choosing a sequence $(\delta_n)_{n\geq 1}$ in $\mathbf{R}_+^\times$ with $\delta_n \downarrow 0$ and putting $\lambda_n := \lambda_n := \lambda|^{\delta_n}$ for all $n \geq 1$ we obtain that

$$\lim_{n\to\infty} \int K(x,a)\lambda_n(dx) = \int K(x,a)\lambda(dx)$$

for each $a \in E'$. Now Property 3.4.8.5 can be employed in order to establish the τ_w-convergence of the sequence $(e_s(\lambda_n))_{n\geq 1}$ towards a measure $\tilde{e}_s(\lambda) \in M^1(E)$ such that

$$\widehat{\tilde{e}_s(\lambda)}(a) = \exp\Big(\int K(x,a)\lambda(dx)\Big)$$

for all $a \in E'$.
(ii) $\Rightarrow$ *(i)*. Let $\lambda \in M^\sigma(E)$ with $\lambda(\{0\}) = 0$ satisfy the conditions *(a)* and *(b)* of *(ii)*. Along with λ also λ^- satisfies *(ii)*, hence the function

$$a \mapsto \exp\Big(\int (\cos\langle x,a\rangle - 1)(\lambda + \lambda^-)(dx)\Big)$$

on E' is the Fourier transform of the measure $\tilde{e}_s(\lambda) * \tilde{e}_s(\lambda^-) \in M^1(E)$. This proves *(i)*. ■

Definition 3.4.10 *The measure* $\tilde{e}_s(\lambda) \in M^1(E)$ *introduced in (ii) of Theorem 3.4.9 is called the* ***generalized exponential of the Lévy measure*** λ.

Clearly, $\tilde{e}_s(\lambda) = \tilde{e}(\lambda)$ *whenever* $\lambda \in L_s(E)$.

Properties 3.4.11 *of the generalized exponential mapping*

3.4.11.1 *The mapping* $\tilde{e}_s$ *from* $L(E)$ *into* $M^1(E)$ *is an involutive semigroup homomorphism, i.e.*

(a) $L(E) + L(E) \subset L(E)$, and

$$\tilde{e}_s(\lambda_1 + \lambda_2) = \tilde{e}_s(\lambda_1) * \tilde{e}_s(\lambda_2)$$

whenever $\lambda_1, \lambda_2 \in L(E)$.

(b) $L(E)^- \subset L(E)$, *and*

$$\tilde{e}_s(\lambda^-) = \tilde{e}_s(\lambda)^-$$

whenever $\lambda \in L(E)$.

3.4.11.2 $\tilde{e}_s(L(E)) \subset EM(E)$, *i.e. for every* $\lambda \in L(E)$ *the generalized exponential* $\tilde{e}_s(\lambda)$ of λ *is (continuously) embeddable with (continuous) embedding semigroup* $(\tilde{e}_s(\lambda)_t)_{t\geq 1}$ *given by*

$$\tilde{e}_s(\lambda)_t := \tilde{e}_s(t\lambda)$$

for all $t \in \mathbf{R}$.

3.4.11.3 *For any* $\lambda_1, \lambda_2 \in L(E)$ *with* $\lambda_1 \leq \lambda_2$ *we have*

$$\tilde{e}_s(\lambda_1) \prec \tilde{e}_s(\lambda_2)\,.$$

3.4.11.4 *Let* $(\lambda_n)_{n\geq 1}$ *be a sequence* in $M^b(E)$ *with* $\lambda_n \uparrow \lambda$. *Then*

$$\tilde{e}_s(\lambda_n) = e_s(\lambda_n) \xrightarrow{\tau_w} \tilde{e}_s(\lambda)\,.$$

3.4.11.5 *For any sequence* $(\lambda_n)_{n\geq 1}$ *in* $L(E)$ *the sequence* $(\tilde{e}_s(\lambda_n))_{n\geq 1}$ *is* τ_w*-relatively compact provided it is relatively shift compact.*

3.4.11.6 *For any sequence* $(\lambda_n)_{n\geq 1}$ *in* $L(E)$ *such that* $(\tilde{e}_s(\lambda_n))_{n\geq 1}$ *is relatively shift compact the sequence* $(\lambda_n|^{\delta})_{n\geq 1}$ *is* τ_w*-relatively compact whenever* $\delta > 0$.

While Properties 3.4.11.1 to 3.4.11.4 are obvious (in particular with respect to their analogs in the symmetric case), we need only note that Property 3.4.11.6 follows from Theorem 3.3.3 together with Property 2.2.16.5 and that Property 3.4.11.5 is a consequence of Theorem 3.3.4 in whose proof the boundedness of the Lévy measures has not been used.

After all these preparations we are ready to approach the Lévy-Khintchine decomposition. We start with the uniqueness of the decomposition whose proof relies on a modification of Lemma 3.3.10 and Theorem 3.3.11.

Given $\lambda \in L(E)$ and $a \in E'$ we introduce the measure λ^a by

$$\lambda^a(B) := \int_B (1 - \cos\langle x, a\rangle)\lambda(dx)$$

for all $B \in \mathfrak{B}(E)$. From Properties 3.4.8.2 and 3.4.8.4 we infer that $\lambda^a \in M^b(E)$.

Lemma 3.4.12 *Let $\lambda_1, \lambda_2 \in L(E)$ satisfying $\lambda_1^a = \lambda_2^a$ for all $a \in E'$. Then $\lambda_1 = \lambda_2$.*

Proof. The assumption yields the equality

$$(\lambda_1|^\delta)^a = (\lambda_2|^\delta)^a$$

for each $\delta > 0$ which, once

$$\lambda_1|^\delta = \lambda_2|^\delta$$

has been shown, implies that $\lambda_1 = \lambda_2$. It suffices therefore to perform the proof of the assertion for $\lambda_1, \lambda_2 \in L(E) \cap M^b(E)$. But in this case the proof of Lemma 3.3.10 takes care of the remaining reasoning. ■

Theorem 3.4.13 *The generalized exponential mapping $\tilde{e}_s : L(E) \to M^1(E)$ is injective. Moreover, let $\lambda_1, \lambda_2 \in L(E)$ and $x_0 \in E$ such that*

$$\tilde{e}_s(\lambda_1) = \tilde{e}_s(\lambda_2) * \varepsilon_{x_0} .$$

Then $\lambda_1 = \lambda_2$ and $x_0 = 0$.

Proof. We modify the arguments applied in the proof of Theorem 3.3.11 in the obvious manner. From the assumption we have that

$$\widehat{\tilde{e}_s(\lambda_1)}(a) = \widehat{\tilde{e}_s(\lambda_2)}(a) e^{i\langle x_0, a\rangle}$$

for all $a \in E'$. On the other hand

$$K(x, a+b) + K(x, a-b) - 2K(x, a) = 2e^{i\langle x, a\rangle}(\cos\langle x, b\rangle - 1)$$

whenever $x \in E, a, b \in E'$. Consequently

$$\exp\Big(2\int e^{i\langle x,a\rangle}(\cos\langle x, b\rangle - 1)\lambda_1(dx)\Big)$$
$$= \exp\Big(2\int e^{i\langle x,a\rangle}(\cos\langle x, b\rangle - 1)\lambda_2(dx)\Big),$$

hence

$$\widehat{\lambda_1^b}(a) - \widehat{\lambda_2^b}(a) = i\pi k(a,b)$$

with $k(a,b) \in \mathbf{Z}$ for $a, b \in E'$. But since $k(a,b)$ turns out to be 0 for all $b \in E'$ we obtain $\lambda_1^b = \lambda_2^b$ for all $b \in E'$, which by Lemma 3.4.12 implies that $\lambda_1 = \lambda_2$ and clearly $x_0 = 0$. ■

Lemma 3.4.14 *Let* $\lambda \in L(E)$. *Then*

$$\lim_{t \to 0} t^2 \int K(x, \frac{a}{t}) \lambda(dx) = 0$$

whenever $a \in E'$.

Proof. For each $x \in U_1^c$ we have $|K(x,a)| \le 2$, hence

$$\lim_{t \to 0} t^2 \int_{U_1^c} |K(x, \frac{a}{t})| \lambda(dx) \le 2 \lim_{t \to 0} t^2 \lambda(U_1^c) = 0 \ (a \in E') .$$

As a consequence of this it suffices to study the integral

$$\int_{U_1} K(x, \frac{a}{t}) \lambda(dx)$$

for $a \in E'$. If $x \in U_1$ we have

$$|K(x, \frac{a}{t})| \le \frac{\langle x, a \rangle^2}{2t^2}$$

and therefore

$$|t^2 K(x, \frac{a}{t})| \le \frac{\langle x, a \rangle^2}{2}$$

for all $x \in E, a \in E'$ and $t > 0$. On the other hand

$$\lim_{t \to 0} t^2 K(x, \frac{a}{t}) = 0$$

for all $x \in E$ and $a \in E'$. Property 3.4.8.4 enables us to apply the dominated convergence theorem, and the assertion has been proved. ■

Theorem 3.4.15 *(Uniqueness of the canonical decomposition)* *Let $\varrho_1, \varrho_2 \in G(E)$, $\lambda_1, \lambda_2 \in L(E)$ and $x_1, x_2 \in E$ such that*

$$\varrho_1 * \tilde{e}_s(\lambda_1) * \varepsilon_{x_1} = \varrho_2 * \tilde{e}_s(\lambda_2) * \varepsilon_{x_2} .$$

Then $\varrho_1 = \varrho_2$, $\lambda_1 = \lambda_2$ and $x_1 = x_2$.

Proof. By Theorem 2.4.7 there exists for $j = 1, 2$ a mapping $q_j : E' \to \mathbf{R}_+$ with the property

$$q_j(ta) = t^2 q_j(a)$$

whenever $t \in \mathbf{R}$ such that $\hat{\varrho}_j(a) = \exp(-q_j(a))$ for all $a \in E'$. From the assumption we deduce the equality

$$\hat{\varrho}_1(a)\tilde{e}_s(\lambda_1)\hat{\ }(a)\hat{\varepsilon}_{x_1}(a) = \hat{\varrho}_2(a)\tilde{e}_s(\lambda_2)\hat{\ }(a)\hat{\varepsilon}_{x_2}(a)$$

and by taking logarithms the equality

$$\begin{aligned} -q_1(a) + \int K(x,a)\lambda_1(dx) + i\langle x_1, a\rangle \\ = -q_2(a) + \int K(x,a)\lambda_2(dx) + i\langle x_2, a\rangle \end{aligned}$$

valid for all $a \in E'$. For given $t > 0$ we now replace $a \in E'$ by $\frac{a}{t}$ and multiply both sides of the above equality by t^2. Taking limits for $t \to 0$ the limit relationship of Lemma 3.4.14 implies $q_1(a) = q_2(a)$, hence $\hat{\varrho}_1(a) = \hat{\varrho}_2(a)$ for all $a \in E'$ and therefore by the uniqueness of the Fourier transform that $\varrho_1 = \varrho_2$. Finally Theorem 3.4.13 implies that $\lambda_1 = \lambda_2$ and $x_1 = x_2$. ■

Remark 3.4.16 *If $\tilde{e}_s(\lambda) * \varepsilon_{x_0} \in G(E)$ for some $\lambda \in L(E)$ and $x_0 \in E$, then $\lambda = 0$ and $x_0 = 0$.*

Theorem 3.4.17 *Let $(\lambda_n)_{n\geq 1}$ be a sequence in $L(E)$ with*

$$\tilde{e}_s(\lambda_n) \xrightarrow{\tau_w} \mu \in M^1(E) .$$

Moreover, let for every $\delta \in \mathbf{R}_+^\times$ *there exists an* $n(\delta) \in \mathbf{N}$ *such that* $\lambda_n(U_\delta^c) = 0$ *for all* $n \geq n(\delta)$. *Then*

(i) $\mu \in G(E)$.

(ii) $\hat{\mu}(a) = \exp(-\frac{1}{2}\lim_{n\to\infty}\int \langle x, a\rangle^2 \lambda_n(dx))$ *whenever* $a \in E'$.

Proof. From Theorem 3.3.12 *(iii)* (obviously valid also for nonsymmetric Lévy measures) we infer that

$$\alpha := \sup\Big\{ \int_{U_1} \langle x, a\rangle^2 \lambda_n(dx) : a \in V_1, n \in \mathbf{N}\Big\} < \infty .$$

For all $a \in E'$ and $n \geq n(\delta)$ we therefore obtain that

$$\int_E |\langle x, a\rangle|^3 \lambda_n(dx) = \int_{U_\delta} |\langle x, a\rangle|^3 \lambda_n(dx) \leq \alpha\, \delta\, \|a\|^3 .$$

This shows that

$$\lim_{n\to\infty} \int_E |\langle x, a\rangle|^3 \lambda_n(dx) = 0 \tag{1}$$

whenever $a \in E'$. Along with K we now consider the kernel M defined by

$$M(x,a) := K(x,a) + \frac{1}{2}\langle x, a\rangle^2$$

for all $x \in E, a \in E'$. It follows from the Taylor expansion of the exponential function that (1) implies

$$\lim_{n\to\infty} \int_E M(x,a)\lambda_n(dx) = 0$$

for all $a \in E'$. Now we employ Theorems 2.3.7, 2.3.3 and Corollary 2.1.16 in order to obtain that

$$\begin{aligned}
\operatorname{Log} \hat{\mu}(a) &= \lim_{n\to\infty} \int K(x,a)\lambda_n(dx) \\
&= \lim_{n\to\infty} \Big\{ \int M(x,a)\lambda_n(dx) - \frac{1}{2}\int \langle x, a\rangle^2 \lambda_n(dx)\Big\} \\
&= \lim_{n\to\infty} \Big(- \frac{1}{2}\int \langle x, a\rangle^2 \lambda_n(dx)\Big)
\end{aligned}$$

whenever $a \in E'$. The mapping $q : E' \to \mathbf{R}$ defined by

$$q(a) := \frac{1}{2} \lim_{n\to\infty} \int \langle x, a\rangle^2 \lambda_n(dx)$$

for all $a \in E'$ obviously has the property that

$$q(ta) = t^2 q(a)$$

valid for all $a \in E', t \in \mathbf{R}$. Consequently by Theorem 2.4.7 $\mu \in G(E)$. ■

Theorem 3.4.18 *Let $(\lambda_n)_{n\geq 1}$ be a sequence in $L(E)$ satisfying*

$$\tilde{e}_s(\lambda_n) \xrightarrow{\tau_w} \mu \in M^1(E)\,.$$

Then there exist measures $\lambda \in L(E), \varrho \in G(E)$ and an element $z \in E$ such that

$$\mu = \tilde{e}_s(\lambda) * \varrho * \varepsilon_z\,.$$

In particular,

(i) for every $\delta \in C(\lambda)$ we have that

$$\lambda_n|^\delta \xrightarrow{\tau_w} \lambda|^\delta,$$

(ii)

$$-2 \ \mathrm{Log}\ \hat{\varrho}(a) = \lim_{\delta\downarrow 0} \limsup_{n\to\infty} \int_{U_\delta} \langle x, a\rangle^2 \lambda_n(dx) = \lim_{\delta\downarrow 0} \liminf_{n\to\infty} \int_{U_\delta} \langle x, a\rangle^2 \lambda_n(dx)$$

holds for all $a \in E'$, and

(iii) z admits a representation

$$z = \int_{U_1\cap U_\delta^c} x\ \lambda(dx) - \lim_{n\to\infty} \int_{U_1\cap U_\delta^c} x\ \lambda_n(dx)$$

for $\delta \in C(\lambda) \cap]0,1[$ *which also equals*

$$\lim_{\substack{\delta\downarrow 0\\ \delta\in C(\lambda)}} \lim_{n\to\infty} \int_{U_\delta\cap U_1^c} x\,\lambda_n(dx)\,.$$

As a supplement we note that $z = 0$ *if* $1 \in C(\lambda)$.

Proof. 1. (Construction of λ). From the assumption we conclude that the sequence $(\tilde{e}_s(\lambda_n))_{n\geq 1}$ is τ_w-relatively compact, hence by Property 3.4.11.6 also $(\lambda_n|^\delta)_{n\geq 1}$ is τ_w-relatively compact (in $M^b(E)$). Next we show that there exist a subsequence $(\lambda_{n'}|^\delta)$ of $(\lambda_n|^\delta)_{n\geq 1}$ and a measure $\lambda \in M^\sigma(E)$ satisfying $\lambda(\{0\}) = 0$ and

$$\lambda_{n'}|^\delta \xrightarrow{\tau_w} \lambda|^\delta$$

for each $\delta \in C(\lambda)$. But then Theorem 3.4.3 implies that $(x(\lambda_n|^\delta))_{n\geq 1}$ whenever $\delta \in C(\lambda) \cap]0,1[$, and we may assume without loss of generality that

$$x(\lambda_{n'}|^\delta) \longrightarrow x_\delta$$

for some $x_\delta \in E$. From Theorem 3.4.6 we now deduce that

$$e_s(\lambda_{n'}|^\delta) \xrightarrow{\tau_w} e_s(\lambda|^\delta) * \varepsilon_{z_\delta},$$

where

$$\begin{aligned} z_\delta &:= x_\delta - x(\lambda|^\delta) \\ &= \lim_{n'\to\infty}\Big(\int_{U_1\cap U_\delta^c} x\,\lambda(dx) - \int_{U_1\cap U_\delta^c} x\,\lambda_{n'}(dx)\Big). \end{aligned}$$

In view of Corollary 3.4.4 z_δ is independent of δ and hence can be named z. We obtain

$$e_s(\lambda_{n'}|^\delta) \xrightarrow{\tau_w} e_s(\lambda|^\delta) * \varepsilon_z$$

whenever $\delta \in C(\lambda) \cap]0,1[$. Since

$$e_s(\lambda_{n'}|^\delta) * e_s(\lambda_{n'}|_\delta) \xrightarrow{\tau_w} \mu,$$

there exists a measure $\varrho_\lambda \in M^1(E)$ satisfying

$$e_s(\lambda_{n'}|^\delta) \xrightarrow{\tau_w} \varrho_\delta,$$

and from Corollary 2.2.4 we infer that

$$\mu = e_s(\lambda|^\delta) * \varrho_\delta * \varepsilon_z$$

as long as $\delta \in C(\lambda) \cap]0,1[$. For any sequence $(\delta_j)_{j\geq 1}$ in $C(\lambda)$ with $\delta_j \downarrow 0$ the sequence $(e_s(\lambda|^{\delta_j}))_{j\geq 1}$ is relatively shift compact by Theorem 2.2.7, hence τ_w-relatively compact by Property 3.4.11.5, and by the characterization theorem 3.4.9 we obtain that $\lambda \in L(E)$ and that

$$e_s(\lambda|^{\delta_j}) \xrightarrow{\tau_w} \tilde{e}_s(\lambda)$$

as $j \to \infty$. Applying Corollary 2.2.4 once more we achieve the τ_w-convergence of the sequence $(\varrho_j)_{j\geq 1}$ towards a measure $\varrho \in M^1(E)$ and the representation

$$\mu = \tilde{e}_s(\lambda) * \varrho * \varepsilon_z\,.$$

In fact, since

$$e_s(\lambda_{n_j}|^{\delta_j}) \xrightarrow{\tau_w} \varrho$$

as $j \to \infty$ for a suitable subsequence $(\lambda_{n_j}|^{\delta_j})$ we deduce from Theorem 3.4.17 that $\varrho \in G(E)$ and hence the desired representation of μ.

Now we show the remaining statements of the theorem.

(i). For an arbitrarily chosen $\delta \in C(\lambda)$ there exists a subsequence $(\lambda_{n_k})_{k\geq 1}$ of $(\lambda_n)_{n\geq 1}$ such that

$$\lambda_{n_k}|^\delta \xrightarrow{\tau_w} \lambda_\delta \in M^b(E)$$

as $k \to \infty$. From Property 3.4.11.6 we infer that $(\lambda_n|^\delta)_{n\geq 1}$ is τ_w-relatively compact. Now we proceed in analogy to part 1 of this proof with $(e_s(\lambda_{n_k}))_{k\geq 1}$ in place of $(e_s(\lambda_n))_{n\geq 1}$ and achieve the representation

$$\mu = \tilde{e}_s(\lambda^{(1)}) * \varrho^{(1)} * \varepsilon_{z^{(1)}}\,.$$

But then the uniqueness theorem 3.4.15 implies that $\lambda^{(1)} = \lambda$, and again considering subsequences we obtain that $\lambda_\delta = \lambda|^\delta$ as the unique accumulation point of the τ_w-relatively compact sequence $(\lambda_n|^\delta)_{n\geq 1}$.
(iii). The proof runs in analogy to that of *(i)*. We fix $\delta \in C(\lambda) \cap \,]0,1[$, observe that by Theorem 3.4.3 the sequence $(x(\lambda_n|^\delta) - x(\lambda^\delta))_{n\geq 1}$ is relatively compact in E, and passing to suitable subsequences together with an application of the uniqueness theorem 3.4.15 yields

$$\mu = \tilde{e}_s(\lambda) * \varrho * \varepsilon_z$$

with

$$z := \lim_{n\to\infty} x(\lambda_n|^\delta) - x(\lambda|^\delta)$$

as asserted.
(ii). For each $\delta \in C(\lambda) \cap \,]0,1[$ we have the limit relationships

$$\lambda_n|^\delta \xrightarrow{\tau_w} \lambda|^\delta$$

and

$$x(\lambda_n|^\delta) \longrightarrow z + x(\lambda|^\delta)\,.$$

Hence we apply Theorem 3.4.6 and obtain

$$e_s(\lambda_n|^\delta) \xrightarrow{\tau_w} e_s(\lambda|^\delta) * \varepsilon_z,$$

consequently with the help of Corollary 2.2.4

$$e_s(\lambda_n|^\delta) \xrightarrow{\tau_w} \varrho_\delta,$$

where ϱ_δ has been introduced in part 1. of this proof. Then Corollary 2.1.16 yields

$$\operatorname{Log} \hat{\varrho}_\delta(a) = \lim_{n\to\infty} \int\limits_{U_\delta} K(x,a)\lambda_n(dx)$$

for all $a \in E'$. Given $\delta > 0$ and $a \in E'$ we choose a subsequence $(\lambda_{n_k})_{k\geq 1}$ of $(\lambda_n)_{n\geq 1}$ such that

$$\lim_{k\to\infty} \int\limits_{U_\delta} \langle x,a\rangle^2 \lambda_{n_k}(dx) = \limsup_{n\to\infty} \int\limits_{U_\delta} \langle x,a\rangle^2 \lambda_n(dx)\,.$$

Clearly,

$$r(\delta, a) := \lim_{k\to\infty} \int\limits_{U_\delta} \Big(K(x,a) + \frac{1}{2}\langle x,a\rangle \Big) \lambda_{n_k}(dx)$$

is a well-defined element of $\mathbf{C}$. But then

$$\operatorname{Log}\, \hat{\varrho}_\delta(a) = r(\delta, a) - \frac{1}{2} \limsup_{n\to\infty} \int\limits_{U_\delta} \langle x,a\rangle^2 \lambda_n(dx)\,.$$

Moreover, employing Theorem 3.3.4 a Taylor expansion argument yields the estimate

$$|r(\delta, a)| \le c(a)\delta$$

with a constant $c(a) \ge 0$, for all sufficiently small $\delta > 0$. Taking a sequence $(\delta_j)_{j\ge1}$ in $C(\lambda)$ with $\delta_j > 0$ and $\delta_j \downarrow 0$ we obtain

$$\begin{aligned} \operatorname{Log}\, \hat{\varrho}(a) &= \lim_{j\to\infty} \operatorname{Log}\, \hat{\varrho}_{\delta_j}(a) \\ &= -\frac{1}{2} \lim_{j\to\infty} \limsup_{n\to\infty} \int\limits_{U_{\delta_j}} \langle x,a\rangle^2 \lambda_n(dx). \end{aligned}$$

But obviously the function

$$\delta \mapsto \limsup_{n\to\infty} \int\limits_{U_\delta} \langle x,a\rangle^2 \lambda_n(dx)$$

is increasing, thus $(\delta_j)_{j\ge1}$ can be chosen in $\mathbf{R}$ arbitrarily, and the first equality in *(ii)* has been established. The second one is proved similarly. ■

In the special case of bounded Lévy measures the preceding theorem can be improved slightly.

Theorem 3.4.19 *Let* $(\lambda_n)_{n\ge1}$ *be a sequence in* $L(E) \cap M^b(E)$ *such that*

$$e(\lambda_n) \xrightarrow{\tau_w} \mu \in M^1(E)\,.$$

Then there exist uniquely $\lambda \in L(E)$, $\varrho \in G(E)$ *and* $x_0 \in E$ *satisfying*

$$\mu = \tilde{e}_s(\lambda) * \varrho * \varepsilon_{x_0} .$$

Moreover, λ *and* ϱ *admit the representations given in Theorem 3.4.18, and*

$$x_0 = \lim_{\substack{\delta \downarrow 1 \\ \delta \in C(\lambda)}} \lim_{n \to \infty} \int_{U_\delta} x \, \lambda_n(dx) .$$

Proof. Since

$$e_s(\lambda_n) * \varepsilon_{-x(\lambda_n)} \xrightarrow{\tau_w} \mu,$$

the sequences $(e_s(\lambda_n))_{n \geq 1}$ and $(x(\lambda_n))$ are τ_w-relatively compact and relatively compact respectively. Indeed, one just applies Theorem 2.2.3 and Fact 3.4.5.1. Let $(\lambda_{n'})$ be a subsequence of (λ_n) such that

$$e_s(\lambda_{n'}) \xrightarrow{\tau_w} \nu \in M^1(E) .$$

Then Theorem 3.4.6 implies that

$$x(\lambda_{n'}) \longrightarrow z_0 \in E,$$

and one obtains the equality $\mu = \nu * \varepsilon_{-z_0}$. On the other hand we infer from Theorem 3.4.18 that

$$\mu = \tilde{e}_s(\lambda) * \varrho * \varepsilon_{z - z_0},$$

where $\lambda \in L(E)$ with

$$\lambda|^\delta := \tau_w - \lim_{n' \to \infty} \lambda_{n'}|^\delta$$

for $\delta \in C(\lambda)$ and

$$z := \lim_{n' \to \infty} x(\lambda_{n'}|^\delta) - x(\lambda|^\delta)$$

for $\delta \in C(\lambda) \cap]0, 1[$. By Theorem 3.3.3 the sequence $(\lambda_n|^\delta)_{n \geq 1}$ is τ_w-relatively compact, thus by the uniqueness theorem 3.4.15 we have

$$\lambda_n|^\delta \xrightarrow{\tau_w} \lambda|^\delta$$

as $n \to \infty$ (*(i)* of Theorem 3.4.18).

Moreover, the sequence $(x(\lambda_n|^\delta))_{n\geq 1}$ is relatively compact by Theorem 3.4.3, hence the sequence $(x(\lambda_n|^\delta) - x(\lambda_n))_{n\geq 1}$ has the same property. Again applying the uniqueness theorem 3.4.15 we obtain

$$\begin{aligned} x_0 &:= z - z_0 \\ &= \lim_{n\to\infty} (x(\lambda_n|^\delta) - x(\lambda_n)) - x(\lambda|^\delta) \end{aligned}$$

whenever $\delta \in C(\lambda) \cap]0,1[$. An application of Corollary 3.4.4 provides us with the limit representation of x_0 (*(iii)* of Theorem 3.4.18).

Finally we prove the representation of $-\operatorname{Log}\hat{\varrho}$ (*(ii)* of Theorem 3.4.18). Let $\delta \in C(\lambda) \cap]0,1[$. By assumption

$$e(\lambda_n|^\delta) * e(\lambda_n|_\delta) \xrightarrow{\tau_w} \mu\,.$$

But Theorem 2.4.12*(ii)* implies that

$$e(\lambda_n|^\delta) \xrightarrow{\tau_w} e(\lambda|^\delta),$$

thus

$$e(\lambda_n|_\delta) \xrightarrow{\tau_w} \nu_\delta \in M^1(E)$$

and

$$\mu = \nu_\delta * e(\lambda|^\delta)$$

(by Corollary 2.2.4). From

$$x(\lambda_n|_\delta) = x(\lambda_n) - x(\lambda_n|^\delta)$$

we conclude that

$$x(\lambda_n|_\delta) \longrightarrow -x_0 - x(\lambda|^\delta)$$

hence that

$$e_s(\lambda_n|_\delta) \xrightarrow{\tau_w} \nu_\delta * \varepsilon_{-x_0 - x(\lambda|^\delta)}\,.$$

Now Theorems 3.4.18 and 3.4.15 applied to $(e_s(\lambda_n|_\delta))_{n\geq 1}$ instead of $(e_s(\lambda_n))_{n\geq 1}$ yield the equality

$$\nu_\delta * \varepsilon_{-x_0-x(\lambda|^\delta)} = e_s(\lambda|_\delta) * \varrho,$$

hence the assertion. ■

Theorem 3.4.20 *(Lévy-Khintchine decomposition of infinitely divisible measures)*

Let $\mu \in I(E)$ with the sequence $(\mu_{\frac{1}{n}})_{n\geq 1}$ of its n-th roots. Then there exist measures $\lambda \in L(E)$, $\varrho \in G(E)$ and an element x_0 of E such that

$$\mu = \tilde{e}_s(\lambda) * \varrho * \varepsilon_{x_0}.$$

Moreover, λ, ϱ and x_0 are uniquely determined and obtained as follows:

(i)
$$(n\mu_{\frac{1}{n}})|^\delta \xrightarrow{\tau_w} \lambda|^\delta$$

for all $\delta \in C(\lambda)$.

(ii)
$$-2 \operatorname{Log} \hat{\varrho}(a) = \lim_{\delta\downarrow 0} \limsup_{n\to\infty} n \int_{U_\delta} \langle x, a\rangle^2 \mu_{\frac{1}{n}}(dx) = \lim_{\delta\downarrow 0} \liminf_{n\to\infty} n \int_{U_\delta} \langle x, a\rangle^2 \mu_{\frac{1}{n}}(dx)$$

for all $a \in E'$.

(iii)
$$x_0 = \lim_{\substack{\delta\downarrow 1\\ \delta\in C(\lambda)}} \lim_{n\to\infty} n \int_{U_\delta} x\, \mu_{\frac{1}{n}}(dx).$$

Proof. From the uniqueness of the roots $\mu_{\frac{1}{n}}$ of μ (Theorem 2.3.5) we infer that for every $n \in \mathbf{N}$ the measure

$$\lambda_n := n(\mu_{\frac{1}{n}} - \mu_{\frac{1}{n}}(\{0\})\varepsilon_0)$$

belongs to $L(E)$. But Property 2.4.10.2 tells us that

$$e_s(n\mu_{\frac{1}{n}}) = e_s(\lambda_n) * e_s(n\mu_{\frac{1}{n}}(\{0\})\varepsilon_0) = e_s(\lambda_n)$$

for all $n \in \mathbf{N}$, and Theorem 2.4.11 implies that

$$e_s(\lambda_n) \xrightarrow{\tau_w} \mu\,.$$

In view of

$$\lambda_n|^\delta = (n\mu_{\frac{1}{n}})|^\delta,$$

$$\int_{U_\delta} \langle x, a\rangle^2 \lambda_n(dx) = n \int_{U_\delta} \langle x, a\rangle^2 \mu_{\frac{1}{n}}(dx),$$

and

$$\int_{U_\delta} x\, \lambda_n(dx) = n \int_{U_\delta} x\, \mu_{\frac{1}{n}}(dx)$$

valid for all $a \in E', \delta \in \mathbf{R}_+^\times$ $(n \in \mathbf{N})$ Theorem 3.4.19 implies the assertion. ■

Remark 3.4.21 *In short Theorem 3.4.20 says that any $\mu \in I(E)$ admits a* ***Lévy-Khintchine representation*** *of the form*

$$\hat{\mu}(a) = \exp\left\{i\langle x_0, a\rangle - \frac{1}{2}\langle Ra, a\rangle + \int \left(e^{i\langle x,y\rangle} - 1 - i\langle x, a\rangle 1_{U_1}(x)\right)\lambda(dx)\right\}$$

valid for all $a \in E'$, where $x_0 \in E$, R is a symmetric linear mapping $E' \to E$ and λ a Lévy measure.

One observes that μ is symmetric if and only if λ is symmetric, and in this case $x_0 = 0$.

On the other hand, if λ satisfies the condition

$$\int_{U_1} \|x\|^2 \lambda(dx) < \infty,$$

then the kernel

$$(x, a) \mapsto e^{i\langle x,a\rangle} - 1 - i\langle x, a\rangle 1_{U_1}$$

(formerly abbreviated by K) can be replaced by the classical more familiar one

$$(x, a) \mapsto e^{i\langle x,a\rangle} - 1 - \frac{i\langle x, a\rangle}{1 + \|x\|^2}\,.$$

In this case the measure $\nu \in I(E)$ given by

$$\hat{\nu}(a) := \exp\Big\{ -\int \Big(e^{i\langle x,a\rangle} - 1 - \frac{i\langle x,a\rangle}{1+\|x\|^2} \Big) \lambda(dx) \Big\}$$

for all $a \in E'$ is a translate of $\tilde{e}_s(\lambda)$.

4

Harmonic Analysis of Convolution Semigroups

4.1 Convolution of Radon measures

We start with an adaptation of the concept of measure given in Chapter 1 to locally compact spaces E. ***Complex Borel measures*** μ on E are introduced as complex-valued σ-additive set functions μ on the Borel-σ-algebra $\mathfrak{B}(E)$ of E having the property that $\mu(B)$ is finite for each relatively compact subset B of E. The totality of complex Borel measures will be abbreviated by $M_0(E)$. Notice that measures in $M_{0,+}(G)$ may take on the value ∞. For every $\mu \in M_0(E)$ one introduces the total variation $|\mu|$ of μ by

$$|\mu|(B) := \sup\Big\{ \sum_{i\in I} |\mu(B_i)| : |I| < \infty,\ \ B_i \in \mathfrak{B}(E) \text{ for all } i \in I,\ \bigcup_{i\in I} B_i = B \Big\},$$

whenever $B \in \mathfrak{B}(E)$. It is shown that $|\mu| \in M_0(E)$ and, of course, $|\mu| \geq 0$. $\mu \in M_0(E)$ is said to be regular if for every $B \in \mathfrak{B}(E)$

$$\begin{aligned} |\mu|(B) &= \sup\{|\mu|(K) : K \in \mathcal{K}(E),\ \ K \subset B\} \\ &= \sup\{|\mu|(O) : O \in \mathcal{O}(E),\ \ O \supset B\}. \end{aligned}$$

The symbol $M(E)$ will serve as a short form for the set of all regular complex Borel measures in E. Finally, for $\mu \in M_0(E)$ we set

$$\|\mu\| := |\mu|(E)$$

and recognize that the set

$$M^b(E) := \{\mu \in M(E) : \|\mu\| < \infty\}$$

of all ***finite regular*** (complex Borel) ***measures*** in E forms a normed vector space over $\mathbf{C}$, with $(\mu, \nu) \mapsto \mu + \nu$ and $\mu \mapsto \alpha\mu$ as vector space operations and $\|\cdot\|$ as the underlying norm. The following sequence of implications

$$M^1(E) \subset M^b_+(E) \subset M^b_{\mathbf{R}}(E) \subset M^b_{\mathbf{C}}(E) := M^b(E)$$

starting with the set

$$M^1(E) := \{\mu \in M^b_+(E) : \|\mu\| = 1\}$$

of probability measures speaks for itself. As for the function spaces applied in the sequel we have the sequence of implications of vector spaces

$$C^c(E) \subset C^0(E) \subset C^b(E) \subset C(E),$$

where the corresponding symbols stand for the complex continuous functions on E which have compact support, vanish at infinity, are bounded and just continuous respectively. While $C(E)$ carries the topology τ_{co} of compact convergence and $C^c(E)$ the canonical inductive limit topology, $C^0(E)$ and $C^b(E)$ are furnished with the topology of uniform convergence. We note that $C^c(E)^- = C^0(E)$. For any vector space B over $\mathbf{C}$ the symbols $L^b(B)$ or $L_+(B)$ will denote the sets of bounded or positive linear functionals on B respectively.

In the above described extended set-up of measure theory the Riesz representation theorem takes on a more general form valid beyond the metric case.

Theorem 4.1.1 *There exists an isometric isomorphism*

$$\mu \mapsto L_\mu$$

from $M^b(E)$ *onto* $C^0(E)^* := L^b(C^0(E))$ *given by*

$$L_\mu(f) := \int f d\mu$$

for all $f \in C^0(E)$.

In fact, to every $L \in L^b(C^0(E))$ *there corresponds a unique* $\mu \in M^b(E)$ *satisfying* $L = L_\mu$, *and*

$$\sup\{|L(f)| : f \in C^0(E), \|f\| = 1\} = \|\mu\| .$$

In particular, $(M^b(E), \|\cdot\|)$ *is a Banach space over* $\mathbf{C}$.

It should be noted that a similar correspondence is available for the sets $L_+(C^c(E))$ *and* $M_+(E)$ *instead of* $L^b(C^0(E))$ *and* $M^b(E)$ *respectively.*

This correspondence justifies Bourbaki's introduction of measures on locally compact spaces E as continuous linear functionals on $C^c(E)$. In other words, $M(E)$ will be interpreted as the topological dual $C^c(E)'$ of $C^c(E)$, and $M(E)$ will appear as the set of ***Radon measures*** on E.

4.1.2 For any $\nu \in M_{0,+}(E)$ and $p \in]0, \infty[$ one introduces the Lebesgue space $L^p(E, \nu)$ of p-times ν-integrable complex-valued functions on E and notes that $L^p(E, \nu)$ is a Banach space provided $p \geq 1$, a Hilbert space for $p = 2$, and that

$$L^p(E, \nu) = C^c(E)^-$$

if $\nu \in M_+(E)$, i.e. if ν is regular. Here the closure of $C^c(E)$ is taken in the p-norm topology of $L^p(E, \nu)$.

For $\mu \in M^b(E)$ and $\nu \in M_{0,+}(E)$ the Radon-Nikodym equivalence is available: μ is ν-(absolutely) continuous if and only if there exists a function $f \in L^1(E, \nu)$ such that $\mu = f \cdot \nu$. In particular, $L^1(E, \nu)$ is isometrically embedded into $M^b(E)$, i.e.

$$\|\mu\| = \|f\|,$$

for every $\mu \in M^b(E)$ of the form $\mu := f \cdot \nu$ with $f \in L^1(E, \nu)$.

4.1.3 On $M(E)$ the ***vague topology*** τ_v is introduced as the topology $\sigma(C^c(E)', C^c(E))$ for the dual pair $(C^c(E)', C^c(E))$ arising from the normed vector space $C^c(E)$. On $M^b(E)$ the vague topology can be compared with the ***weak topology*** considered as the topology $\sigma(C^b(E)', C^b(E))$ for the dual pair arising from $C^b(E)$. In fact, τ_w is finer than τ_v.

Proposition *For any net $(\mu_\alpha)_{\alpha \in A}$ in $M^b_+(E)$ and any measure $\mu \in M^b_+(E)$ the following statements are equivalent:*

(i) $\tau_w - \lim_\alpha \mu_\alpha = \mu$.

(ii) $\tau_v - \lim_\alpha \mu_\alpha = \mu$ *and* $\lim_\alpha \|\mu_\alpha\| = \|\mu\|$.

Proof. It suffices to show the implication *(ii)* $\Rightarrow$ *(i)*. Let *(ii)* be satisfied for a net $(\mu_\alpha)_{\alpha \in A}$ and a measure μ in $M^b_+(E)$. We take a function $f \in C^b(E)$ and fix $\varepsilon > 0$. Since μ is regular, there exists a set $K \in \mathcal{K}(E)$ such that $\mu(K^c) < \varepsilon$. Now choose a function $g \in C^c(E)$ satisfying $0 \leq h \leq 1$ and $h(x) = 1$ for all $x \in K$. By assumption we have

$$\begin{aligned} \lim_\alpha \int (1 - g) d\mu_\alpha &= \int (1 - g) d\mu \\ &\leq \mu(K^c) < \varepsilon \end{aligned}$$

and

$$\lim_\alpha \int f g \, d\mu_\alpha = \int f g \, d\mu \, .$$

Consequently, there exists an $\alpha_0 \in A$ such that for all $\alpha \in A$ with $\alpha \geq \alpha_0$ the inequalities

$$\int (1 - g) d\mu_\alpha < \varepsilon$$

and

$$\Big|\int fg\,d\mu_\alpha - \int fg\,d\mu\Big| < \varepsilon$$

hold. But this implies for all $\alpha \in A$ with $\alpha \geq \alpha_0$ that

$$\Big|\int f d\mu_\alpha - \int f d\mu| \leq \Big|\int fgd\mu_\alpha - \int fgd\mu\Big|$$
$$+\Big|\int f(1-g)d\mu_\alpha\Big| + \Big|\int f(1-g)d\mu\Big| \leq \varepsilon(1+2\|f\|),$$

hence that

$$\tau_w - \lim_\alpha \mu_\alpha = \mu\,. \qquad \blacksquare$$

It follows from the Proposition that on $M^1(E)$ the topologies τ_w and τ_v coincide.

Further topological properties of concern the τ_v-metrizability of $M_+(E)$ which holds if and only if E admits a countable basis of its topology, and the τ_v-compactness of $M^1(E)$ which is equivalent to the compactness of E.

From now on let $E := G$ be a locally compact Abelian group. For every $a \in G$ the group translation $x \mapsto x + a$ can be extended to functions f on G by

$$T_a f(x) := f_a(x) := f(x-a)$$

for all $x \in G$ and to measures μ on G by

$$\begin{aligned} T_a(\mu)(f) :&= \int f_{-a}(x)\mu(dx) \\ &= \int f(x+a)\mu(dx) \end{aligned}$$

for all $f \in C^c(G)$.

Definition 4.1.4 *A measure $\mu \in M_+(G)$ is called a* ***Haar measure*** *on G if $\mu \neq 0$ and if μ is* ***translation invariant*** *in the sense that*

$$\mu(T_a f) = \mu(f)$$

for all $f \in C^c(G)$ *and* $a \in G$.

Theorem 4.1.5 *On any locally compact Abelian group* G *there exists a Haar measure.*

Proof. 1. Producing the crucial functional.
In the following the space $C^c(G)$ is assumed to contain only functions $\neq 0$. For $f, \varphi \in C^c_+(G)$ we define

$$(f : \varphi) := \inf \left\{ \sum_{j=1}^{n} c_j : f \leq \sum_{j=1}^{n} c_j \, \varphi_{x_j} \text{ for } c_1, \ldots, c_n \in \mathbf{R}_+, \right. \\ \left. x_1, \ldots, x_n \in G, n \geq 1 \right\}$$

and derive the following properties

(1) $(f : \varphi) = (T_y f : \varphi)$ for all $y \in G$.

(2) $(f_1 + f_2 : \varphi) \leq (f_1 : \varphi) + (f_2 : \varphi)$.

(3) $(cf : \varphi) = c(f : \varphi)$ for any $c \in \mathbf{R}_+^\times$.

(4) $(f_1 : \varphi) \leq (f_2 : \varphi)$ whenever $f_1 \leq f_2$.

(5) $(f : \varphi) \geq \frac{\|f\|}{\|\varphi\|}$.

(6) $(f : \varphi) \leq (f : \psi)(\psi : \varphi)$ for any $\psi \in C^c_+(G)$.

In these statements f, f_1, f_2 and φ are functions in $C^c_+(G)$.

In order to justify the above assertions it suffices to note that since the compact support of f can be covered by a finite number N of translates of the set

$$\left\{ x \in G : \varphi(x) > \frac{1}{2}\|\varphi\| \right\},$$

$$(f : \varphi) \leq 2N \frac{\|f\|}{\|\varphi\|}$$

and hence $(f:\varphi)$ is well-defined. While properties (1) to (5) are easily verified, property (6) follows from the additional observation that for

$$f \leq \sum_{j=1}^{n} c_j\, \varphi_{x_j}$$

and

$$g \leq \sum_{k=1}^{m} d_k\, \varphi_{y_k}$$

with the obvious meaning of the summands involved,

$$f \leq \sum_{j=1}^{n} \sum_{k=1}^{m} c_j\, d_k\, \varphi_{x_j+y_k}$$

holds.

Now, we fix $f_0 \in C_+^c(G)$ and define

$$I_\varphi(f) := \frac{(f:\varphi)}{(f_0:\varphi)}$$

whenever $f, \varphi \in C_+^c(G)$. From properties (1) to (4) we deduce that I_φ is invariant, subadditive, homogeneous, and increasing. Moreover, by property (6) we obtain that

(7) $\frac{1}{(f_0:f)} \leq I_\varphi(f) \leq (f:f_0)$.

2. A preparative inequality.
We show that given $f_1, f_2 \in C_+^c(G)$ and $\varepsilon > 0$ there exists a $V \in \mathfrak{V}(0)$ such that

$$I_\varphi(f_1) + I_\varphi(f_2) \leq I_\varphi(f_1 + f_2) + \varepsilon$$

whenever $\operatorname{supp}(\varphi) \subset V$.

In fact, let $g \in C_+^c(G)$ such that $g(x) = 1$ for all $x \in \operatorname{supp}(f_1 + f_2)$, and let $\delta > 0$. Moreover, let $h := f_1 + f_2 + \delta g$ and for $1 = 1, 2$

$$h_i = \frac{f_i}{h}$$

with the convention that $h_i = 0$ whenever $f_i = 0$. Then $h_i \in C_+^c(G)$, hence there exists a $V \in \mathfrak{V}(0)$ such that

$$|h_i(x) - h_i(y)| < \delta$$

for $x - y \in V (i = 1, 2)$. Now we take $\varphi \in C_+^c(G)$ with $\operatorname{supp}(\varphi) \subset V$. If

$$h \leq \sum_{j=1}^{n} c_j \ \varphi_{x_j}$$

with the summands as described above, then

$$\begin{aligned} f_i(x) &= h(x)h_i(x) \\ &\leq \sum_{j=1}^{h} c_j \varphi(x - x_j) h_i(x) \\ &\leq \sum_{j=1}^{n} c_j \varphi(x - x_j)(h_i(x_j) + \delta), \end{aligned}$$

since $|h_i(x) - h_i(x_j)| < \delta$ whenever $x - x_j \in \operatorname{supp}(\varphi)$. But we have $h_1 + h_2 \leq 1$, hence obtain

$$\begin{aligned} (f_1 : \varphi) + (f_2 : \varphi) &\leq \sum_{j=1}^{n} c_j(h_1(x_j) + \delta) + \sum_{j=1}^{n} c_j(h_2(x_j) + \delta) \\ &\leq \sum_{j=1}^{n} (1 + 2\delta), \end{aligned}$$

and taking the infimum over all sums of the form $\sum_{j=1}^{n} c_j$, properties (2) and (3) imply that

$$\begin{aligned} I_\varphi(f_1) + I_\varphi(f_2) &\leq (1 + 2\delta) I_\varphi(h) \\ &\leq (1 + 2\delta)(I_\varphi(f_1 + f_2) + \delta I_\varphi(g)) \end{aligned}$$

holds. Finally, property (7) yields

$$2\delta(f_1 + f_2 : f_0) + \delta(1 + 2\delta)(g : f_0) < \varepsilon$$

once δ is chosen sufficiently small, and this proves the assertion.

3. The completion of the proof.

For each $f \in C_+^c(G)$ let M_f denote the interval $\left[\frac{1}{(f_0:f)}, (f : f_0)\right]$ arising from property (7), and put $M := \Pi_{f\in C_+^c(G)} M_f$. Clearly, M is a compact space which by property (7) contains all mappings I_φ for $\varphi \in C_+^c(G)$. For each $N \in \mathfrak{V}(0)$ let $c(V)$ denote the closure in M of the set of those I_φ for which $\operatorname{supp}(\varphi) \subset V$. From

$$\bigcap_{j=1}^{n} c(V_j) \supset c\left(\bigcap_{j=1}^{n} V_j\right)$$

for neighborhoods $V_j \in \mathfrak{V}(0)$ we conclude that the system of sets $c(V)$ $(V \in \mathfrak{V}(0))$ has the finite intersection property, hence the compactness of M secures the existence of an element I of M with $I \in c(V)$ for all $V \in \mathfrak{V}(0)$ which implies that every neighborhood of I in M contains mappings I_φ with arbitrarily small $\operatorname{supp}(\varphi)$. This means that for any $V \in \mathfrak{V}(0)$, any $\varepsilon > 0$, and all $f_1, ..., f_n \in C_+^c(G)$ there exists a $\varphi \in C_+^c(G)$ satisfying $\operatorname{supp}(\varphi) \subset V$ and

$$|I(f_j) - I_\varphi(f_j)| < \varepsilon$$

for all $j = 1, ..., n$. From properties (1) to (3) and part 2. of this proof we deduce that I is translation invariant, additive and homogeneous.

Now, any $f \in C_+^c(G)$ admits a representation $f = g - h$ with $g, h \in C_+^c(G)$. If, in addition $f = g' - h'$ with $g', h' \in C_+^c(G)$, then $g + h' = h + g'$, hence

$$I(g) + I(h') = I(h) + I(g'),$$

and

$$I(f) := I(g) - I(g)$$

yields a well-defined extension of I to a positive linear functional on $C^c(G)$ which by the Riesz Theorem 4.1.1 is a Haar measure. ■

Theorem 4.1.6 *For two Haar measures μ and ν on G there exists a constant $c \in \mathbf{R}_+^*$ such that*

$$\nu = c\ \mu\,.$$

Proof. Let $g \in C^c(G)$ with $\int g d\mu = 1$. Then, putting

$$c := \int_G g(-x)\nu(dx)$$

we obtain for any $f \in C^c(G)$ that

$$\begin{aligned}
\int_G f d\nu &= \int_G g(y)\mu(dy) \int_G f(x)\nu(dx) \\
&= \int_G g(y)\mu(dy) \int_G f(x+y)\nu(dx) \\
&= \int_G \left(\int_G g(y) f(x+y)\mu(dy) \right) \nu(dx) \\
&= \int_G \left(\int_G g(y-x) f(y)\mu(dy) \right) \nu(dx) \\
&= \int_G f(y)\mu(dy) \int_G g(y-x)\nu(dx) \\
&= c \int_G f d\mu,
\end{aligned}$$

hence that $\nu = c\ \mu$. In the above chain of equalities the Fubini theorem was applicable since the integrands of the double integrals belong to $C^c(G \times G)$. ■

Convention 4.1.7 *Since by Theorem 4.1.6 Haar measure is unique up to a multiplicative constant, one talks about* **the** *Haar measure of G and denotes it by $\omega = \omega_G$.*

Properties 4.1.8

4.1.8.1 *ω is positive on non-empty open subsets of G, i.e.*
$$\omega(O) > 0 \text{ for all } O \in \mathcal{O}(G), \quad O \neq \emptyset.$$

4.1.8.2 *ω is **inverse invariant**, i.e.*
$$\omega(-B) = \omega(B) \text{ for all } B \in \mathfrak{B}(G).$$

Definition 4.1.9 *A pair (μ, ν) of measures in $M(G)$ is said to be **convolvable** if the integral*

$$\int\limits_{G\times G} f(x+y)\mu \otimes \nu(d(x,y))$$

exists in $\mathbf{C}$ for every $f \in C^c(G)$. In this case the mapping

$$f \mapsto \int\limits_{G\times G} f(x+y)\mu \otimes \nu(d(x,y))$$

*is a continuous linear functional on $C^c(G)$, hence a measure in $M(G)(\cong C^c(G)')$; it is called the **convolution** of μ and ν and is denoted by $\mu * \nu$.*

In the case of convolvability the convolution viewed as a mapping

$$(\mu, \nu) \mapsto \mu * \nu$$

from $M(G) \times M(G)$ into $M(G)$ yields a commutative and associative operation.

The following result subsuming various useful properties of the convolution is easily proved.

Theorem 4.1.10 *Any pair (μ, ν) in $M^b(G) \times M^b(G)$ is convolvable, and together with convolution $\mu * \nu$ and norm $\|\cdot\|$ thc space $M^b(G)$ becomes a commutative Banach algebra with ε_0 as (convolution) unit element. In particular,*

$$\|\mu * \nu\| \leq \|\mu\|\|\nu\|$$

whenever $\mu, \nu \in M^b(G)$.

Moreover, the Banach algebra $M^b(G)$ *is involutive with respect to the involution* $\mu \mapsto \mu^\sim$ *given by*

$$\mu^\sim(f) := \overline{\mu(f^\sim)}$$

for all $f \in C^c(G)$, *where* $f^\sim = \overline{f^*}$ *with* $f^*(x) := f(-x)$ *for all* $x \in G$. *Here, the properties defining the involution read as follows*

$$\mu^{\sim\sim} = \mu,$$

$$(\mu * \nu)^\sim = \mu^\sim * \nu^\sim, and$$

$$\|\mu^\sim\| = \|\mu\|,$$

whenever $\mu, \nu \in M^b(G)$.

For measures $\mu, \nu \in M^b_+$ *the* ***support formula***

$$\text{supp}\,(\mu * \nu) = (\text{supp}\,(\mu) + \text{supp}\,(\nu))^-$$

holds

4.1.11 Applying the convolution to measures of the form $\mu := f \cdot \omega_G$ for $f \in L^1(G, \omega_G)$ one obtains a convolution in $L^1(G, \omega_G)$ which for $f, g \in L^1(G, \omega_G)$ is given by

$$f * g(x) = \int f(x - y) g(y) \omega_G(dy)$$

whenever $x \in G$.

Clearly, together with this convolution and the norm $\|\cdot\|_1$ the space $L^1(G, \omega_G)$ becomes a closed ideal of $M^b(G)$ and therefore also a commutative Banach algebra with involution.

$M^b(G)$ and $L^1(G, \omega_G)$ are called the ***measure algebra*** and the ***group algebra*** of G respectively.

While $M^b(G)$ has a unit $\varepsilon_0, L^1(G, \omega_G)$ has a unit only if G is discrete.

For functions f, g in an arbitrary Lebesgue space $L^p(G, \omega_G)$ ($p \in [1, \infty]$) the convolution $f * g$ is introduced by

$$f * g(x) := \int_G f(x-y)g(y)\omega_G(dy)$$

provided that

$$\int_G |f(x-y)g(y)|\omega_G(dy) < \infty$$

for all $x \in G$.

Proposition 4.1.12 *For each $f \in L^p(G, \omega_G)$ ($p \in [1, \infty[$) the mapping $x \mapsto f_x$ from G into $L^p(G, \omega_G)$ is uniformly continuous.*

Proof. Let $f \in L^p(G, \omega_G)$ and let $\varepsilon > 0$. Since $C^c(G)$ is dense in $L^p(G, \omega_G)$, there exists a function $g \in C^c(G)$ such that

$$\|g - f\|_p < \frac{\varepsilon}{3} \,.$$

Let $K := \operatorname{supp}(g)$. From the uniform continuity of g we deduce the existence of a neighborhood $V \in \mathfrak{V}(0)$ such that

$$\|g - g_x\| < \frac{\varepsilon}{3\omega_G(K)^{\frac{1}{p}}}$$

for all $x \in V$. Therefore

$$\|g - g_x\|_p < \frac{\varepsilon}{3}$$

and hence

$$\|f - f_x\|_p \leq \|f - g\|_p + \|g - g_x\|_p + \|g_x - f_x\|_p < \varepsilon$$

whenever $x \in V$. Finally we note that $f_x - f_y = (f - f_{y-x})_x$, so that

$$\|f_x - f_y\|_p < \varepsilon$$

whenever $y - x \in V$. ■

Proposition 4.1.13 *Let $\mathfrak{U}$ be a neighborhood system of $0 \in G$. For each $U \in \mathfrak{U}$ let ψ_U be a measurable function on G with compact support* $\operatorname{supp}(\psi_U) \subset U$ *such that $\psi_U \geq 0$, $\psi_U^* = \psi_U$ and $\int \psi_U d\omega_G = 1$. Then*

$$\|f * \psi_U - f\|_p \to 0 \text{ as } U \to \{0\}$$

for all $f \in L^p(G, \omega_G)$, $p \in [1, \infty[$.

The family $\{\psi_U : U \in \mathfrak{U}\}$ is said to be an ***approximate identity*** in $L^p(G, \omega_G)$.

Proof. By Proposition 4.1.12 we choose $U \in \mathfrak{U}$ such that

$$\|f - f_y\|_p \to 0 \text{ as } U \to \{0\}$$

whenever $y \in G$. But for each $h \in L^q(G, \omega_G)$ (where q is conjugate to p) Fubini's theorem and Hölder's inequality imply

$$\left| \int_G (f * \psi_U - f) h \, d\omega_G \right| \leq \|h\|_q \int_G \|f - f_y\|_p \psi_U(y) \omega_G(dy)$$

and therefore

$$\begin{aligned} \|f * \psi_U - f\|_p &\leq \int_U \|f - f_y\|_p \psi_U(y) \omega_G(dy) \\ &\leq \sup_{y \in U} \|f - f_y\|_p \to 0 \text{ as } U \to \{0\}. \end{aligned}$$ ■

4.2 Duality of locally compact Abelian groups

Generalizing classical Fourier theory for Euclidean spaces two settings can be chosen in order to establish the appropriate analysis: locally convex vector spaces E, the structure underlying Chapters 1 to 3, and locally compact Abelian groups G, the structure to remain the basis of discussion in the subsequent chapters 4 to 6. The counterpart of the linear dual of E will be the group of continuous characters of G.

Definition 4.2.1 *Any homomorphism $\chi : G \to \mathbf{T}$ is called a* ***character*** *of G.*

The set $G^\wedge$ of all continuous characters of G forms a group under addition in the sense that for $\chi, \rho \in G^\wedge$ the sum is given by

$$(\chi + \rho)(x) := \chi(x)\rho(x),$$

the inverse $-\chi$ of χ by

$$(-\chi)(x) := \overline{\chi(x)},$$

and the neutral element 0 by

$$0(x) := 1$$

whenever $x \in G$.

In view of the duality between G and $G^\wedge$ to be discussed at a later stage, $G^\wedge$ is called the ***dual*** *of G.*

In the sequel we shall show that $G^\wedge$ can be furnished with a topology that makes it a locally compact group so that G and $G^\wedge$ belong to the same category of objects and consequently $G^{\wedge\wedge} := (G^\wedge)^\wedge$ can be formed.

Theorem 4.2.2 *There exists a one-to-one correspondence between the dual $G^\wedge$ of G and the space $M(L^1(G, \omega_G))$ of all nonvanishing multiplicative linear functionals on the group algebra $L^1(G, \omega_G)$ of G given as a mapping*

$$\chi \mapsto \tau_\chi$$

with

$$\tau_\chi(f) := \int_G f\overline{\chi} d\omega_G$$

for all $f \in L^1(G, \omega_G)$.

Proof. 1. The mapping $f \mapsto \tau_\chi(f)$ is obviously linear. Moreover it is multiplicative, as follows from the following sequence of equalities valid for all $f, g \in L^1(G, \omega_G)$ and $\chi \in G^\wedge$:

$$\begin{aligned}\tau_\chi(f * g) &= \int (f * g)(x)\overline{\chi(x)}\, \omega_G(dx) \\ &= \int \overline{\chi(x)}\, \omega_G(dx) \int f_y(x) g(y)\, \omega_G(dy) \\ &= \int g(y)\overline{\chi(y)}\, \omega_G(dy) \int f_y(x)\chi(-x+y)\, \omega_G(dx) \\ &= \tau_\chi(f)\tau_\chi(g).\end{aligned}$$

The nonvanishing of the mapping $f \mapsto \tau_\chi(f)$ follows from $\tau_\chi(f) \neq 0$ for some $f \in L^1(G, \omega_G)$, since $|\chi(x)| = 1$ for all $x \in G$. One concludes that $\tau_\chi \in M(L^1(G, \omega_G))$

2. Now, let $\tau \in M(L^1(G, \omega_G))$. Since $L^1(G, \omega_G) \cong L^\infty(G, \omega_G)^*$ and τ is a bounded linear functional with $\|\tau\| = 1$, there exists a function $\varphi \in L^\infty(G, \omega_G)$ with $\|\varphi\| = 1$ such that

$$\tau(f) = \int_G f\varphi\ d\omega_G$$

for all $f \in L^1(G, \omega_G)$. For $f, g \in L^1(G, \omega_G)$ we have

$$\begin{aligned}\int \tau(f) g\varphi\ d\omega_G &= \tau(f)\tau(g) \\ &= \tau(f * g) = \int (f * g)\varphi\ d\omega_G \\ &= \int g d\omega_G \int f_y(x)\varphi(x)\omega_G(dx) \\ &= \int g(y)\tau(f_y)\omega_G(dy),\end{aligned}$$

consequently

$$\tau(f)\varphi(y) = \tau(f_y) \tag{1}$$

for ω_G-a.a. $y \in G$. It follows from Proposition 4.1.12 that the continuity of τ implies the continuity of $y \mapsto \tau(f_y)$ for each $f \in L^1(G, \omega_G)$. Choosing $f \in L^1(G, \omega_G)$ such that $\tau(f) \neq 0$ we deduce from (1) that φ is ω_G-a.e. continuous. Hence φ can be assumed to be continuous, and (1) holds for all $y \in G$.

Now we replace y by $x + y$ and then f by f_x in (1) in order to obtain

$$\begin{aligned}\tau(f)\varphi(x+y) &= \tau(f_{x+y}) = \tau((f_x)_y) \\ &= \tau(f_x)\varphi(y) = \tau(f)\varphi(x)\varphi(y)\end{aligned}$$

and consequently

$$\varphi(x+y) = \varphi(x)\varphi(y)$$

for all $x, y \in G$. In particular, $\varphi^* = \varphi^{-1}$. Since $|\varphi| \leq 1$ it follows that $|\varphi| = 1$, hence that $\varphi \in G^\wedge$ and so $\tau = \tau_\varphi$.

3. As for the uniqueness of φ we just note that

$$\tau_\chi(f) = \tau_\rho(f)$$

for $\chi, \rho \in G^\wedge$ and all $f \in L^1(G, \omega_G)$ implies that

$$\chi(x) = \rho(x)$$

for ω_G-a.a. $x \in G$, but since χ, ρ are continuous functions, even for all $x \in G$. ■

4.2.3 For every $f \in L^1(G, \omega_G)$ the function $\hat{f} = \mathcal{F}(f)$ defined by

$$\hat{f}(\chi) := \tau_\chi(f)$$

for all $\chi \in G^\wedge$ is said to be the ***Fourier transform*** of f.

In terms of Gelfand's theory as described in Appendix C 8 $\hat{f}$ is the Gelfand transform of f and $f \mapsto \hat{f}$ the Gelfand mapping $\mathcal{F}$ on $L^1(G, \omega_G)$. The space $M(L^1(G, \omega_G))$ identified by Theorem 4.2.2 with the dual $G^\wedge$ of G can be interpreted as the Gelfand space

$\Delta(L^1(G,\omega_G))$ of $L^1(G,\omega_G)$. Since $\Delta(L^1(G,\omega_G))$ is a locally compact space with respect to the weak topology induced by the set

$$A(G^\wedge) = L^1(G,\omega_G)^\wedge,$$

$G^\wedge$ is also a locally compact *space*. This topology τ_g, sometimes named the Gelfand topology on $G^\wedge$, admits a neighborhood system of $\chi_0 \in G^\wedge$ of the form

$$V_{f_1,\dots,f_n;\varepsilon}(\chi_0) := \{\chi \in G^\wedge : |\widehat{f}_i(\chi) - \widehat{f}_i(\chi_0)| < \varepsilon \text{ for all } i = 1,\dots,n\},$$

where $f_1,\dots,f_n \in L^1(G,\omega_G)$ and $\varepsilon > 0$.

From the Gelfand theory we obtain the following essential

Properties *of* $\mathcal{F}$.

4.2.3.1 *$A(G^\wedge)$ is a selfadjoint subalgebra of $C^0(G^\wedge)$ which separates $G^\wedge$, hence*

$$A(G^\wedge)^- = C^0(G^\wedge).$$

4.2.3.2 *$\mathcal{F}$ is a norm-decreasing involutive homomorphism of the group algebra $L^1(G,\omega_G)$ into $C^0(G^\wedge)$.*

Theorem 4.2.4 *The dual $G^\wedge$ of a locally compact group G is again a locally compact group.*

Proof. Since we already know that $G^\wedge$ is an Abelian group and a locally compact space with respect to the Gelfand topology τ_g it remains to show that the mapping

$$(\chi,\rho) \mapsto \chi - \rho$$

from $G^\wedge \times G^\wedge$ into $G^\wedge$ is continuous.

1. We prove that the mapping

$$(x, \chi) \mapsto \chi(x)$$

from $G \times G^\wedge$ into $\mathbf{C}$ is continuous.

First of all we note that for every $f \in L^1(G, \omega_G)$ and $x \in G$

$$\widehat{f_x}(\chi) = \hat{f}(\chi)\overline{\chi(x)}$$

holds whenever $\chi \in G^\wedge$. Thus it suffices to show that

$$(x, \chi) \mapsto \widehat{f_x}(\chi)$$

is continuous on $G \times G^\wedge$ for every $f \in L^1(G, \omega_G)$. So, take $x_0 \in G, \chi_0 \in G^\wedge$ and $\varepsilon > 0$. There are neighborhoods $V \in \mathfrak{V}_G(x_0)$ and $W \in \mathfrak{V}_{G^\wedge}(\chi_0)$ such that

$$\|f_x - f_{x_0}\|_1 < \varepsilon$$

as well as

$$|\widehat{f_{x_0}}(\chi) - \widehat{f_{x_0}}(\chi_0)| < \varepsilon$$

whenever $x \in V$, $\chi \in W$. This follows from Proposition 4.1.12 and from the continuity of f_{x_0} respectively. But since

$$|\widehat{f_{x_0}}(\chi) - \widehat{f_{x_0}}(\chi)| \leq \|f_x - f_{x_0}\|_1,$$

we obtain

$$|\widehat{f_x}(\chi) - \widehat{f_{x_0}}(\chi_0)| < 2\varepsilon$$

whenever $x \in V$ and $\chi \in W$, hence the assertion.

In the following two parts of the proof we shall employ the compact open topology τ_{co} in $G^\wedge$.

2. For compact $K \subset G$ and $r > 0$ the set

$$V_{K,r} := \{\chi \in G^\wedge : |\chi(x) - 1| < r \text{ for all } x \in K\}$$

is an open subset of $G^\wedge$.

In fact, we choose a compact subset K of G, an $r > 0$ and a $\chi_0 \in V_{K,r}$. Part 1. of this proof implies that for every $x_0 \in K$ there exist neighborhoods $V \in \mathfrak{V}_G(x_0)$ and $W \in \mathfrak{V}_{G^\wedge}(\chi_0)$ such that $|\chi(x)-1| < r$ for all $x \in V, \chi \in W$. Since K is compact, finitely many of these sets V cover K, hence the intersection W_0 of the corresponding sets W is a subset of $V_{K,r}$. Since $W_0 \in \mathfrak{V}_{G^\wedge}(\chi_0)$, $V_{K,r}$ is open.

3. The family consisting of the sets $V_{K,r}$ and their translates generate the topology $\tau_{G^\wedge}$ of $G^\wedge$.

In order to see this we pick a neighborhood V of $\chi_0 \in G^\wedge$ and show that $\chi_0 + V_{K,r} \subset V$ for some choice of $K \in \mathcal{K}(G)$ and $r > 0$. Without loss of generality let $\chi_0 = 0$. From the definition of the Gelfand topology τ_g in $G^\wedge$ we deduce the existence of functions $f_1, \ldots, f_n \in L^1(G, \omega_G)$ and of $\varepsilon > 0$ such that

$$\bigcap_{i=1}^{n} \{\chi \in G^\wedge : |\widehat{f}_i(\chi) - \widehat{f}_i(0)| < \varepsilon\} \subset V .$$

But now $C^c(G)$ is dense in $L^1(G, \omega_G)$, so we may assume that $f_1, \ldots, f_n$ vanish outside a compact subset K of G. With the choices

$$r < \frac{\varepsilon}{\max_{1 \le i \le n} \|f_i\|_1}$$

and $\chi \in V_{K,r}$ we get

$$\begin{aligned} |\hat{f}_i(\chi) - \hat{f}_i(0)| &\le \int_K |\overline{\chi(x)} - 1| \; |f_i(x)| \omega_G(dx) \\ &\le r \|f_i\|_1 < \varepsilon, \end{aligned}$$

hence that $V_{K,r} \subset V$.

4. The statement of the theorem now follows from the inclusion

$$(\chi + V_{K,\frac{r}{2}}) - (\rho + V_{K,\frac{r}{2}}) \subset \chi - \rho + V_{K,r}$$

valid for all $\chi, \rho \in G^\wedge$ and any $V_{K,r} \in \mathfrak{V}_{G^\wedge}(0)$. ■

4.2.5 From the proof of Theorem 4.2.4 we learn that on $G^\wedge \cong \Delta(L^1(G,\omega_G))$ the topologies τ_g and τ_{co} coincide. Therefore we may consider $G^\wedge$ as a locally compact Abelian *group* furnished with either of the topologies τ_g or τ_{co}. $G^\wedge$ will be called the ***character group*** or the dual (group) of G.

4.2.6 Since the norm of the spaces $L^p(G,\omega_G)$ for $p \in [1,\infty[$ is invariant (with respect to translation in G) we have

$$\|f * g\|_p \leq \|f\|\|g\|_p$$

whenever $f \in L^1(G,\omega_G)$ and $g \in L^p(G,\omega_G)$, i.e. $L^1(G,\omega_G)$ operates linearly by means of $g \mapsto f * g$ on $L^p(G,\omega_G)$.

Let $\|f\|_T$ denote the norm of the operator defined on $L^2(G,\omega_G)$ by $f \in L^1(G,\omega_G)$. Then, clearly,

$$\|f\|_T \leq \|f\|,$$

and

$$\|f * g\|_2 \leq \|f\|_T\|g\|_2$$

for all $f \in L^1(G,\omega_G)$ and $g \in L^2(G,\omega_G)$.

The completion of $L^1(G,\omega_G)$ (or $C^c(G)$) with respect to the operator norm $\|\cdot\|_T$ is called the ***extended group algebra*** of G and will be denoted by $\Lambda(G)$.

Properties of $\Lambda(G)$.

4.2.6.1 *$\Lambda(G)$ is a commutative Banach algebra containing $L^1(G,\omega_G)$ as a subalgebra.*

4.2.6.2 *$\Lambda(G)$ is an algebra of normal operators on the Hilbert space $L^2(G,\omega_G)$ and hence a commutative C^*-algebra.*

4.2.6.3 *$\Lambda(G)$ admits a unit element if and only if G is discrete.*

4.2.6.4 *For any $f \in \Lambda(G)$ and $\chi \in G^\wedge$ we have*

$$\|\chi f\|_T = \|f\|_T \,.$$

4.2.7 For the Gelfand mapping $\mathcal{F} : f \mapsto \hat{f}$ on $\Lambda(G)$ we quote the following

Properties

4.2.7.1 *$\mathcal{F}(\Lambda(G))$ is a subalgebra of $C^0(\Delta(\Lambda(G)))$.*

4.2.7.2 *$\mathcal{F}$ is a norm and involution preserving isomorphism from $\Lambda(G)$ onto $C^0(\Delta(\Lambda(G)))$.*

4.2.7.3 $\Delta(\Lambda(G)) \cong G^\wedge$

For the proof of 4.2.7.3 we take $f \in L^1(G, \omega_G)$ with $f \neq 0$. By Property 4.2.7.2 there exists $\tau' \in \Delta(\Lambda(G))$ such that $\tau'(f) \neq 0$ for all $f \in \Lambda(G)$. Now apply Theorem 4.2.2 in order to obtain $\chi_1 \in G^\wedge$ such that

$$\tau'(f) = \int_G f\overline{\chi}_1 d\omega_G \,.$$

For any $\chi \in G^\wedge$ we put

$$\tau(f) := \tau'((\chi_1 - \chi)f) = \int f\overline{\chi} d\omega_G$$

whenever $f \in L^1(G, \omega_G)$. Since

$$\begin{aligned}|\tau(f)| &= |\tau'((\chi_1 - \chi)f)| \\ &\leq \|(\chi_1 - \chi)f\|_T = \|f\|_T\end{aligned}$$

for all $f \in L^1(G, \omega_G), \tau$ can be uniquely extended to $\Lambda(G)$. Clearly, τ appears to be an element of $\Delta(\Lambda(G))$.

4.2.8 Let $\Lambda_\infty(G)$ and $\Lambda_2(G)$ denote the subsets of classes of $\|\cdot\|_T$-Cauchy sequences that contain at least one $\|\cdot\|$- or $\|\cdot\|_2$-Cauchy sequence respectively. In the sequel $\Lambda_\infty(G)$ and $\Lambda_2(G)$ will be viewed as subsets of $\Lambda(G)$. $\Lambda_2(G)$ will also be considered as a subset of $L^2(G, \omega_G)$.

Properties *of the spaces* $\Lambda_\infty(G))$ *and* $\Lambda_2(G)$.

4.2.8.1 $\Lambda(G) * C^c(G) \subset \Lambda_2(G)$.

4.2.8.2 $\Lambda_2(G) * \Lambda_2(G) \subset \Lambda_\infty(G)$.

4.2.8.3 $\Lambda(G) * C^c(G) * C^c(G) \subset \Lambda_\infty(G)$.

4.2.8.4 *If* $f \in \Lambda(G)$ *such that* $\hat{f}$ *is real or* $\hat{f} \geq 0$, *then* $f(0)$ *is real or* $f(0) \geq 0$ *respectively.*

The proof of Property 4.2.8.1 follows for $f \in \Lambda(G)$ of the form

$$f = \|\cdot\|_T - \lim_{n \to \infty} f_n$$

with a sequence $(f_n)_{n \geq 1}$ in $L^1(G, \omega_G)$ from the inequalities

$$\|f_n * g - f_m * g\|_2 \leq \|f_n - f_m\|_T \|g\|_2$$

and

$$\|f_n * g - f_m * g\|_T \leq \|f_n - f_m\|_T \|g\|_T$$

valid for $n, m \geq 1$.

Similarly, Property 4.2.8.2 is implied by the analogous estimates

$$\begin{aligned}\|f_n * g_n - f_m * g_m\| &\leq \|(f_n - f_m) * g_n\| + \|f_m * (g_n - g_m)\| \\ &\leq \|f_n - f_m\|_2 \|g_n\|_2 + \|f_m\|_2 \|g_n - g_m\|_2\end{aligned}$$

and

$$\|f_n * g_n - f_m * g_m\|_T \leq \|f_n - f_m\|_T \|g_n\|_T + \|f_m\|_T \|g_n - g_m\|_T .$$

Property 4.2.8.3 follows from Properties 4.2.8.1 and 4.2.8.2 with the help of $C^c(G) \subset \Lambda_2(G)$.

Finally, as for Property 4.2.8.4, we assume $\hat{f} \geq 0$. Since by Property 4.2.7.2 the Gelfand isomorphism $f \mapsto \hat{f}$ from $\Lambda(G)$ onto $C^0(G^\wedge)$ is involution invariant, ${}_+\sqrt{\hat{f}} = \hat{g}$ yields a hermitian function g with $g * g = f$. This implies that

$$f(0) = g * g(0) = \int g(-y)g(y)\omega_G(dy) = \int_G \overline{g^\sim(y)}g(y)\omega_G(dy)$$

$$= \int \overline{g(y)}g(y)\omega_G(dy) = \int_G |g(y)|^2\omega_G(dy) \geq 0.$$

The case of $f \in \Lambda(G)$ such that $\hat{f}$ is real follows from the decomposition $f := g - h$, where $g, h \in \Lambda(G)$ with $\hat{g}, \hat{h} \geq 0$.

The aim of the following discussion is the proof of Pontryagin's fundamental theorem stating that any locally compact Abelian group G can be identified as a topological group with its double dual $G^{\wedge\wedge}$. On the way we shall establish two useful tools of harmonic analysis: the inversion formula for Fourier transforms and the Plancherel isomorphism.

Lemma 4.2.9 *Let $f \in \Lambda(G)$ such that $\hat{f} \in C_+^c(G^\wedge)$, and let $\varepsilon > 0$. There exist functions $f_1, f_2 \in \Lambda_\infty(G)$ satisfying*

(i) $\hat{f}_1, \hat{f}_2 \in C^c(G^\wedge)$,

(ii) $\hat{f}_1 \geq \hat{f} \geq \hat{f}_2$, *and*

(iii) $f_1(0) - f_2(0) < \varepsilon$.

Proof. (*i*) Let $C := \operatorname{supp}(\hat{f})$. There exists a neighborhood $U \in \mathfrak{V}_G(0)$ such that

$$|\hat{h}(\chi) - 1| < \varepsilon$$

for all $h \in C_+^c(G, U)$ with $\int_G h d\omega_G = 1$ and for all $\chi \in C$. In order to see this one just chooses

$$U := \{x \in G : |\chi(x) - 1| < \varepsilon \text{ for all } \chi \in C\}.$$

Consequently there exists a $g_0 \in C^c_+(G)$ with $\hat{g}_0(\chi) \geq 1$ for all $\chi \in C$, and for each $\delta > 0$ there is a $g \in C^c_+(G)$ satisfying

$$1 + \delta \geq \hat{g}(\chi) \geq 1 - \delta$$

whenever $\chi \in C$. Now we define

$$f_1 := f * (g + \delta g_0)$$

and

$$f_2 := f * (g - \delta g_0) .$$

Since there exist $h, h_0 \in C^c(G)$ such that $g_0 = h_0 * h_0^\sim$ and $g = h * h^\sim$, we obtain that $f_1, f_2 \in \Lambda_\infty(G)$. And clearly, $\hat{f}_1, \hat{f}_2 \in C^c(G^\wedge)$.

(ii) follows from

$$\hat{f}_1 = \hat{f}\hat{g} + \delta \hat{f}\hat{g}_0 \geq \hat{f}(1 - \delta + \delta) = \hat{f}$$

and

$$\hat{f} = \hat{f}(1 + \delta - \delta) \geq \hat{f}\hat{g} - \delta \hat{f}\hat{g}_0 = \hat{f}_2 .$$

(iii) Obviously,

$$f_1(0) - f_2(0) = 2\delta f * g_0(0) .$$

But

$$(f * g_0)^\wedge = \hat{f}\hat{g}_0 \geq 0$$

implies $f * g_0(0) \geq 0$. A proper choice of δ yields the assertion. ■

Theorem 4.2.10 *(Inversion) Let $f \in \Lambda_\infty(G)$ with $\hat{f} \in C^c(G^\wedge)$.*

Then a Haar measure $\omega_{G^\wedge}$ on $G^\wedge$ can be chosen such that for all $x \in G$ the inversion formula

$$f(x) = \int_{G^\wedge} \hat{f}(\chi)\chi(x)\omega_{G^\wedge}(dx)$$

holds.

Proof. For every $f \in \Lambda(G)$ with $\hat{f} \in C^c_+(G^\wedge)$ we introduce

$$\begin{aligned} F(\hat{f}) : &= \sup\{g(0) : \hat{g} \leq \hat{f},\ f \in \Lambda_\infty(G)\} \\ &= \inf\{h(0) : \hat{h} \geq \hat{f},\ h \in \Lambda_\infty(G)\}, \end{aligned}$$

where the equalities defining the mapping $F : C^c_+(G^\wedge) \to \mathbf{C}$ are justified by Lemma 4.2.9. Obviously F is additive and positive homogeneous on $C^c_+(G^\wedge)$. Since Re $C^c(G^\wedge)$ is a vector lattice, F can be extended to a linear functional on Re $C^c(G^\wedge)$ and hence on $C^c(G^\wedge)$.

Obviously F is positive, since for every $f \in \Lambda_\infty(G)$ with $\hat{f} \geq 0$ we have $F(\hat{f}) = f(0) \geq 0$.

Moreover, $F \neq 0$. In fact, there exist an $f \in \Lambda(G)$, $f \neq 0$ with $\hat{f} \in C^c(G^\wedge)$ and a $g \in C^c(G)$ with $h := f * g \in \Lambda_2(G)$ and $h \neq 0$. But then $\ell := h * h^\sim \in \Lambda_\infty(G)$ and therefore

$$F(\hat{\ell}) = \ell(0) = h * h^\sim(0) = \|h\|_2^2 \neq 0\,.$$

In order to show the translation invariance of F it suffices to note that for every $\chi \in G^\wedge$

$$\hat{f}_\chi = (\chi f)^\wedge$$

and

$$\chi(0) f(0) = f(0)\,.$$

But then there exists a Haar measure $\omega_{G^\wedge} \in M_+(G^\wedge)$ satisfying

$$F(\hat{f}) = \int\limits_{G^\wedge} \hat{f}(\chi)\omega_{G^\wedge}(d\chi)$$

for all $f \in \Lambda(G)$ with $\hat{f} \in C^c(G^\wedge)$.

For $f \in \Lambda_\infty(G)$ with $\hat{f} \in C^c(G^\wedge)$ we have

$$F(\hat{f}) = f(0),$$

hence

$$\begin{aligned} f(x) = f_{-x}(0) &= F(\widehat{f_{-x}}) \\ &= \int\limits_{G^\wedge} \widehat{f_{-x}}(\chi)\omega_{G^\wedge}(d\chi) \\ &= \int\limits_{G^\wedge} \hat{f}(\chi)\chi(x)\omega_{G^\wedge}(d\chi), \end{aligned}$$

whenever $x \in G$. ■

Theorem 4.2.11 *(Plancherel)*

(i) $\Lambda_2(G)$ *is dense in* $L^2(G, \omega_G)$.

(ii) The mapping $f \mapsto \hat{f}$ *from* $\Lambda_2(G)$ *in* $C^0(G^\wedge)$ *is an isometry onto a dense subset of* $L^2(G^\wedge, \omega_{G^\wedge})$ *which can be extended uniquely to an isometry from* $L^2(G, \omega_G)$ *onto* $L^2(G, \omega_{G^\wedge})$.

(iii) For $f, g \in L^2(G, \omega_G)$ *we have*

$$\int f\overline{g}\ d\omega_G = \int \hat{f}\hat{\overline{g}}\ d\omega_{G^\wedge}\,.$$

Proof. From Properties 4.2.7.2 and 4.2.7.3 we deduce the existence of a dense subset M of $\Lambda(G)$ with $M^\wedge = C^c(G^\wedge)$.

1. We show that the set

$$N := \{f * g : f \in M,\ \ g \in C^c(G)\}$$

is dense in $L^2(G, \omega_G)$. Since $N \subset \Lambda_2(G)$, we therefore obtain that $\Lambda_2(G)$ is dense in $L^2(G, \omega_G)$.

For the proof let $f \in L^2(G, \omega_G)$ and let $\varepsilon > 0$. First of all there is a $g \in C^c(G)$ with $\|f - g\| < \varepsilon$, and for this g there exists an $h \in C^c(G)$ with $\|g * h - g\|_2 < \varepsilon$. Next, for h there exists a $k \in M$ such that $\|h - k\|_T < \varepsilon$. Our assertion now follows from the inequality

$$\begin{aligned}\|k * g - f\|_2 &\leq \|k * g - h * g\|_2 + \|h * g - g\|_2 + \|g - f\|_2 \\ &< \|k - h\|_T \|g\|_2 + \varepsilon + \varepsilon \\ &< \varepsilon(\varepsilon + \|f\|_2 + 2).\end{aligned}$$

2. For every $f \in N$ we obtain that $\hat{f} \in C^c(G^\wedge)$ and $f * f^\sim \in \Lambda_\infty(G)$. Applying the inversion formula 4.2.10 the equalities

$$\begin{aligned}\|f\|_2^2 &= \int_G f\overline{f}\ d\omega_G = f * f^\sim(0) \\ &= \int_{G^\wedge} (f * f^\sim)^\wedge \omega_{G^\wedge} = \int_{G^\wedge} \hat{f}\hat{\overline{f}}\ d\omega_{G^\wedge} = \|\hat{f}\|_2^2\end{aligned}$$

show that $f \mapsto \hat{f}$ is an isometry on N. But this isometry extends to an isometry from $\Lambda_2(G)$ into $L^2(G^\wedge, \omega_{G^\wedge})$.

3. For the statements (*i*) and (*ii*) it remains to be shown that $\Lambda_2(G)^\wedge$ is dense in $L^2(G^\wedge, \omega_G)$.

In fact, let $\psi \in L^2(G^\wedge, \omega_G)$ with $\int_{G^\wedge} \varphi\hat{\psi}\, d\omega = 0$ for all $\varphi \in \Lambda_2(G)^\wedge$. Then

$$\int_{G^\wedge} \check{\chi}\varphi\overline{\psi}\, d\omega_{G^\wedge} = 0$$

whenever $\check{\chi} \in G^{\wedge\wedge}$, thus $(\varphi\overline{\psi})^\wedge = 0$ and hence $\varphi\overline{\psi} = 0$ $\omega_{G^\wedge}$-a.e.. Since for each $\chi_0 \in G^\wedge$ there is a $\varphi \in \Lambda_2(G)^\wedge$ such that $\varphi \neq 0$ in some neighborhood of $\chi_0, \psi = 0$ ω_G-a.e., the asserted density property has been shown.

4. For the proof of (*iii*) it suffices to refer to the well-known identity

$$4f\overline{g} = |f+g|^2 - |f-g|^2 + i|f+ig|^2 - i|f-ig|^2$$

valid for all $f, g \in L^2(G, \omega_G)$. ■

Theorem 4.2.12 *The group algebra $L^1(G, \omega_G)$ is regular, i.e. to every proper closed subset C of $G^\wedge$ and to every $\chi \in C^c$ there exists an $f \in L^1(G, \omega_G)$ such that $\hat{f}(C) = 0$ and $\hat{f}(\chi) \neq 0$.*

Proof. Let C be a proper closed subset of $G^\wedge$. We shall show that there exists a function $\psi \in L^1(G, \omega_G)^\wedge$ satisfying $\psi(C) = 0$ and $\psi(\chi) \neq 0$ for $\chi \in U := C^c$. Let $\chi = \chi_1 + \chi_2$ with $\chi_1, \chi_2 \in G^\wedge$. Then there are open sets U_1 and U_2 with $\chi_1 \in U_1,\ \chi_2 \in U_2$ and $U_1 + U_2 \subset U$. Now we choose functions $f, g \in L^2(G, \omega_G)$ with the properties $\hat{f}, \hat{g} \in C_+^c(G^\wedge),\ \hat{f}(\chi_1) \neq 0,\ \hat{g}(\chi_2) \neq 0,\ \hat{f}(U_1^c) = \hat{g}(U_2^c) = 0$. Applying (*iii*) of Theorem 4.2.11 we have

$$(fg)^\wedge = \hat{f} * \hat{g},$$

hence $\psi := (fg)^\wedge$ satisfies the desired conditions: $\psi \in L^1(G, \omega_G)^\wedge$, $\psi(C) = \psi(U^c) = 0$ and $\psi(\chi) \neq 0$. ■

Theorem 4.2.13 *(Pontryagin) Let G be a locally compact Abelian group with dual and double dual groups $G^\wedge$ and $G^{\wedge\wedge}$ respectively. For every $x \in G$ let $\Omega_x : G^\wedge \to \mathbf{C}$ be defined by*

$$\Omega_x(\chi) := \chi(x)$$

for all $\chi \in G^\wedge$. Then the homomorphism $\Omega : G \to G^{\wedge\wedge}$ defined by

$$\Omega(x) := \Omega_x$$

for all $x \in G$ is a topological isomorphism.

In short: $G^{\wedge\wedge} \cong G$.

Proof. 1. Ω as a mapping from (G, τ) into $(G^{\wedge\wedge}, \tau_{co})$ is continuous.

In fact, it is sufficient to show that Ω is continuous at $0 \in G$. Let C be a compact subset of $G^\wedge$ and $\varepsilon > 0$. Then the set

$$V_{C,\varepsilon} := \{\check{\chi} \in G^{\wedge\wedge} : |\check{\chi}(\chi) - 1| < \varepsilon \text{ for all } \chi \in C\}$$

is a neighborhood $W \in \mathfrak{V}_{G^{\wedge\wedge}}(0)$. By Part 1. of the proof of Theorem 4.2.4 the mapping $(x, \chi) \mapsto \chi(x)$ from $G \times G^\wedge$ into $\mathbf{C}$ is continuous. Since C is compact, the set

$$V := \{x \in G : |\chi(x) - 1| < \varepsilon \text{ for all } \chi \in C\}$$

belongs to $\mathfrak{V}_G(0)$, and $\Omega(V) \subset W$.

2. Ω is open.

It is to be shown that to each $U \in \mathfrak{V}_G(0)$ there exist a compact subset C of $G^\wedge$ and an $\varepsilon > 0$ such that

$$|\chi(x) - 1| < \varepsilon$$

for all $\chi \in C$ implies that $x \in U$.

In fact, we choose $g \in L^2(G, \omega_G)$ with $\|g\|_2 = 1$ such that

$$\|g - g_x\| < 1$$

implies $x \in U$. In order to see this, pick $V \in \mathfrak{V}_G(0)$ with $V^2 \subset U$. Moreover, take $g \in L^2_+(G, \omega_G)$ with $\operatorname{supp}(g) \subset V$ and $\|g\|_2 = 1$. If $x \notin V$ then $\operatorname{supp}(g) \cap \operatorname{supp}(g_x) = \emptyset$, hence

$$\|g - g_x\|_2 \geq \|g\|_2 = 1,$$

and we have a contradiction.

Next, let $f \in L^1_+(G, \omega_G)$ with $\|f\|_1 = 1$ satisfying the inequalities

$$\|f * g - g\|_2 < \frac{1}{3}$$

and

$$\|f * g_x - g_x\|_2 < \frac{1}{3}.$$

The triangle inequality immediately implies that

$$\|f * g - f * g_x\|_2 \leq \frac{1}{3}$$

yields $x \in U$. But since

$$\begin{aligned}\|f * g - f * g_x\|_2 &= \|f * g - f_x * g\|_2 \\ &= \|(f - f_x) * g\|_2 \\ &\leq \|f - f_x\|_T,\end{aligned}$$

we may continue the reduction procedure and obtain that

$$\|f - f_x\|_T \leq \frac{1}{3}$$

yields $x \in U$.

Moreover, we have that

$$\|f - f_x\|_T = \sup\{|\tau(f - f_x)| : \tau \in \Delta(\Lambda(G))\}$$

Thus, the inequality

$$|\tau(f - f_x)| < \frac{1}{3}$$

valid for all $\tau \in \Delta(\Lambda(G))$ implies $x \in U$. But

$$|\tau(f - f_x)| = |\tau(f)||\chi_\tau(x) - 1|,$$

where χ_τ denotes the character associated with τ by Property 4.2.7.3. Now applying Property 4.2.7.1 we obtain that $\hat{f} := \tau(f) \in C^0(G^\wedge)$, thus there exists a compact subset C of $G^\wedge$ such that $\chi_\tau \notin C$ implies that

$$|\tau(f)| < \frac{1}{6}.$$

Now, choose $\varepsilon := \frac{1}{3\|f\|_T}$, and let $x \in G$ be such that

$$|\chi(x) - 1| < \varepsilon$$

for all $\chi \in C$. Then

$$|\tau(f - f_x)| < \frac{1}{3}$$

holds for all $\tau \in \Delta(\Lambda(G))$ and consequently $x \in U$.

3. $\Omega(G)$ is dense in $G^{\wedge\wedge}$.

Once we have shown this it will be clear that $\Omega(G)$ as the image of a locally compact and hence complete group G is itself complete and therefore closed in $G^{\wedge\wedge}$. The proof of the theorem will then be terminated.

Assume that $\Omega(G)$ is not dense in $G^{\wedge\wedge}$. Then by Theorem 4.2.12 there exists a $\varphi \in L^1(G^\wedge, \omega_{G^\wedge})$ with $\varphi \neq 0$ satisfying $\hat{\varphi}(\Omega(x)) = 0$ for all $x \in G$. Since $\varphi \neq 0$, there is a $g \in \Delta(G)$ such that

$$\int_{G^\wedge} \varphi(\chi)\hat{g}(\chi)\ \omega_{G^\wedge}(d\chi) \neq 0.$$

But $C^c(G)$ is dense in $\Delta(G)$, hence there also exists an $h \in C^c(G)$ such that

$$\int_{G^\wedge} \varphi(\chi)\hat{h}(\chi)\omega_{G^\wedge}(d\chi) \neq 0.$$

On the other hand, by assumption we have

$$\begin{aligned}\int\limits_{G^\wedge} \varphi(\chi)\hat{h}(\chi)\omega_{G^\wedge}(d\chi) &= \int\limits_{G^\wedge} \varphi(\chi)\Big(\int\limits_{G} h(x)\overline{\chi(x)}\omega_G(dx)\Big)\omega_{G^\wedge}(d\chi)\\ &= \int\limits_{G} h(x)\Big(\int\limits_{G^\wedge} \varphi(\chi)\overline{\Omega(x)}(\chi)\omega_{G^\wedge}(d\chi)\Big)\omega_G(dx)\\ &= \int\limits_{G} h(x)\hat{\varphi}(\Omega(x))\omega_G(dx) = 0,\end{aligned}$$

which is the desired contradiction. ■

We draw a few consequences from Pontryagin's duality theorem 4.2.13.

Theorem 4.2.14

(i) G is compact if and only if $G^\wedge$ is discrete.

(ii) G is discrete if and only if $G^\wedge$ is compact.

Proof. An application of Theorem 4.2.13 reduces the proofs of (i) and (ii) to those of (i') and (ii') below.

(i') If G is compact, then $G^\wedge$ is discrete.

The only subgroup of $\mathbf{T}$ whose elements z satisfy $|z-1| < \sqrt{3}$ is the trivial one. Therefore the set

$$V_{G,\sqrt{3}} := \{\chi \in G^\wedge : |\chi(x) - 1| < \sqrt{3} \text{ for all } x \in G\} = \{0\}$$

is a neighborhood in $\mathfrak{V}_{G^\wedge}(0)$. But this means that $G^\wedge$ is discrete.

(ii') If G is discrete, then $G^\wedge$ is compact.

It follows from the assumption that the function f_0 on G defined by $f_0(0) = 1$ and $f_0(x) = 0$ for all $x \in G$ with $x \neq 0$ is a unit of the group algebra $L^1(G, \omega_G)$. Consequently, by Preparation C 4.2 $\Delta(L^1(G, \omega_G))$ and by Theorem 4.2.2 also $G^\wedge$ is compact. ■

Definition 4.2.15 The ***Fourier***(-Stieltjes)***transform*** $\hat{\mu}$ of a measure $\mu \in M^b(G)$ is given by

$$\hat{\mu}(\chi) := \int_G \overline{\chi(x)} \mu(dx)$$

for all $\chi \in G^\wedge$.

Properties 4.2.16 *of the* ***Fourier mapping***

$$\mathcal{F} = \mathcal{F}_G : M^b(G) \to C^b(G^\wedge)$$

given by

$$\mathcal{F}(\mu) := \hat{\mu}$$

for all $\mu \in M^b(G)$.

4.2.16.1 $\mathcal{F}$ *maps* $M^b(G)$ *into the space* $C^u(G^\wedge)$ *of uniformly continuous bounded functions on* $G^\wedge$.

4.2.16.2 $\mathcal{F}$ *is a norm-decreasing homomorphism of involutive algebras.*

4.2.16.3 $\mathcal{F}$ *is injective.*

By the isometric embedding of the group algebra $L^1(G, \omega_G)$ into the measure algebra $M^b(G)$ analogous properties remain valid for the restriction of $\mathcal{F}$ to $L^1(G, \omega_G)$. In particular $M^b(G)$ and $L^1(G, \omega_G)$ are semisimple Banach algebras.

We content ourselves with the proof of 4.2.16.3. By Pontryagin's Theorem 4.2.13 it suffices to show the injectivity of the ***inverse Fourier mapping***

$$\check{\mathcal{F}} : M^b(G^\wedge) \to C^b(G)$$

defined by

$$\check{\mathcal{F}}(\mu) := \check{\mu}$$

for all $\mu \in M^b(G^\wedge)$, where

$$\check{\mu}(x) := \overline{\mathcal{F}}_{G^\wedge}(\mu)(x) := \int_{G^\wedge} \chi(x)\mu(d\chi)$$

whenever $x \in G$.

So, let $\mu \in M^b(G^\wedge)$ such that $\check{\mu} = 0$. For every $f \in L^1(G, \omega_G)$ we have

$$\begin{aligned}\int_{G^\wedge} \hat{f}\,d\mu &= \int_{G^\wedge} \left(\int_G \overline{\chi(x)} f(x)\omega_G(dx) \right) \mu(d\chi) \\ &= \int_G f(x)\omega_G(dx) \int_{G^\wedge} \overline{\chi(x)}\mu(d\chi) = 0.\end{aligned}$$

Since $A(G^\wedge)$ is dense in $C^0(G^\wedge)$ by Property 4.2.3.1 we obtain that

$$\int_G g\,d\mu = 0$$

for every $g \in C^0(G^\wedge)$ which implies that $\mu = 0$. ■

4.2.17 *The following* ***functorial properties*** *of the duality of locally compact Abelian groups will be useful for its application to harmonic analysis.*

Let G and H be locally compact Abelian groups and let $\varphi : G \to H$ be a continuous homomorphism. For every $\chi \in H^\wedge$ let

$$\varphi^\wedge(\chi) := \chi \circ \varphi \in G^\wedge .$$

Therefore, with the notation of Theorem 4.2.13,

$$\Omega_x(\varphi^\wedge(\chi)) = \Omega_{\varphi(x)}(\chi),$$

whenever $x \in G, \chi \in H^\wedge$. Moreover, $\varphi^\wedge$ is a continuous homomorphism $H^\wedge \to G^\wedge$.

Indeed, the duality $^\wedge$ is a contravariant functor in the category of locally compact Abelian groups (together with continuous homomorphisms as morphisms).

For any subset M of G let

$$M^\perp := \{\chi \in G^\wedge : \chi(x) = 1 \text{ for all } x \in M\}$$

be the ***annihilator*** or orthogonal complement of M.

Clearly $M^\perp$ is a closed subgroup of $G^\wedge$. From Pontryagin's theorem 4.2.13 we infer that $M^{\perp\perp} := (M^\perp)^\perp$ is a closed subgroup of G. For $N \subset M$ one has $M^\perp \subset N^\perp$, and $M \subset M^{\perp\perp}$ which implies

$$M^\perp \supset (M^{\perp\perp})^\perp = (M^\perp)^{\perp\perp} \supset M^\perp,$$

hence $M^\perp = M^{\perp\perp\perp}$.

4.2.17.1 *The closed subgroup $[M]^-$ generated by a subset M of G coincides with $M^{\perp\perp}$.*

It is clear that $G_1 := [M]^- \subset M^{\perp\perp}$. For the remaining inclusion we consider the canonical projection π from G onto G/G_1 and take an element $x \notin G_1$. Since $G^\wedge$ separates G (by the Pontryagin theorem 4.2.13) there exists $\chi \in (G/G_1)^\wedge$ such that $\chi \circ \pi(x) \neq 1$. Hence $\chi \circ \pi \in G^\wedge$ with $\chi \circ \pi(y) = 1$ for all $y \in M$ but $\chi \circ \pi(x) \neq 1$, so $x \notin M^{\perp\perp}$.

As an immediate consequence of this property we note that

4.2.17.2 *the mapping*

$$H \longmapsto H^\perp$$

is a bijection from the class of closed subgroups of G onto the class of closed subgroup of $G^\wedge$

4.2.17.3 *$(G/H)^\wedge$ and $H^\perp$ are (canonically) isomorphic locally compact Abelian groups.*

Again we define for every $\chi' \in (G/H)^\wedge$ the character $\chi \in G^\wedge$ by

$$\chi(y) := \chi'(yH)$$

whenever $y \in G$. The mapping $\chi' \mapsto \chi$ from $(G/H)^\wedge$ into $G^\wedge$ serves as the desired topological isomorphism.

It follows

4.2.17.4 *that $H^\wedge$ is isomorphic to $G^\wedge/H^\perp$.*

One just has to observe the identifications

$$G^\wedge/H^\perp \cong (G^\wedge/H^\perp)^{\wedge\wedge} \cong (H^{\perp\perp})^\wedge = H^\wedge\,.$$

4.2.17.5 *H is a compact subgroup of G if and only if $H^\perp$ is open in $G^\wedge$.*

This statement follows from Property 4.2.17.4 together with Theorem 4.2.14

***Examples* 4.2.18** *of locally compact Abelian groups and their duals.*

4.2.18.1 $(\mathbf{R}^d)^\wedge \cong \mathbf{R}^d$, *so $\mathbf{R}^d$ is **self dual** for any $d \geq 1$.*

4.2.18.2 $(\mathbf{T}^d)^\wedge \cong \mathbf{Z}^d$ *for $d \geq 1$.*

4.2.18.3 $(\mathbf{Z}^d)^\wedge \cong \mathbf{T}^d$ *for $d \geq 1$.*

In order to establish these identifications the characters of the underlying groups have to be exhibited, and the Gelfand topology has to be recognized as the natural topology in each case.

In this context we mention the evident fact that

4.2.18.4 *for locally compact Abelian groups $G_1, ..., G_n$ the identification*

$$\left(\prod_{i=n}^{n} G_i\right)^\wedge \cong \prod_{i=1}^{n} G_i^\wedge$$

holds.

This property helps to see further special dualities once one accepts the validity of the following fundamental structure results.

Theorem 4.2.19 Let G be a compactly generated locally compact Abelian group.

Then

$$G \cong \mathbf{R}^d \times \mathbf{Z}^e \times K,$$

where $d, e \geq 0$ and K is a compact Abelian group.

From this theorem follows without difficulty

Theorem 4.2.20 *(Pontryagin, van Kampen) For every locally compact Abelian group G there exists an open subgroup G_1 of G of the form*

$$G_1 \cong \mathbf{R}^d \times K,$$

where $d \geq 0$ and K denotes a compact Abelian group.

If, in addition, G is connected, then

$$G \cong \mathbf{R}^d \times K,$$

where $d \geq 0$ and K is a connected compact Abelian group.

For a proof of these assertions the reader is referred for example to Guichardet [15].

4.3 Positive definite functions

The constituents of this title are important tools of the harmonic analysis of locally compact Abelian groups. For the first definition and its basic properties we note that a matrix $A = (a_{ij}) \in \mathbf{M}(n \times n, \mathbf{C})$ is said to be ***positive hermitian*** if for all $c_1, \ldots, c_n \in \mathbf{C}$

$$\sum_{i,j=1}^{n} a_{ij} c_i \overline{c}_j \geq 0\,.$$

If $B = (b_{ij}) \in \mathbf{M}(n \times n, \mathbf{C})$ is another positive hermitian matrix, then so is the product $AB = (d_{ij})$, where $d_{ij} := a_{ij} b_{ij}$ for all $i, j = 1, \ldots, n$.

Now let G be a locally compact group.

Definition 4.3.1 *A complex-valued function φ on G is called **positive definite** if for all $n \geq 1$ and all $x_1, \ldots, x_n \in G$ the matrix $(\varphi(x_i - x_j)) \in \mathbf{M}(n \times n, \mathbf{C})$ is positive hermitian.*

The totality of positive definite functions on G will be abbreviated by $PD(G)$. At a later stage we shall exclusively employ the set $CPD(G) := PD(G) \cap C(G)$.

Properties 4.3.2 *of a function $\varphi \in PD(G)$.*

4.3.2.1 *$\varphi^{\sim} = \varphi$ and $|\varphi| \leq \varphi(0)$, in particular φ is bounded, and*

$$\sup_{x \in G} |\varphi(x)| = \varphi(0)\,.$$

4.3.2.2 *For any $x, y \in G$ we have*

$$|\varphi(x) - \varphi(y)|^2 \leq 2\varphi(0)(\varphi(0) - \operatorname{Re} \varphi(x - y))$$

4.3.2.3 *If $\varphi(0) = 1$, then*

$$|\varphi(x + y) - \varphi(x)\varphi(y)|^2 \leq (1 - |\varphi(x)|^2)(1 - |\varphi(y)|^2)$$

4.3.2.4 *If* Re φ *is lower semicontinuous at 0, then* φ *is uniformly continuous.*

For proofs of these properties we consider specially chosen positive hermitian matrices.

Clearly, $\varphi(0) \geq 0$. Moreover, for every $x \in G$ the matrix

$$\begin{pmatrix} \varphi(0) & \varphi(-x) \\ \varphi(x) & \varphi(0) \end{pmatrix}$$

is positive hermitian. This fact takes care of Property 4.3.2.1.

Next, we consider the case $n = 3$, i.e. for any $x, y \in G$ we look at the matrix

$$\begin{pmatrix} \varphi(0) & \overline{\varphi(x)} & \overline{\varphi(y)} \\ \varphi(x) & \varphi(0) & \varphi(x-y) \\ \varphi(y) & \overline{\varphi(x-y)} & \varphi(0) \end{pmatrix}$$

under the assumption that $\varphi(x) \neq \varphi(y)$ we choose for $\lambda \in \mathbf{R}$ the complex numbers $c_1 := 1, c_2 := \lambda|\varphi(x) - \varphi(y)|(\varphi(x) - \varphi(y))^{-1}$, and $c_3 := -c_2$. The positive hermitian property of the above matrix yields the inequality

$$\varphi(0)(1 + 2\lambda^2) + 2\lambda|\varphi(x) - \varphi(y)| - 2\lambda^2 \operatorname{Re} \varphi(x-y) \geq 0$$

valid for all $\lambda \in \mathbf{R}$, and since the discriminant of the polynomial in λ occurring in this inequality is ≤ 0, we obtain Property 4.3.2.2.

In order to show Property 4.3.2.3 we observe that a positive hermitian matrix of the form

$$\begin{pmatrix} 1 & z_1 & z_2 \\ \overline{z_1} & 1 & z_3 \\ \overline{z_2} & \overline{z_3} & 1 \end{pmatrix}$$

has a determinant ≥ 0, hence

$$1 + z_1\overline{z}_2 z_3 + \overline{z}_1 z_2 \overline{z}_3 \geq |z_1|^2 + |z_2|^2 + |z_3|^2$$

or equivalently,

$$|z_3 - \overline{z}_1 z_2|^2 \leq (1 - |z_1|^2)(1 - |z_2|^2)\,.$$

Applying this inequality to the matrix in terms of φ and employing Property 4.3.2.1 implies the assertion.

Concerning Property 4.3.2.4 we note that from the assumption follows that Re φ is continuous at 0. This is a consequence of the equality

$$\begin{aligned}\{x \in G : \operatorname{Re} \varphi(x) \in]\varphi(0) - \varepsilon, \varphi(0) + \varepsilon[\} \\ = \{x \in G : \operatorname{Re} \varphi(x) > \varphi(0) - \varepsilon\}\end{aligned}$$

together with Re $\varphi \leq |\varphi| \leq \varphi(0)$ (Property 4.3.2.1). But then Property 4.3.2.2 provides the remaining argument.

4.3.2.5 *Let H be an open subgroup of G and let $\varphi \in CPD(H)$. Then the function φ_0 defined by*

$$\varphi_0(x) := \begin{cases} \varphi(x) & \text{if } x \in H \\ 0 & \text{otherwise} \end{cases}$$

belongs to $CPD(G)$.

For the proof of the positive-definiteness of φ_0 we pick a finite set in G and for each $x \in F$, let $c_x \in \mathbf{C}$. The set F intersects only finitely many distinct cosets $w_1H, \ldots, w_nH$ of H. With the notation $F_k := F \cap (w_kH)$ for $k = 1, \ldots, n$ we then obtain

$$\sum_{x \in F} \sum_{y \in F} \varphi_0(x-y) c_x \overline{c_y} = \sum_{k=1}^{n} \sum_{x \in F_k} \sum_{y \in F_k} \varphi(x-y) c_x \overline{c_y},$$

where

$$\begin{aligned}&\sum_{x \in F_k} \sum_{y \in F_k} \varphi(x-y) c_x \overline{c_y} \\ &\quad = \sum_{x \in F_k} \sum_{y \in F_k} \varphi((x - w_k) - (y - w_k)) c_x \overline{c_y} \\ &\quad = \sum_{u \in F_k - w_k} \sum_{v \in F_k - w_k} \varphi(u - v) c_{u+w_k} \overline{c}_{v+w_k} \geq 0.\end{aligned}$$

Clearly, φ_0 extends φ continuously.

Properties 4.3.3 *of the set* $PD(G)$

4.3.3.1 *The set* $PD(G)$ *is closed under formation of complex conjugates and real parts.*

4.3.3.2 *The constant function* ≥ 0 *on* G *belongs to* $PD(G)$.

4.3.3.3 $PD(G)$ *is closed under formation of products.*

4.3.3.4 $PD(G)$ *is a convex cone closed with respect to the topology* τ_p, $CPD(G)$ *is a convex cone closed with respect to* τ_{co} (in $C(G)$).

Only Property 4.3.3.3 requires an argument, and this has been quoted at the beginning of the section.

Examples 4.3.4 *Besides the constant function* ≥ 0

4.3.4.1 *all characters* χ *of* G *are elements of* $PD(G)$.

This follows from the inequalities

$$\sum_{i,j=1}^{n} \chi(x_i - x_j)c_i\bar{c}_j = |\sum_{i=1}^{n} \chi(x_i)c_i|^2 \geq 0$$

valid for all $n \geq 1$, $x_1, \ldots, x_n \in G$ and $c_1, \ldots, c_n \in \mathbf{C}$.

On the other hand,

4.3.4.2 *any positive definite function* $\varphi : G \to \mathbf{T}$ *is a character of* G, *as is evident from Property 4.3.2.3.*

4.3.4.3 *The Fourier transform of a measure* $\mu \in M_+^b(G^\wedge)$ *belongs to* $CPD(G)$.

In fact, for all $n \geq 1$, $x_1, \ldots, x_n \in G$ and $c_1, \ldots, c_n \in \mathbf{C}$ we have as in 4.3.4.1

$$\sum_{i,j=1}^{n} \hat{\mu}(x_i - x_j)c_i\bar{c}_j = \left|\sum_{i=1}^{n} \hat{\mu}(x_i)c_i\right|^2 \geq 0,$$

hence $\hat{\mu} \in PD(G)$. The continuity of $\hat{\mu}$ follows from Proposition 4.2.16.1.

The converse of this property is the statement of

Theorem 4.3.5 *(Bochner) For every $\varphi \in CPD(G)$ there exists a unique measure $\beta := \beta_\varphi \in M_+^b(G^\wedge)$ such that*

$$\check{\mathcal{F}}(\beta) = \check{\beta} = \varphi\,.$$

β is said to be the **Bochner measure** *of φ. It satisfies $\|\beta\| = \varphi(0)$.*

Proof. 1. We first show that $\varphi \in CPD(G)$ is of positive type in the sense that it satisfies

$$\int (f^* * f)\varphi \geq 0$$

for all $f \in C^c(G)$ and, since $C^c(G)$ is dense in $L^1(G, \omega_G)$, also for all $f \in L^1(G, \omega_G)$. For a given $f \in C^c(G)$ the function F on $G \times G$ defined by

$$F(x, y) := f(x)\overline{f(x)}\varphi(x - y)$$

for all $(x, y) \in G \times G$ belongs to $C^c(G \times G)$, hence is uniformly continuous. For $K := \operatorname{supp} f$ we have $\operatorname{supp} F \subset K \times K$, and $K \times K$ can be covered by finitely many open sets $U \times U$ such that the variation of F on each of these sets is less than a prescribed $\varepsilon > 0$. By neglecting overlaps we therefore obtain a partition $\{E_1, \ldots, E_n\}$ of K and $x_l \in E_l$ $(l = 1, \ldots, n)$ such that $|F(x, y) - F(x_i, x_j)| < \varepsilon$ whenever $(x, y) \in E_i \times E_j$ $(i, j, \ldots, n)$. But then

$$\begin{aligned}
&\int (f^* * f)\varphi \, d\omega_G \\
&\quad = \int\int F(x, y)\omega_G(dx)\omega_G(dy) \\
&\quad = \sum_{i,j=1}^{n} \int_{E_i}\int_{E_j} F(x, y)\omega_G(dx)\omega_G(dy) \\
&\quad = \sum_{i,j=1}^{n} F(x_i, x_j)\omega_G(E_i)\omega_G(E_j) + R \\
&\quad = \sum_{i,j=1}^{n} f(x_i)\omega_G(E_i)\overline{f(x_j)\omega_G(E_j)}\varphi(x_i - x_j) + R \qquad (1)
\end{aligned}$$

where

$$|R| = \left| \sum_{i,j=1}^{n} \int_{E_i} \int_{E_j} (F(x,y) - F(x_i,x_j))\omega_G(dx)\omega_G(dy) \right| < \varepsilon\, \omega_G(K)^2 .$$

Since by the positive definiteness of φ

$$\sum_{i,j=1}^{n} f(x_i)\omega_G(E_i)\overline{f(x_j)\omega_G(E_j)}\varphi(x_i - x_j) \geq 0$$

and ε was chosen arbitrarily, the assertion has been proved.

2. Next we construct a linear functional

$$\hat{f} \mapsto \int \varphi f d\omega_G$$

on the Fourier algebra $A(G^\wedge) = L^1(G,\omega_G)^\wedge$. Without loss of generality we assume that $\varphi(0) = 1$. From the Schwarz inequality we deduce that the positive Hermitian (sesquilinear) form

$$(f,g) \mapsto [f,g]_\varphi := \int \varphi(f^* * g)d\omega_G$$

on $L^1(G,\omega_G)$ satisfies

$$|[f,g]_\varphi|^2 \leq [f,f]_\varphi [g,g]_\varphi$$

for all $f,g \in L^1(G,\omega_G)$. Letting g run through an approximate identity $\{\psi_U : U \in \mathfrak{U}\}$ in $L^1(G,\omega_G)$ with

$$\psi_U^* * f \longrightarrow f \text{ (in } L^1(G,\omega_G))$$

(Proposition 4.1.13) we obtain that

$$[\psi_U, f] \to \int \varphi f \, d\omega_G \quad \text{as} \quad U \longrightarrow \{0\} .$$

But along with $\{\psi_U : U \in \mathfrak{U}\}$ also $\{\psi_U^* * \psi_U : U \in \mathfrak{U}\}$ is an approximate identity in $L^1(G,\omega_G)$. In fact, if $\operatorname{supp}\psi_U \subset U$ then $\operatorname{supp}(\psi_U^* * \psi_U) \subset U - U$, and

$$\int \psi_U^* * \psi_U \, d\omega_G = \left|\int \psi_U \, d\omega_G\right|^2 = 1 .$$

It follows that

$$\left|\int \varphi f d\omega_G\right|^2 \leq [f,f]_\varphi$$

for all $f \in L^1(G,\omega_G)$. For the function $h := f^* * f$ we have $h^* = h$. Applying the above inequality to the functions $f, h, h^{(2)} := h * h, h^{(3)} := h * h * h, \ldots$ we obtain

$$\begin{aligned}\left|\int \varphi f d\omega_G\right| &\leq \left|\int \varphi h \, d\omega_G\right|^{\frac{1}{2}} \\ &\leq \left|\int \varphi h^{(2)} d\omega_G\right|^{\frac{1}{4}} \\ &\leq \ldots \\ &\leq \left|\int \varphi h^{(n)} d\omega_G\right|^{\frac{1}{2^{n+1}}} \\ &\leq \|h^{(2^n)}\|_1^{\frac{1}{2^{n+1}}},\end{aligned}$$

where $\|\varphi\|_\infty = \varphi(0) = 1$ has been applied. Now we infer from Theorem C 6 that

$$\lim_{n\to\infty} \|h^{(2^n)}\|_1^{\frac{1}{2^{n+1}}} = \|\hat{h}\|_\infty^{\frac{1}{2}} = \||\hat{f}|^2\|_\infty^{\frac{1}{2}} = \|\hat{f}\|_\infty .$$

Thus the mapping $f \mapsto \int \varphi f d\omega_G$ induces a linear functional

$$\hat{f} \mapsto \int \varphi f d\omega_G$$

on $A(G^\wedge)$.

3. Since by Property 4.2.3.1 $A(G^\wedge)$ is dense in $C^0(G^\wedge)$ this linear functionals extends to a linear functional F on $C^0(G^\wedge)$ with $\|F\| \leq 1$.

We now apply the Riesz representation theorem 4.1.1 in order to obtain a measure $\nu \in M^b(G^\wedge)$ with $\|\nu\| \leq 1$ satisfying

$$F(f) = \int \hat{f} d\nu = \int \int f(x)\overline{\chi(x)}\nu(d\chi)\omega_G(dx)$$

for all $f \in L^1(G, \omega_G)$ or equivalently

$$\varphi(x) = \int \chi(x)\nu^*(d\chi)$$

for all $x \in G$, hence $\varphi = \check{\beta}$ for $\beta := \nu^*$. From $1 = \varphi(0) = \beta(G^\wedge) \leq \|\beta\| \leq 1$ we conclude that $\beta(G^\wedge) = \|\beta\| \geq 0$, so that $\beta \in M_+^b(G^\wedge)$. The desired representation of φ has been established. ■

Remark 4.3.6 *Replacing G by $G^\wedge$ and applying the Pontryagin theorem 4.2.13 Theorem 4.3.5 can be rephrased as follows: The Fourier mapping $\mathcal{F} := \mathcal{F}_G$ is a bijection of the cone $M_+^b(G)$ onto the cone $CPD(G^\wedge)$. In particular, $\mathcal{F}$ maps the convex set $M^1(G)$ onto the convex set $\{\varphi \in CPD(G^\wedge) : \varphi(0) = 1\}$.*

We may now deepen our knowledge about this bijection by proving that

Theorem 4.3.7 *$\mathcal{F}$ is a homeomorphism of the cone $(M_+^b(G), \tau_w)$ onto the cone $(CPD(G^\wedge), \tau_{co})$.*

Proof. Let $(\mu_\alpha)_{\alpha \in A}$ be a net in $M_+^b(G)$ such that

$$\mu_\alpha \xrightarrow{\tau_w} \mu \in M_+^b(G).$$

Then, clearly,

$$\hat{\mu}_\alpha(\chi) \longrightarrow \hat{\mu}(\chi)$$

for all $\chi \in G^\wedge$.

1. At first we show that for each $\varepsilon > 0$ there exists a $V \in \mathfrak{V}_{G^\wedge}(0)$ and there exists an $\alpha_0 \in A$ such that for all $\alpha \geq \alpha_0$ and for all $\chi_1, \chi_2 \in G^\wedge$ with $\chi_1 - \chi_2 \in V$ the inequality

$$|\hat{\mu}_\alpha(\chi_1) - \hat{\mu}_\alpha(\chi_2)| \leq \varepsilon$$

holds.

In fact, let $\varepsilon > 0$ be given, choose $\delta > 0$ such that $\delta(3 + \|\mu\|) \leq \varepsilon$ and then pick $\varphi \in C_+^c(G)$ with $0 \leq \varphi \leq 1$ and $\int (1-\varphi) d\mu < \delta$. Since $\mu_n \xrightarrow{\tau_w} \mu$, there is an $\alpha_0 \in A$ such that the inequalities

$$\|\mu_\alpha\| < \|\mu\| + 1$$

and

$$\int (1-\varphi) d\mu_\alpha < \delta$$

are satisfied for all $\alpha \geq \alpha_0$. Now let V be a neighborhood in $\mathfrak{V}_{G^\wedge}(0)$ of the form $V := V_{\operatorname{supp}\varphi, \delta}$. For $\alpha \in A$ and $\chi_1, \chi_2 \in G^\wedge$ with $\alpha \geq \alpha_0, \chi_1 - \chi_2 \in V$ we obtain

$$\begin{aligned}
|\hat{\mu}_\alpha(\chi_1) - \hat{\mu}_\alpha(\chi_2)| &\leq \int |\chi_1(x) - \chi_2(x)| \mu_\alpha(dx) \\
&\leq \int |1 - (\chi_1 - \chi_2)(x)| \varphi(x) \mu_\alpha(dx) \\
&\quad + \int |1 - (\chi_1 - \chi_2)(x)| (1 - \varphi(x)) \mu_\alpha(dx) \\
&\leq \delta \int \varphi(x) \mu_\alpha(dx) + 2 \int |1 - \varphi(x)| \mu_\alpha(dx) \\
&\leq \delta(\|\mu\| + 1) + 2\delta \leq \varepsilon.
\end{aligned}$$

2. Next we show that

$$\hat{\mu}_\alpha \xrightarrow{\tau_{co}} \hat{\mu}\,.$$

Let K be a compact subset of $G^\wedge$, and let $\varepsilon > 0$. We choose α_0 and V as above, and by taking the limits along α it follows that

$$|\hat{\mu}(\chi_1) - \hat{\mu}(\chi_2)| \leq \varepsilon$$

whenever $\chi_1, \chi_2 \in G^\wedge$ satisfy $\chi_1 - \chi_2 \in V$. Since K is compact, there exist $\chi_1, \ldots, \chi_n \in K$ such that $K \subset \bigcup_{i=1}^{n} (\chi_i + V)$, hence there exist $\alpha_1, \ldots, \alpha_n \in A$ with

$$|\hat{\mu}_\alpha(\chi_i) - \hat{\mu}(\chi_i)| \leq \varepsilon$$

for all $\alpha \geq \alpha_i \quad (i = 1, \ldots, n)$. Let $\alpha^* \in A$ be chosen such that $\alpha^* > \alpha_i$ for $i = 0, 1, \ldots, n$. Then, for $\chi \in \chi_i + V$ and $\alpha \geq \alpha^*$ we obtain the estimate

$$
\begin{aligned}
&|\hat{\mu}_\alpha(\chi) - \hat{\mu}(\chi)| \\
&\qquad \leq |\hat{\mu}_\alpha(\chi) - \hat{\mu}_\alpha(\chi_i)| + |\hat{\mu}_\alpha(\chi_i) - \hat{\mu}(\chi_i)| + |\hat{\mu}(\chi_i) - \hat{\mu}(\chi)| \\
&\qquad \leq 3\,\varepsilon
\end{aligned}
$$

and hence that

$$\sup_{\chi \in K} |\hat{\mu}_\alpha(\chi) - \hat{\mu}(\chi)| \leq 3\,\varepsilon$$

which is the desired statement.

3. We now suppose that

$$\hat{\mu}_\alpha \xrightarrow{\tau_{co}} \hat{\mu}$$

and show that

$$\mu_\alpha \xrightarrow{\tau_w} \mu\,.$$

From the hypothesis follows that

$$\lim_\alpha \|\mu_\alpha\| = \lim_\alpha \hat{\mu}_\alpha(0) = \hat{\mu}(0) = \|\mu\|\,.$$

Therefore, as a consequence of Proposition 4.1.3 it suffices to verify the limit relationship

$$\mu_\alpha \xrightarrow{\tau_v} \mu\,.$$

For $\varphi \in C^c_+(G)$ and $\varepsilon > 0$ we choose $f \in C^c(G^\wedge)$ such that

$$\|\varphi - \mathcal{F}_{G^\wedge} f\| < \varepsilon\,.$$

But then

$$
\begin{aligned}
\left| \int \varphi d\mu_\alpha - \int \varphi d\mu \right| \quad &\leq \left| \int (\varphi - \mathcal{F}_{G^\wedge} f) d\mu_\alpha \right| \\
+ \left| \int \mathcal{F}_{G^\wedge} f \, d\mu_\alpha - \int \mathcal{F}_{G^\wedge} f \, d\mu \right| &+ \left| \int (\mathcal{F}_{G^\wedge} f - \varphi) d\mu \right| \\
\leq \ \varepsilon(\|\mu_\alpha\| + \|\mu\|) &+ \int |\hat{\mu}_\alpha(\chi) - \hat{\mu}(\chi)| \, |f(\chi)| \omega_{G^\wedge}(dx),
\end{aligned}
$$

hence by assumption

$$\limsup_{\alpha} \left| \int \varphi \, d\mu_\alpha - \int \varphi \, d\mu \right| \leq 2\varepsilon \|\mu\|$$

which implies the assertion. ■

Theorem 4.3.8 *(Sequential continuity of the Fourier transform)*

Let $(\mu_n)_{n\geq 1}$ be a sequence of measures in $M_+^b(G)$, and let φ be a complex-valued function on $G^\wedge$ which is continuous at $0 \in G^\wedge$ such that

$$\hat{\mu}_n(\chi) \longrightarrow \varphi(\chi)$$

for all $\chi \in G^\wedge$. Then there exists a measure $\mu \in M_+^b(G)$ with $\hat{\mu} = \varphi$ such that

$$\mu_n \xrightarrow{\tau_w} \mu \,.$$

Proof. Clearly, $\hat{\mu} \in CPD(G^\wedge)$ for all $n \geq 1$. From Property 4.3.3.4 we infer that

$$\varphi = \lim_{n\to\infty} \hat{\mu}_n \in PD(G^\wedge),$$

and Property 4.3.2.4 yields that $\varphi \in C(G^\wedge)$, hence $\varphi \in CPD(G^\wedge)$. Now we apply the Bochner theorem 4.3.5 and obtain a measure $\mu \in M_+^b(G)$ satisfying $\hat{\mu} = \varphi$. It remains to be shown that for all $\psi \in C_+^c(G^\wedge)$

$$\lim_{n\to\infty} \int \psi d\mu_n = \int \psi d\mu \,.$$

As in the proof of part 3 of Theorem 4.3.7 we establish the inequality

$$\left| \int \psi d\mu_n - \int \psi d\mu \right|$$

$$\leq \varepsilon(\|\mu_n\| + \|\mu\|) + \int_G |\hat{\mu}_n(\chi) - \hat{\mu}(\chi)||f(x)|\omega_{G^\wedge}(dx).$$

But the dominated convergence theorem implies

$$\lim_{n\to\infty} \int |\hat{\mu}_n(\chi) - \hat{\mu}(\chi)| \, |f(\chi)|\omega_{G^\wedge}(d\chi) = 0,$$

hence the assertion. ■

4.4 Positive definite measures

We proceed to the study of Fourier transforms of not necessarily bounded positive definite measures on a locally compact Abelian group G. Our next aim will be to prove an analog of Theorem 4.3.7 for nonnegative positive definite measures on G.

Some measure-theoretical supplements will facilitate the comprehension.

Definition 4.4.1 *A measure $\mu \in M(G)$ is said to be* ***shift bounded*** *if*

$$\mu * C^c(G) \subset C^b(G).$$

It is easily seen that μ is shift bounded if and only if the set $\{T_a(\mu) : a \in G\}$ of translates of μ is τ_v-bounded.

Properties 4.4.2

4.4.2.1 *Along with μ also μ^* and $\mu^\sim$ are shift bounded measures.*

4.4.2.2 *If μ is shift bounded, also $|\mu|$ is shift bounded.*

4.4.2.3 *Any pair (μ, ν) consisting of a shift bounded measure μ and a bounded measure ν is convolvable.*

4.4.2.4 *A measure $\mu \in M_+(G)$ is shift bounded if and only if for each set $K \in \mathcal{K}(G)$ the function*

$$x \mapsto \mu(K + x)$$

is bounded.

Only Properties 4.4.2.2 and 4.4.2.3 deserve an argument. Concerning Property 4.4.2.2 we look at the inductive limit representation

$$C^c(G) = \varinjlim_{K \in \mathcal{K}(G)} C^c(G, K)$$

in the sense of Appendix B 4. Suppose that $f \in C^c_+(G, K)$. Then for any $g \in C^c(G)$ with $|g| \le f$ we have

$$\|\mu * g\|_\infty \le C_K \|g\|_\infty \le C_K \|f\|_\infty,$$

where C_K is a constant ≥ 0. In particular we obtain for all $x \in G$ that

$$\begin{aligned}|\mu| * f(x) &= \int (f^*)_x \, d|\mu| \\ &= \sup_{|g| \leq f} \left| \int (g^*)_x \, d\mu \right| \\ &= \sup_{|g| \leq f} |\mu * g(x)| \\ &\leq C_K \|f\|_\infty\end{aligned}$$

which says that $|\mu|$ is shift bounded.

As for Property 4.4.2.3 we just note that for each $f \in C_+^c(G)$ we have

$$\int \Big(\int f(x+y)|\mu|(dx) \Big) |\nu|(dy) = \int (|\mu^*| * f)(y)|\nu|(dy) < \infty,$$

since $|\mu^*| * f$ is a bounded function on G.

Definition 4.4.3 *A measure $\mu \in M(G)$ is said to* ***vanish at infinity*** *if*

$$\mu * C^c(G) \subset C^0(G).$$

With the obvious notation $M^{sb}(G)$ and $M^\infty(G)$ for the measures defined in 4.4.1 and 4.4.3 respectively we note that

$$M^b(G) \subset M^\infty(G) \subset M^{sb}(G).$$

Examples 4.4.4

4.4.4.1 *The Haar measure ω_G of G belongs to $M^{sb}(G)$, and*

4.4.4.2 *$\omega_G \in M^\infty(G)$ if and only if G is compact.*

Properties 4.4.5

4.4.5.1 *For $\mu \in M^{sb}(G)$ the linear mapping $f \mapsto \mu * f$ from $C^c(G)$ into $C^b(G)$ is continuous.*

4.4.5.2 *Moreover,*

$$\mu * C^c(G) \subset C^u(G)$$

whenever $\mu \in M^{sb}(G)$.

While Property 4.4.5.1 is an easy consequence of Appendix B 6 (closed graph theorem), Property 4.4.5.2 requires a proof. Given $\varepsilon > 0$ and a compact symmetric neighborhood $V_0 \in \mathfrak{V}_G(0)$. Then Property 4.4.2.4 implies that

$$\alpha := \sup_{x \in G} |\mu|(V_0 - \text{supp }(f) + x) < \infty .$$

Since $f \in C^u(G)$, there exists a $V \in \mathfrak{V}_G(0)$ with $V \subset V_0$ such that

$$|f(x) - f(y)| \leq \frac{\varepsilon}{\alpha}$$

for all $x, y \in G$ with $x - y \in V$. But for such x, y it follows that

$$\begin{aligned} |\mu * f(x) - \mu * f(y)| &\leq \int |f(x-z) - f(y-z)||\mu|(dz) \\ &\leq \frac{\varepsilon}{\alpha}|\mu|(V_0 - \text{supp }(f) + x) \leq \varepsilon. \end{aligned}$$

As a motivation for the notion of positive definite measures on G we note that a function $\varphi \in C(G)$ is positive definite if and only if it is of ***positive type*** in the sense of the inequality

$$\int \varphi(f * f^{\sim}) d\omega_G \geq 0$$

valid for all $f \in C^c(G)$ (See part 1. of the proof of Theorem 4.3.5).

Definition 4.4.6 *A measure $\mu \in M(G)$ is called* **positive definite** *if for all $f \in C^c(G)$ one has*

$$\int f * f^{\sim}\, d\mu \geq 0\,.$$

The set of positive definite measures on G will be denoted by $M_p(G)$.

Obviously $M_p(G)$ is a τ_v-closed cone in $M(G)$ which is stable under formation of reflections and complex conjugates.

As first

Examples 4.4.7 *of positive definite measures we mention ε_0 and ω_G.*

Facts 4.4.8

4.4.8.1 *For a function $\varphi \in C(G)$ such that $\varphi \cdot \omega_G \in M_p(G)$ it is necessary and sufficient that $\varphi \in PD(G)$.*

4.4.8.2 *If for $\mu \in M(G)$ the pair $(\mu, \mu^{\sim})$ is convolvable, then $\mu * \mu^{\sim} \in M_p(G)$.*

4.4.8.3 *If $\varphi \in CPD(G)$ with Bochner measure $\beta \in M_+^b(G^\wedge)$, then for any $f \in C^c(G)$ the function $\varphi * f * f^{\sim}$ belongs to $CPD(G)$ and has Bochner measure $|\hat{f}|^2 \cdot \beta$.*

In order to see this we show that the inverse Fourier transform of the measure $|\hat{f}|^2 \cdot \beta \in M_+^b(G)$ is $\varphi * f * f^{\sim}$, and this in turn follows from the subsequent computation valid for all $x \in G$:

$$\begin{aligned}
\overline{\mathcal{F}}_{G^\wedge}(|\hat{f}|^2 \cdot \beta) &= \int \Big(\chi(x) \int \overline{\chi(y)} f * f^{\sim}(y)\omega_G(dy)\Big)\beta(d\chi)\\
&= \int \Big(\int \chi(y) f * f^{\sim}(x-y)\omega_G(dy)\Big)\beta(d\chi)\\
&= \int \varphi(y) f * f^{\sim}(x-y)\omega_G(dy)\\
&= \varphi * f * f^{\sim}(x).
\end{aligned}$$

With this statement as a motivation we proceed by looking at the positive definiteness of functions $\mu * f * f^{\sim}$ with measures μ instead of functions φ.

4.4.9 Proposition *Let* $\mu \in M(G)$.

(i) $\mu \in M_p(G)$ *if and only if* $\mu * f * f^{\sim} \in CPD(G)$ *for all* $f \in C^c(G)$.

(ii) If $\mu \in M_{p,+}(G) := M_p(G) \cap M_+(G)$, *then* $\mu \in M^{sb}(G)$.

Proof. *(i)* Let $\mu \in M_p(G)$ and let $f, g \in C^c(G)$. Then $\mu * f * f^{\sim} \in C(G)$, and

$$\begin{aligned}\int \mu * f * f^{\sim}(g * g^{\sim}) d\omega_G &= \int (f^* * g) * (f^* * g)^{\sim} d\mu \\ &= \left| \int f^* * g \, d\mu \right|^2 \geq 0.\end{aligned}$$

The statement preceding Definition 4.4.6 implies that $\mu * f * f^{\sim} \in CPD(G)$.

Conversely, if $\mu * f * f^{\sim} \in CPD(G)$ for all $f \in C^c(G)$, then $\mu * f * f^{\sim}(0) \geq 0$ which says that

$$\int f^* * (f^*)^{\sim} d\mu \geq 0$$

for all $f \in C^c(G)$, hence that $\mu \in M_p(G)$.

(ii) Let $\mu \in M_{p,+}(G)$ and $f \in C_+^c(G)$. There exists a function $g \in C_+^c(G)$ such that $f \leq g * g^{\sim}$. It follows that

$$\mu * f \leq \mu * g * g^{\sim},$$

where $\mu * g * g^{\sim} \in CPD(G)$ (by *(i)*) and hence bounded. Consequently $\mu * f \in C^b(G)$, i.e. $\mu \in M^{sb}(G)$. ■

Theorem 4.4.10 *(Existence of generalized Bochner measure) Let* $\mu \in M_p(G)$. *There exists a unique measure* $\beta \in M_+(G^{\wedge})$ *such that for all* $f \in C^c(G)$ *the following conditions hold:*

(i) $\int |\hat{f}|^2 d|\beta| < \infty$.

(ii) $\mu * f * f^{\sim}(x) = \int \chi(x)|\hat{f}(\chi)|^2\beta(d\chi)$ *whenever* $x \in G$.

β *is called the* ***generalized Bochner measure associated with*** μ.

Proof. We first note that by Proposition 4.4.9 *(i)* $\mu * f * f^{\sim} \in CPD(G)$ for all $f \in C^c(G)$. Now Theorem 4.3.5 provides us with a Bochner measure $\beta_f \in M_+^b(G^\wedge)$ satisfying

$$\mu * f * f^{\sim} = \check{\beta}_f\,.$$

Since for $f, g \in C^c(G)$

$$(|\hat{f}|^2\beta_g)^\vee = \mu * f * f^{\sim} * g * g^{\sim} = (|\hat{g}|^2\beta_f)^\vee,$$

the uniqueness of the Bochner measure yields

$$|\hat{f}|^2\beta_g = |\hat{g}|^2\beta_f. \tag{2}$$

But for every $\beta \in M_+(G^\wedge)$ satisfying *(i)* and *(ii)* of the theorem we must have

$$|\hat{f}|^2 = \beta_f$$

whenever $f \in C^c(G)$, so we define β accordingly.

In order to show that β is well-defined we verify that the integral $\int\limits_{G^\wedge} h\,d\beta$ is uniquely determined for every $h \in C^c(G^\wedge)$. Indeed, choose $g \in C^c(G)$ with $\hat{g} \neq 0$ on $\operatorname{supp} h$ (by applying Property 4.2.3.1) such that

$$\int\limits_{G^\wedge} h\,d\beta = \int\limits_{G^\wedge} \frac{h}{|\hat{g}|^2}\,d\beta_g,$$

where $\frac{h}{|\hat{g}|^2}$ denotes the function in $C^c(G^\wedge)$ given as

$$\chi \mapsto \begin{cases} \frac{h(\chi)}{|\hat{g}(\chi)|^2} & \text{if } \hat{g}(\chi) \neq 0 \\ 0 & \text{otherwise.} \end{cases}$$

It now follows from (2) that $\int\limits_{G^\wedge} h\,d\beta$ is independent of the choice of g.

Clearly, $h \mapsto \int\limits_{G^\wedge} h \, d\beta$ is a positive linear functional on $C^c(G^\wedge)$, hence by the Riesz representation theorem 4.1.1 $\beta \in M_+(G^\wedge)$. It remains to show that β satisfies the conditions *(i)* and *(ii)* of the theorem. For this it suffices to establish the equality

$$\beta_f = |\hat{f}|^2 \cdot \beta$$

whenever $f \in C^c(G)$. Indeed let $h \in C^c(G^\wedge)$, choose again $g \in C^c(G)$ with $\hat{g} \neq 0$ on $\operatorname{supp}(h)$ and apply again (2) together with the definition of β. Then the inequalities

$$\begin{aligned}\int\limits_G |\hat{f}|^2 h \, d\beta &= \int\limits_{G^\wedge} \frac{|\hat{f}|^2 h}{|\hat{g}|^2} \, d\beta_g \\ &= \int\limits_{G^\wedge} \frac{h}{|\hat{g}|^2} \, |\hat{g}|^2 d\beta_f \\ &= \int\limits_{G^\wedge} h \, d\beta_f\end{aligned}$$

imply the assertion. ■

Theorem 4.4.11 *For any measure $\mu \in M(G)$ the following statements are equivalent:*

(i) $\mu \in M_p(G)$.

(ii) There exists a measure $\sigma \in M_+(G^\wedge)$ such that

$$\int f * f^\sim d\mu = \int |\overline{\mathcal{F}}_G f|^2 d\sigma \tag{3}$$

whenever $f \in C^c(G)$.

If (i) is fulfilled, then σ is the generalized Bochner measure β_μ associated with μ.

Proof. *(ii)* $\Longrightarrow$ *(i)*. Let μ be a measure in $M(G)$ for which there exists $\sigma \in M_+(G^\wedge)$ satisfying the equality (3) valid for all $f \in C^c(G)$.

Then obviously μ belongs to $M_p(G)$. We now fix $f \in C^c(G)$ and $x \in G$. With f replaced by $f^\sim$ *(ii)* yields

$$\int |\hat{f}|^2 d\sigma = \int f^* * \overline{f}\, d\mu < \infty,$$

i.e. condition *(i)* of Theorem 4.4.10. Moreover, polarization of the equality in *(ii)* implies

$$\int f * g^\sim d\mu = \int \mathcal{F}_G f\, \overline{\mathcal{F}_G g}\, d\sigma$$

whenever $g \in C^c(G)$. Replacing f by $\overline{f}_x$ and g by $\overline{f}$ shows that condition *(ii)* of Theorem 4.4.10 is fulfilled and therefore $\sigma = \beta$.
(ii) $\Longrightarrow$ *(i)*. If conversely $\mu \in M_p(G)$ then the generalized Bochner measure β associated with μ (by Theorem 4.4.10) satisfies the equality (3) for each $f \in C^c(G)$. One needs only replace f by f^* and specialize condition *(ii)* of Theorem 4.4.10 to $x = 0$. ■

Corollary 4.4.12 *The mapping $\mu \mapsto \beta_\mu$ from $M_p(G)$ into $M_+(G^\wedge)$ established in the theorem (and envisaged to serve as a generalization of the Fourier mapping) is injective.*

Proof. Suppose that measures $\mu, \nu \in M_p(G)$ admit the same generalized Bochner measure β. By the theorem we have that

$$\int f * f^\sim d\mu = \int f * f^\sim d\nu$$

for all $f \in C^c(G)$, and by polarization we immediately obtain

$$\int f * g^\sim d\mu = \int f * g^\sim d\nu$$

whenever $f, g \in C^c(G)$. Letting g run through an approximate identity in $C^c(G)$ (See Proposition 4.1.13) we achieve that

$$\int f d\mu = \int f d\nu$$

holds for all $f \in C^c(G)$, i.e. $\mu = \nu$. ■

Consequences 4.4.13 *of the proceeding discussion*

4.4.13.1 *If $\mu \in M_p(G)$ then $\beta_\mu \in M^{sb}(G^\wedge)$. If, in addition, $\mu \ll \omega_G$ then $\beta_\mu \in M^\infty(G)$.*

We show the first statement. Clearly,

$$\chi(f * f^\sim) = (\chi f) * (\chi f)^\sim$$

for all $\chi \in G^\wedge, f \in C^c(G)$. Replacing f by $\overline{\chi} f$ in (3) this implies

$$\begin{aligned}\int \overline{\chi(x)} f * f^\sim(x)\mu(dx) &= \int |\hat{f}(\chi - \varrho)|^2 \beta_\mu(d\varrho) \\ &= \beta_\mu * |\hat{f}|^2(\chi).\end{aligned}$$

We observe that

$$\beta_\mu * |\hat{f}|^2 = ((f * f^\sim) \cdot \mu)^\wedge$$

is a bounded function, since $(f * f^\sim) \cdot \mu \in M^b(G)$. But for every function $\psi \in C^c_+(G^\wedge)$ there exists a function $f \in C^c(G)$ with $\psi \leq |\hat{f}|^2$ (which is a consequence of Property 4.2.3.2), hence $\beta_\mu * \psi \in C^b(G^\wedge)$ for all $\psi \in C^c_+(G^\wedge)$.

4.4.13.2 *Let $\mu \in M_p(G)$ with generalized Bochner measure β_μ, and let $\varphi \in CPD(G)$ with Bochner measure β_φ. Then $\varphi \cdot \mu \in M_p(G), (\beta_\mu, \beta_\varphi)$ is a convolvable pair, and*

$$\beta_\mu * \beta_\varphi = \beta_{\varphi \cdot \mu}\,.$$

From 4.4.13.1 we infer that $\beta_\mu \in M^{sb}(G^\wedge)$. For every $f \in C^c(G)$ we have

$$\begin{aligned}\int \varphi(x) f * f^\sim(x)\mu(dx) &= \int \Big(f * f^\sim(x) \int \chi(x)\beta_\varphi(d\chi)\Big)\mu(dx) \\ &= \int \Big(\int \chi f * (\chi f)^\sim d\mu\Big)\beta_\varphi(d\chi) \\ &= \int \Big(\int |\overline{\mathcal{F}}_G(\chi f)|^2 d\beta_\mu\Big)\beta_\varphi(d\chi) \\ &= \int\int |\overline{\mathcal{F}}_G f|^2(\chi + \varrho)\beta_\mu(d\varrho)\beta_\varphi(d\chi)\,,\end{aligned}$$

and since the last double integral is ≥ 0, $\varphi{\cdot}\mu \in M_p(G)$. The rest follows from Theorem 4.4.11.

4.4.13.3 *Given $\mu \in M_p(G)$ with generalized Bochner measure β_μ we have that $T_\chi(\beta_\mu) = \beta_\mu$ for all $\chi \in G^\wedge$ if and only if $\chi = 1$ on* supp (μ) *for all $\chi \in G^\wedge$.*

In fact, for $\chi \in G^\wedge$, $\beta_\chi = \varepsilon_\chi$, hence $\beta_{\chi\cdot\mu} = \varepsilon_\chi * \beta_\mu = T_\chi\beta_\mu$ by Consequence 4.4.13.2. But now Corollary 4.4.12 applies and yields that $T_\chi(\beta_\mu) = \beta_\mu$ if and only if $\chi{\cdot}\mu = \mu$ holds. This statement, however, is equivalent to $\chi = 1$ on supp (μ).

4.4.13.4 *Since $\varepsilon_0 \in M_p(G)$, Consequence 4.4.13.3 implies that*

$$\beta_{\varepsilon_0} = \omega_{G^\wedge},$$

hence by Theorem 4.4.10 that

$$f * f^\sim(x) = \int \chi(x)|\hat{f}(\chi)|^2\beta_\mu(d\chi)$$

for all $f \in C^c(G)$ and $x \in G$. In particular, for $x = 0$ we obtain

$$\int |f(x)|^2\omega_G(dx) = \int |\hat{f}(\chi)|^2\beta_\mu(d\chi)$$

and hence that $\beta_\mu = \omega_{G^\wedge}$.

We have regained the classical version of the Plancherel theorem.

Theorem 4.4.14 *For any measure $\mu \in M^b(G)$ the following statements are equivalent:*

(i) $\mu \in M_p(G)$.

(ii) $\hat{\mu}(\chi) \geq 0$ for all $\chi \in G^\wedge$.

If any of these equivalent conditions is satisfied then

$$\beta_\mu = \hat{\mu}{\cdot}\omega_{G^\wedge}.$$

Proof. *(i)* $\Longrightarrow$ *(ii)*. For $\mu \in M^b(G)$ we have $|\hat{\mu}(\chi)| \leq \|\mu\|$, hence

$$\int |\hat{f}(\chi)|^2 |\hat{\mu}(\chi)| \omega_{G^\wedge}(d\chi) \leq \|\mu\| \int |\hat{f}(\chi)|^2 \omega_{G^\wedge}(d\chi) < \infty$$

whenever $f \in C^c(G)$. From Consequence 4.4.13.4 (with β_μ replaced by $\omega_{G^\wedge}$) we infer that for all $f \in C^c(G)$ and $x \in G$

$$\mu * f * f^\sim(x) = \int \chi(x) |\hat{f}(\chi)|^2 \hat{\mu}(\chi) \omega_{G^\wedge}(d\chi) \tag{4}$$

holds. Thus the measure $\hat{\mu} \cdot \omega_{G^\wedge}$ fulfills the conditions *(i)* and *(ii)* of Theorem 4.4.10 Since $\mu \in M_p(G), \beta_\mu = \hat{\mu} \cdot \omega_{G^\wedge}$ and therefore $\beta_\mu \geq 0$. It follows that $\hat{\mu} \geq 0$.

(ii) $\Longrightarrow$ *(i)*. If conversely $\hat{\mu} \geq 0$ then (4) implies that

$$\int f * f^\sim d\mu = \int |\overline{\mathcal{F}}_G f(\chi)|^2 \hat{\mu}(\chi) \omega_{G^\wedge}(d\chi) \geq 0$$

for all $f \in C^c(G)$, hence that $\mu \in M_p(G)$. ■

Definition 4.4.15 *For any measure $\mu \in M_p(G)$ with associated Bochner measure $\beta_\mu \in M_+(G^\wedge)$ the* ***generalized Fourier transform*** *of μ is given by*

$$\mathcal{F}(\mu) = \mathcal{F}_G(\mu) := \beta_\mu \,.$$

Clearly, the ***generalized Fourier mapping*** *$\mathcal{F} := \mathcal{F}_G : M_p(G) \to M_+(G^\wedge)$ is additive, positive homogeneous and injective, the latter property following from Corollary 4.4.12.*

In analogy to Theorem 4.3.7 we now prove

Theorem 4.4.16 *$\mathcal{F}_G$ is a homeomorphism of the cone $(M_{p,+}(G), \tau_v)$ onto the cone $(M_{p,+}(G^\wedge), \tau_v)$ with $\mathcal{F}_G^{-1} = \mathcal{F}_{G^\wedge}$.*

Proof. 1. $\mathcal{F}_G$ maps $CPD_+(G)$ into $M_{p,+}(G^\wedge)$, and

$$\text{Res}_{\,CPD_+(G)} \mathcal{F}_{G^\wedge} \mathcal{F}_G = \text{Id}_{CPD_+(G)} \,.$$

In fact, let $\varphi \in CPD_+(G)$ with Bochner measure $\beta := \mathcal{F}_G\varphi$. Then

$$\mathcal{F}_{G^\wedge}\beta(x) = \overline{\mathcal{F}}_{G^\wedge}\beta(-x) = \varphi(-x) = \varphi(x)$$

whenever $x \in G$. Theorem 4.4.14 implies that $\beta \in M_{p,+}(G^\wedge)$, hence the assertion follows.

2. Now we show that

$$\mathcal{F}_G M_{p,+}(G) \subset M_{p,+}(G^\wedge).$$

Let $\mu \in M_{p,+}(G)$. Given an approximate identity $\{\psi_U : U \in \mathfrak{U}\}$ in $C^c(G)$ for which necessarily

$$\psi_U \cdot \omega_G \xrightarrow{\tau_w} \varepsilon_0$$

or

$$\hat{\psi}_U \xrightarrow{\tau_{co}} 1 \quad \text{as} \quad U \longrightarrow \{0\}$$

holds (Theorem 4.3.7), for every $U \in \mathfrak{U}$ the function $\psi_{U,\mu} := \mu * \psi_U * \psi_U^\sim$ belongs to $CPD_+(G)$, hence

$$\mathcal{F}_G\psi_{U,\mu} = |\widehat{\psi_U}|^2 \cdot \mathcal{F}_G\mu \in M_{p,+}(G^\wedge).$$

This implies that

$$\int g * g^\sim |\widehat{\psi_U}|^2 \, d(\mathcal{F}_G\mu) \geq 0$$

for all $g \in C^c(G^\wedge)$ and all $U \in \mathfrak{U}$. For $U \longrightarrow \{0\}$ this yields

$$\int g * g^\sim d(\mathcal{F}_G\mu) \geq 0$$

for all $g \in C^c(G^\wedge)$, and this shows that $\mathcal{F}_G\mu \in M_{p,+}(G^\wedge)$.

3. Next we prove that $\mathcal{F}_{G^\wedge}\mathcal{F}_G = \mathrm{Id}$ (on $M_{p,+}(G)$). Let $\mu \in M_{p,+}(G)$, $\{\psi_U : U \in \mathfrak{U}\}$ and $\psi_{U,\mu}$ $(U \in \mathfrak{U})$ be as in part 2. of this proof. Applying Consequence 4.4.13.2 with G replaced by $G^\wedge$ we obtain on the one hand

$$\begin{aligned}\mathcal{F}_{G^\wedge}(\mathcal{F}_G\psi_{U,\mu}) &= \mathcal{F}_{G^\wedge}(|\hat{\psi}_U|^2 \cdot \mathcal{F}_G\mu) \\ &= \psi_U * \psi_U^\sim * \mathcal{F}_{G^\wedge}\mathcal{F}_G\mu,\end{aligned}$$

on the other hand with the help of part 1. of this proof

$$\begin{aligned}\mathcal{F}_{G^\wedge}(\mathcal{F}_G\psi_{U,\mu}) &= \psi_{U,\mu}\\ &= \psi_U * \psi_U^{\sim} * \mu,\end{aligned}$$

since $\psi_{U,\mu} \in CPD_+(G)$ for all $U \in \mathfrak{U}$. Consequently

$$\psi_U * \psi_U^{\sim} * \mathcal{F}_{G^\wedge}\mathcal{F}_G\mu = \psi_U * \psi_U^{\sim} * \mu$$

for all $U \in \mathfrak{U}$ and in the limit as $U \to \{0\}$ the desired identity $\mathcal{F}_{G^\wedge}\mathcal{F}_G\mu = \mu$.

4. It remains to be shown that $\mathcal{F}_G$ is a homeomorphism. For the purpose of that proof it is sufficient to verify the continuity of the Fourier mapping $\mathcal{F}_G : M_{p,+}(G) \to M_{p,+}(G^\wedge)$. Employing part 3. of this proof and interchanging the roles of G and $G^\wedge$ yields the final statement.

Let $(\mu_\alpha)_{\alpha\in A}$ be a net in $M_{p,+}(G)$ such that

$$\mu_\alpha \xrightarrow{\tau_v} \mu \in M_{p,+}(G)\,.$$

It is easy to see that this implies

$$\mu_\alpha * f * f^\sim \xrightarrow{\tau_{co}} \mu * f * f^\sim$$

for all $f \in C^c(G)$. By Theorem 4.3.7

$$\begin{aligned}\tau_w - \lim_{\alpha\in A} |\hat{f}|^2\mathcal{F}_G\mu_\alpha &= \tau_w - \lim_{\alpha\in A} \mathcal{F}_G(\mu_\alpha * f * f^\sim)\\ &= \mathcal{F}_G(\mu * f * f^\sim)\\ &= |\hat{f}|^2\mathcal{F}_G\mu.\end{aligned}$$

Now, for any $\psi \in C^c(G^\wedge)$ we choose $f \in C^c(G)$ such that $\hat{f}(\chi) \neq 0$ for all $\chi \in \operatorname{supp}(\psi)$. The function

$$\chi \mapsto h(\chi) := \begin{cases} \frac{\psi(\chi)}{|\hat{f}(\chi)|^2} & \text{if } \hat{f}(\chi) \neq 0 \\ 0 & \text{otherwise} \end{cases}$$

belongs to $C^c(G^\wedge)$. Then

$$\lim_{\alpha \in A} \int h|\hat{f}|^2 d(\mathcal{F}_G\mu_\alpha) = \int h|\hat{f}|^2 d(\mathcal{F}_G\mu)$$

which implies

$$\lim_{\alpha \in A} \int \psi \, d(\mathcal{F}_G\mu_\alpha) = \int \psi \, d(\mathcal{F}_G\mu)$$

and hence that

$$\mathcal{F}_G\mu_\alpha \xrightarrow{\tau_v} \mathcal{F}_G\mu \, .$$

■

On the way of showing that Haar measures ω_H of closed subgroups H of G are positive definite we are starting by studying the invariance set of a measure on G.

Definition 4.4.17 *A measure $\mu \in M(G)$ is said to be a-**invariant** with invariance point $a \in G$ if*

$$\mu * \varepsilon_a = \mu \, .$$

*The totality of all invariance points of μ which obviously is a closed subgroup of G, will be called the **invariance group** of μ.*

Both notions are also employed for functions $f \in C(G)$ by considering f as the measure $f\omega_G$.

The invariance groups of μ and f are denoted by $\mathrm{Inv}\,(\mu)$ and $\mathrm{Inv}\,(f)$ respectively.

*Finally, $\mu \in M(G)$ is said to be H-**invariant** for some subset H of G if $H \subset \mathrm{Inv}\,(\mu)$.*

Clearly, for any closed subgroup H of G Haar measure ω_H of H (viewed as a measure in $M_+(G)$) is H-invariant (with $\mathrm{supp}\,(\omega_H) = H$). If $\mu \in M^b(G)\backslash\{0\}$ then $\mathrm{Inv}\,(\mu)$ is a compact subgroup of G.

In fact, there exists a function $f \in C^c(G)$ such that $g := \mu * f \neq 0$. For any $x_0 \in G$ with $g(x_0) \neq 0$ we obtain the inclusion

$$x_0 + \operatorname{Inv}(\mu) \subset \{x \in G : g(x) = g(x_0)\}.$$

Since $M^b(G) \subset M^\infty(G)$, the set $\{x \in G : g(x) = g(x_0)\}$ is compact, hence $\operatorname{Inv}(\mu)$ is compact.

Properties 4.4.18 *of invariance groups.*

4.4.18.1 *For measures $\mu \in M^b(G)$ we have that*

$$\operatorname{Inv}(\mu) = (\operatorname{supp}(\hat{\mu}))^\perp$$

and

$$\operatorname{Inv}(\hat{\mu}) = (\operatorname{supp}(\mu))^\perp.$$

We only argue in favor of the first equality. By the injectivity of the Fourier mapping (Property 4.2.16.3) $a \in G$ is an invariance point of μ if and only if

$$\overline{\chi(a)}\hat{\mu}(\chi) = \hat{\mu}(\chi)$$

for all $\chi \in G^\wedge$ and this in turn holds if and only if $\chi(a) = 1$ for all $\chi \in \operatorname{supp}(\hat{\mu})$ which says that $a \in (\operatorname{supp}(\hat{\mu}))^\perp$.

4.4.18.2 *For functions $\varphi \in CPD(G)$ with Bochner measure $\beta_\varphi \in M_+^b(G^\wedge)$ we have that*

$$\begin{aligned}\operatorname{Inv}(\varphi) &= \{x \in G : \varphi(x) = \varphi(0)\}\\ &= (\operatorname{supp}(\beta_\varphi))^\perp.\end{aligned}$$

At first we infer from Property 4.4.18.1 that

$$\begin{aligned}\operatorname{Inv}(\varphi) &= (\operatorname{supp}(\beta_\varphi^*))^\perp\\ &= (\operatorname{supp}(\beta_\varphi))^\perp.\end{aligned}$$

Now, any invariance point of φ satisfies $\varphi(x) = \varphi(0)$. On the other hand, if $\varphi(x) = \varphi(0)$ for $x \in G$ then Property 4.3.2.3 implies that

$$\varphi(x + y) = \varphi(y)$$

whenever $y \in G$. But this yields $x \in \operatorname{Inv}(\varphi)$.

4.4.18.3 *For measure* $\mu \in M_p(G)$ *we have that*

$$\mathrm{Inv}\,(\mu) = (\mathrm{supp}\ (\mathcal{F}\mu))^{\perp}$$

and

$$\mathrm{Inv}\,(\mathcal{F}\mu) = (\mathrm{supp}\ (\mu))^{\perp}\,.$$

While the last equality follows from Consequence 4.4.18.3, the first one requires a detailed proof. From formula (ii) of Theorem 4.4.10 modified by polarization we deduce that

$$\mu * \varepsilon_a * f * g^{\sim}(x) = \int \chi(x)\overline{\chi(a)}\hat{f}(\chi)\overline{\hat{g}(\chi)}\mathcal{F}\mu(d\chi) \tag{5}$$

valid for all $f, g \in C^c(G)$ and $a \in G$. If $a \in \ \mathrm{Inv}\,(\mu)$, then this equality implies that the bounded measures $\hat{f}\overline{\hat{g}} \cdot \mathcal{F}\mu$ and $\hat{\varepsilon}_a\hat{f}\overline{\hat{g}} \cdot \mathcal{F}\mu$ have the same inverse Fourier transforms and hence are equal. But then $\chi(a) = 1$ for all $\chi \in \mathrm{supp}\,(\mathcal{F}\mu)$ and hence $a \in (\mathrm{supp}\ (\mathcal{F}\mu))^{\perp}$. If, conversely, $a \in G$ satisfies $\chi(a) = 1$ for all $\chi \in \mathrm{supp}\ (\mathcal{F}\mu)$, then (5) implies that

$$\mu * f * g^{\sim}(x) = \mu * \varepsilon_a * f * g^{\sim}(x)$$

holds for all $f, g \in C^c(G)$ and all $x \in G$. Letting f and g run through an approximate identity in $C^c(G)$ we achieve $\mu = \mu * \varepsilon_a$ or $a \in \mathrm{Inv}\,(\mu)$ as desired.

Now, let H *be a closed subgroup of the given locally compact Abelian group* G, *and let* π *denote the canonical homomorphism from* G *onto the quotient group* G/H. *For every* H*-invariant function* $f \in C(G)$ *there exists exactly one quotient function* $\dot{f} \in C(G/H)$ *such that* $\dot{f} \circ \pi = f$. *Let* ω_H *be a fixed Haar measure of* H. *Since, for any* $f \in C^c(G)$ *the function* $\omega_H * f$ *is* H*-invariant, its quotient function* $(\omega_H * f)^{\cdot}$ *is uniquely determined and an element of* $C^c(G/H)$. *Consequently we obtain a mapping* $\sigma : C^c(G) \to C^c(G/H)$ *defined by*

$$\sigma(f) := (\omega_H * f)^{\cdot}$$

for all $f \in C^c(G)$ *which is linear, positive and continuous (with respect to the topology* τ_{co}*).* The transpose σ^t of σ maps $M(G/H)$ linearly into the set $M(G,H)$ of H-invariant measures in $M(G)$.

Properties 4.4.19 *of the mapping* σ.

4.4.19.1 σ *is a surjection from* $C^c(G)$ *onto* $C^c(G/H)$

4.4.19.2 σ *satisfies the invariance condition*

$$\sigma(T_a f) = T_{\pi(a)}\sigma(f)$$

valid for all $f \in C^c(G), a \in G$.

4.4.19.3 *Given Haar measures* ω_G *and* ω_H *of* G *and* H *respectively there exists a unique Haar measure* $\omega_{G/H}$ *of* (G/H) *such that*

$$\sigma^t(\omega_{G/H}) = \omega_G .$$

4.4.19.4 *Let* $\mu \in M_+(G,H)$*. Then there exists a unique* ***quotient measure*** $\dot{\mu} \in M_+(G/H)$ *(**associated with*** μ*) such that*

$$\sigma^t(\dot{\mu}) = \mu$$

More generally,

4.4.19.5 σ^t *is an isomorphism from* $M(G/H)$ *onto* $M(G,H)$.

While Properties 4.4.19.2 through 4.4.19.4 follow from obvious computations, Property 4.4.19.1 requires a proof. Let $h \in C^c_+(G/H)$. Since π is a proper mapping, there exists a compact set $K \subset G$ with $\pi(K) = \operatorname{supp}(h)$. Now we choose $\psi \in C^c(G)$ with the property that $\psi = 1$ on K and define

$$f(x) := \begin{cases} \frac{\psi(x) h\circ\pi(x)}{\omega_H * \psi(x)} & \text{if } \omega_H * \psi(x) \neq 0 \\ 0 & \text{otherwise.} \end{cases}$$

Then $f \in C^c_+(G)$, hence

$$\omega_H * f(x) = \begin{cases} h\circ\pi(x) & \text{if } \omega_h * \psi(x) \neq 0 \\ 0 & \text{otherwise} \end{cases}$$

and consequently $\sigma(f) = h$.

4.4.19.6 *Given the Haar measure ω_G of G and a fixed Haar measure ω_H of the closed subgroup H of G there exists a Haar measure $\omega_{G/H}$ of G/H satisfying*

$$\sigma^t(\omega_{G/H}) = \omega_G .$$

4.4.19.7 *For a closed subgroup H of G any Haar measure ω_H of H belongs to $M_p(G)$, and the generalized Fourier transform $\mathcal{F}\omega_H$ of ω_H is a Haar measure of the closed subgroup $H^\perp$ of $G^\wedge$.*

In fact, for any $f \in C^c(G)$ we obtain from Property 4.4.19.6 that

$$\begin{aligned}\int f * f^\sim d\omega_H &= \int (\omega_H * \overline{f}) f \, d\omega_G \\ &= \int |\sigma(f)|^2 \, d\omega_{G/H},\end{aligned}$$

hence that $\omega_H \in M_p(G)$. From Property 4.4.18.3 we infer that $\mathcal{F}_G\omega_H$ is $H^\perp$-invariant. The rest is clear. ■

Special Case 4.4.20 *Let K be a compact subgroup of G and let ω_K be the* **normed Haar measure** *of K in the sense that $\omega_K \in M^1(G)$. Then*

$$\hat{\omega}_K = 1_{K^\perp},$$

and

$$1_{K^\wedge} \cdot \omega_{G^\wedge} = \mathcal{F}_G\omega_K$$

is a Haar measure of $K^\perp$.

The normed Haar measure ω_K of a compact subgroup K of G will play an important role in the probabilistic implications of this section to be discussed in Chapter 6.

5

Negative Definite Functions and Convolution Semigroups

5.1 Negative definite functions

In the present section we are going to study a notion dual to positive definiteness in order to obtain an analytic tool for the description of convolution semigroups of measures on the locally compact Abelian group G with dual group $G^\wedge$.

Definition 5.1.1 *A complex-valued function ψ on $G^\wedge$ is called* ***negative definite*** *if for all $n \geq 1$ and for all $\chi_1, \ldots, \chi_n \in G^\wedge$ the matrix*

$$(\psi(\chi_i) + \overline{\psi(\chi_j)} - \psi(\chi_i - \chi_j)) \in \mathbf{M}(n \times n, \mathbf{C})$$

is hermitian.

Let $ND(G^\wedge)$ denote the totality of all negative definite functions on G, and let $CND(G^\wedge) := ND(G^\wedge) \cap C(G^\wedge)$.

Properties 5.1.2 *of a function $\psi \in ND(G^\wedge)$.*

5.1.2.1 $\psi(0) \geq 0$.

5.1.2.2 $\psi^{\sim} = \psi$ *and* Re $\psi \geq \psi(0)$.

5.1.2.3 $\sqrt{|\psi|}$ *is subadditive.*

For proofs we note that Property 5.1.2.2 follows from the fact that for every $\chi \in G^\wedge$ the matrix

$$\begin{pmatrix} \psi(\chi) + \overline{\psi(\chi)} - \psi(0) & \psi(\chi) + \overline{\psi(0)} - \psi(\chi) \\ \psi(0) + \overline{\psi(\chi)} - \psi(-\chi) & \psi(0) + \overline{\psi(0)} - \psi(0) \end{pmatrix} \in \mathbf{M}(2 \times 2, \mathbf{C})$$

is positive hermitian, and that Property 5.1.2.3 is a consequence of the positive hermiteness of the matrix

$$\begin{pmatrix} \psi(\chi) + \overline{\psi(\chi)} - \psi(0) & \psi(\chi) + \overline{\psi(\varrho)} - \psi(\chi - \varrho) \\ \psi(\varrho) + \overline{\psi(\chi)} - \psi(\varrho - \chi) & \psi(\varrho) + \overline{\psi(\varrho)} - \psi(0) \end{pmatrix}$$

$$\in \mathbf{M}(2 \times 2, \mathbf{C})$$

valid for all $\chi, \varrho \in G^\wedge$. In fact, applying that $\psi^{\sim} = \psi$ one obtains the inequalities

$$\begin{aligned} |\psi(\chi) + \overline{\psi(\varrho)} - \psi(\chi - \varrho)|^2 & \\ &\leq (2\,\mathrm{Re}\,\psi(\chi) - \psi(0))(2\,\mathrm{Re}\,\psi(\varrho) - \psi(0)) \\ &\leq 4|\psi(\chi)||\psi(\varrho)|, \end{aligned}$$

hence

$$|\psi(\chi + \varrho)| \leq (\sqrt{|\psi(\chi)|} + \sqrt{|\psi/(\varrho)|})^2,$$

and this yields the subadditivity of $\sqrt{|\psi|}$.

Properties 5.1.3 *of the set* $ND(G^\wedge)$.

5.1.3.1 $ND(G^\wedge)$ *is closed under formation of complex conjugates and real parts.*

5.1.3.2 *The constant function* ≥ 0 *belongs to* $ND(G^\wedge)$.

5.1.3.3 $ND(G^\wedge)$ *is a* τ_p*- closed convex cone,* $CND(G^\wedge)$ *a* τ_{co}*- closed convex cone (in* $C(G^\wedge)$*).*

The proofs of these properties are obvious.

Theorem 5.1.4 *For any complex-valued function ψ on $G^\wedge$ the following statement are equivalent:*

(i) $\psi \in ND(G^\wedge)$.

(ii) *(a)* $\psi(0) \geq 0$,

(b) $\psi^\sim = \psi$, *and*

(c) for all $n \geq 1$, $\chi_1, \ldots, \chi_n \in G^\wedge$ *and* $c_1, \ldots, c_n \in \mathbf{C}$ *with* $\sum_{i=1}^n c_i = 0$ *one has*

$$\sum_{i,j=1}^n \psi(\chi_i - \chi_j) c_i \overline{c_j} \leq 0 .$$

Proof *(i)* $\Rightarrow$ *(ii)*. Given $\psi \in ND(G^\wedge)$ it remains to show *(c)*. Let $n \geq 1$, $\chi_1, \ldots, \chi_n \in G^\wedge$ and $c_1, \ldots, c_n \in \mathbf{C}$ with $\sum_{i=1}^n c_i = 0$. Then

$$\begin{aligned} 0 &\leq \sum_{i,j=1}^n (\psi(\chi_i) + \overline{\psi(\chi_j)} - \psi(\chi_i - \chi_j)) c_i \bar{c}_j \\ &= \sum_{j=1}^n \bar{c}_j \Big(\sum_{i=1}^n \psi(\chi_i) c_i \Big) + \sum_{i=1}^n c_i \Big(\sum_{j=1}^n \overline{\psi(\chi_j) c_j} \Big) \\ &\quad - \sum_{i,j=1}^n \psi(\chi_i - \chi_j) c_i \bar{c}_j \\ &= - \sum_{i,j=1}^n \psi(\chi_i - \chi_j) c_i \bar{c}_j . \end{aligned}$$

(ii) $\Rightarrow$ *(i)*. Let ψ satisfy *(a)* to *(c)* of *(ii)*, and assume given $\chi_1, \ldots, \chi_n \in G^\wedge$, $c_1 \ldots, c_n \in \mathbf{C}$. Considering the sequences $\{0, \chi_1,$

$\ldots, \chi_n\}$ and $\{c, c_1, \ldots, c_n\}$ with $c := -\sum_{i=1}^{n} c_i$ we obtain from *(c)* that

$$\psi(0)|c|^2 + \sum_{i=1}^{n} \psi(\chi_i)c_i\bar{c} + \sum_{j=1}^{n} \psi(-\chi_j)c\bar{c}_j + \sum_{i,j=1}^{n} \psi(\chi_i - \chi_j)c_i\bar{c}_j \leq 0$$

and with the help of *(a)* that

$$\sum_{i,j=1}^{n} (\psi(\chi_i) + \overline{\psi(\chi_j)} - \psi(\chi_i - \chi_j))c_i\bar{c}_j \geq \psi(0)|c|^2 \geq 0\,,$$

i.e. *(i)*. ■

Further **Properties 5.1.5**

5.1.5.1 *If $\psi \in ND(G^\wedge)$, then also $\psi - \psi(0) \in ND(G^\wedge)$.*

In fact, for $\chi_1, \ldots, \chi_n \in G^\wedge$, $c_1, \ldots, c_n \in \mathbf{C}$ with $\sum_{i=1}^{n} c_i = 0$ we obtain that

$$\sum_{i,j=1}^{n} (\psi(\chi_i - \chi_j) - \psi(0))c_i\bar{c}_j = \sum_{i,j=1}^{n} \psi(\chi_i - \chi_j)c_i\bar{c}_j \leq 0$$

by assumption. But the function $\psi - \psi(0)$ clearly satisfies the conditions *(a)* and *(b)* of *(ii)* of Theorem 5.1.4. The theorem implies that it belongs to $ND(G^\wedge)$.

5.1.5.2 *If $\varphi \in PD(G^\wedge)$, then $\varphi(0) - \varphi \in ND(G^\wedge)$.*

Again, we see that for $\chi_1, \ldots, \chi_n \in G^\wedge$ and $c_1, \ldots, c_n \in \mathbf{C}$ with $\sum_{i=1}^{n} c_i = 0$ the positive definiteness of φ implies that

$$\sum_{i,j=1}^{n} (\varphi(0) - \varphi(\chi_i - \chi_j))c_i\bar{c}_j = -\sum_{i,j=1}^{n} \varphi(\chi_i - \chi_j)c_i\bar{c}_j \leq 0\,.$$

Since the function $\varphi(0) - \varphi$ satisfies the conditions *(a)* and *(b)* of *(ii)* of Theorem 5.1.4, $\varphi(0) - \varphi$ turns out to belong to $ND(G^\wedge)$.

The following result describes the connection between the sets $PD(G^\wedge)$ and $ND(G^\wedge)$.

Theorem 5.1.6 *(Schoenberg duality, noncontinuous case) For any complex-valued function ψ on $G^\wedge$ the subsequent statements are equivalent:*

(i) $\psi \in ND(G^\wedge)$.

(ii) *(a)* $\psi(0) \geq 0$ *and*

(b) $\exp(-t\psi) \in PD(G^\wedge)$ *for all* $t > 0$.

Proof *(i) ⇒ (ii).* Given $\psi \in ND(G^\wedge)$ it suffices to show that $\exp(-\psi) \in PD(G^\wedge)$. Let therefore $\chi_1, \ldots, \chi_n \in G^\wedge$. Since the matrix

$$(\psi(\chi_i) + \overline{\psi(\chi_j)} - \psi(\chi_i - \chi_j)) \in \mathbf{M}(n \times n, \mathbf{C})$$

is positive hermitian, also the matrix

$$(\exp(\psi(\chi_i) + \overline{\psi(\chi_j)} - \psi(\chi_i - \chi_j)))$$

is positive hermitian. But then we obtain for $c_1, \ldots, c_n \in \mathbf{C}$ that

$$\begin{aligned}
&\sum_{i,j=1}^{n} \exp(-\psi(\chi_i - \chi_j))c_i\bar{c}_j \\
&\quad = \sum_{i,j=1}^{n} \exp(\psi(\chi_i) + \overline{\psi(\chi_j)} - \psi(\chi_i - \chi_j)) \exp(-\psi(\chi_i)) \\
&\qquad \cdot \exp(-\overline{\psi(\chi_j)})c_i\bar{c}_j \\
&\quad = \sum_{i,j=1}^{n} \exp(\psi(\chi_i) + \overline{\psi(\chi_j)} - \psi(\chi_i - \chi_j))d_i\bar{d}_j \geq 0 ,
\end{aligned}$$

where $d_i := \exp(-\psi(\chi_i))c_i \in \mathbf{C}$ for $i = 1, \ldots, n$, and this is the assertion.

(ii) ⇒ *(i)*. Suppose now that ψ satisfies *(a)* and *(b)* of *(ii)*. From *(a)* we infer that $\exp(-t\psi(0)) \leq 1$ for all $t > 0$, hence by Property 5.1.5.2 that

$$\frac{1}{t}(1 - \exp(-t\psi)) \in ND(G^\wedge)$$

for all $t > 0$. Moreover we have that

$$\psi = \lim_{t \to 0} \frac{1}{t}(1 - \exp(-t\psi))$$

on $G^\wedge$, hence by Property 5.1.3.3 that $\psi \in ND(G^\wedge)$. ■

The Schoenberg duality permits to prove two more important

Properties 5.1.7 *of functions* $\psi \in ND(G^\wedge)$.

5.1.7.1 *(Forming the inverse). If* $\psi(0) > 0$ *then* $\frac{1}{\psi} \in PD(G^\wedge)$.

By Theorem 5.1.6 the function $\exp(-t\psi)$ belongs to $PD(G^\wedge)$ for all $t > 0$. Moreover,

$$|\exp(-t\psi)| \leq \exp(-t\psi(0))$$

for all $t > 0$. Consequently

$$\frac{1}{\psi} = \int_0^\infty \exp(-t\psi)dt \in PD(G^\wedge).$$

5.1.7.2 *(Approximation) There exist sequences* $(a_n)_{n\geq 1}$ *in* $\mathbf{R}_+$ *and* $(\varphi_n)_{n\geq 1}$ *in* $PD(G^\wedge)$ *such that for the sequence* $(\psi_n)_{n\geq 1}$ *with*

$$\psi_n := a_n + \varphi_n(0) - \varphi_n \ (n \geq 1)$$

the limit relationship

$$\psi = \tau_p - \lim_{n \to \infty} \psi_n$$

holds (on $G^\wedge$*).*

In fact, considering the sequence $(\varphi_n)_{n\geq 1}$ of functions

$$\varphi_n := n \exp(-\frac{1}{n}(\psi - \psi(0)))$$

which by Theorem 5.1.6 belong to $PD(G^\wedge)$, and the sequence $(a_n)_{n\geq 1}$ of numbers $a_n := \psi(0) \in \mathbf{R}_+$ we obtain that for any $\chi \in G^\wedge$

$$\psi(\chi) - \psi_n(\chi) = \frac{1}{n}\left\{\frac{(\psi(\chi) - \psi(0))^2}{2!} - \frac{(\psi(\chi) - \psi(0))^3}{3!\, n} + \cdots\right\}$$

hence

$$|\psi(\chi) - \psi_n(\chi)| \leq \frac{1}{n} \exp(|\psi(\chi) - \psi(0)|)$$

holds.

We note that given sequences $(a_n)_{n\geq 1}$ in $\mathbf{R}_+$ and $(\varphi_n)_{n\geq 1}$ in $PD(G^\wedge)$ the sequence $(\psi_n)_{n\geq 1}$ defined in Property 5.1.7.2 clearly belongs to $ND(G^\wedge)$ as follows from Properties 5.1.3 and 5.1.5.2.

Examples 5.1.8 *of functions in* $ND(G^\wedge)$.

5.1.8.1 *Any homomorphism* h *from* $G^\wedge$ *into* $\mathbf{R}$ *belongs to* $ND(G^\wedge)$.

Moreover,

5.1.8.2 *given a real function* h *on* $G^\wedge$ *the function* $\psi := ih$ *belongs to* $ND(G^\wedge)$ *if and only if* h *is a homomorphism.*

In fact, for all $t > 0$ the function

$$\chi \mapsto \varphi_t(\chi) := \exp(-t\psi(\chi))$$

is positive definite with $|\varphi_t| = 1$. But then from Example 4.3.4.2 we conclude that $\varphi_t \in G^\wedge$ and therefore

$$\exp(-tih(\chi + \varrho)) = \exp(-ti(h(\chi) + h(\varrho)))$$

which implies that

$$h(\chi + \varrho) = h(\chi) + h(\varrho)$$

whenever $\chi, \varrho \in G^\wedge$.

5.1.8.3 *Any quadratic form $q \geq 0$ on $G^\wedge$ belongs to $ND(G^\wedge)$.*

We recall that quadratic forms q on $G^\wedge$ defined by

$$q(\chi + \varrho) + q(\chi - \varrho) = 2[q(\chi) + \varrho(\varrho)]$$

for all $\chi, \varrho \in G^\wedge$ have the properties $q(0) = 0, q(\chi) = q(-\chi)$ and $q(n\chi) = n^2 q(\chi)$ whenever $\chi \in G^\wedge$ and $n \in \mathbf{N}$. The mapping $Q : G^\wedge \times G^\wedge \to \mathbf{R}$ given by

$$Q(\chi, \varrho) := q(\chi) + q(\varrho) - q(\chi - \varrho)$$

for all $\chi, \varrho \in G^\wedge$ is non-negative, symmetric and additive in both variables the latter property following from the subsequent sequence of equalities (related to the first variable):

$$\begin{aligned}
&q(\chi) + q(\varrho) - q(\chi - \varrho) + q(\sigma) + q(\varrho) - q(\sigma - \varrho) \\
&= q(\chi) + q(\sigma) + 2q(\varrho) - \frac{1}{2}[q(\chi - \varrho + \sigma - \varrho) + q(\chi - \varrho - (\sigma - \varrho))] \\
&= q(\chi) + q(\sigma) - \frac{1}{2}q(\chi - \sigma) + 2q(\varrho) - \frac{1}{2}[2q(\chi + \sigma - \varrho) + \\
&\quad 2q(\varrho) - q(\chi + \sigma)] \\
&= \frac{1}{2}q(\chi + \sigma) + 2q(\varrho) - q(\chi + \sigma - \varrho) - q(\varrho) + \frac{1}{2}q(\chi + \sigma) \\
&= q(\chi + \sigma) + q(\varrho) - q(\chi + \sigma - \varrho),
\end{aligned}$$

χ, ϱ, σ being taken from $G^\wedge$. But now

$$\sum_{i,j=1}^{n} [q(\chi_i) + \overline{q(\chi_j)} - q(\chi_i - \chi_j)]c_i \bar{c}_j = Q(\chi, \chi) \geq 0$$

for $n \geq 1$, $\chi_1, \ldots, \chi_n \in G^\wedge, c_1, \ldots, c_n \in \mathbf{Z}$ and $\chi := \sum_{i=1}^{n} c_i \chi_i$, and this suffices to see that $q \in ND(G^\wedge)$.

For later application we note that

5.1.8.4 *given a constant* $c \geq 0$, *a homomorphism* $h : G^\wedge \to \mathbf{R}$ *and a quadratic form* $q \geq 0$ *on* $G^\wedge$ *the function*

$$\psi := c + ih + q$$

belongs to $ND(G^\wedge)$.

This follows from Properties 5.1.3.

5.2 Convolution semigroups and resolvents

We are now prepared to discuss the representation of continuous negative definite functions on $G^\wedge$ in terms of measures on G. For this purpose we introduce the subset

$$M_+^{(a)}(G) := \{\mu \in M_+^b(G) : \|\mu\| \le a\}$$

of nonnegative measures on G bounded by $a > 0$.

Definition 5.2.1 *A (one-parameter) family $(\mu_t)_{t>0}$ of measures in $M_+^{(1)}(G)$ is called a* **semigroup** *(of measures)* **in** *$M_+^{(1)}(G)$ if the mapping $t \mapsto \mu_t$ from $\mathbf{R}_+^\times$ into $M_+^{(1)}(G)$ is a homomorphism (of semigroups).*

$(\mu_t)_{t>0}$ is said to be τ_v- or τ_w - **continuous** *if this homomorphism is τ_v- or τ_w-continuous respectively. In the case of τ_v-continuity*

$$\mu_0 := \tau_v - \lim_{t \to 0} \mu_t$$

exists, and μ_0 is an idempotent measure in $M_+^{(1)}(G)$.

$(\mu_t)_{t>0}$ is called a **convolution semigroup in** *$M_+^{(1)}(G)$ if it is τ_v-continuous and if*

$$\mu_0 = \varepsilon_0 \, .$$

We then write $(\mu_t)_{t\ge 0}$ instead of $(\mu_t)_{t>0}$.

Remark 5.2.2 *For a convolution semigroup $(\mu_t)_{t\ge 0}$ in $M_+^{(1)}(G)$ the τ_v- and τ_w-continuities coincide.*

In fact, let $(\mu_t)_{t\ge 0}$ be τ_v-continuous. Then for $f \in C^c(G)$ with $0 \le f \le 1$ and $f(0) = 1$ one sees that

$$\begin{aligned} 1 = f(0) &= \lim_{t \to 0} \int f d\mu_t \\ &\le \liminf_{t \to 0} \|\mu_t\| \\ &\le \limsup_{t \to 0} \|\mu_t\| \le 1, \end{aligned}$$

hence by Proposition 4.1.3 that

$$\tau_w - \lim_{t\to 0} \mu_t = \varepsilon_0 \,.$$

For $t, t_0 > 0$ and $\chi \in G^\wedge$ we have that

$$|\hat{\mu}_t(\chi) - \hat{\mu}_{t_0}(\chi)| \leq |\hat{\mu}_{|t-t_0|}(\chi) - 1|$$

holds. The above limit relationship together with a double application of Theorem 4.3.7 for $t \to t_0$ yields the assertion.

By $\mathbf{S}(G)$ and $\mathbf{C}\,\mathbf{S}(G)$ we shall denote the set of semigroups and convolution semigroups in $M_+^{(1)}(G)$ respectively.

Theorem 5.2.3 *(Schoenberg correspondence) There is a one-to-one correspondence*

$$(\mu_t)_{t\geq 0} \mapsto \psi$$

between the sets $\mathbf{C}\,\mathbf{S}\,(G)$ *and* $CND(G^\wedge)$ *given by*

$$\hat{\mu}_t = \exp(-t\psi)$$

for all $t \geq 0$.

In the situation of this correspondence $(\mu_t)_{t\geq 0}$ *and* ψ *are said to be associated (with each other). Symbolically we shall write*

$$(\mu_t)_{t\geq 0} \longleftrightarrow \psi \,.$$

Proof. 1. Let $(\mu_t)_{t\geq 0} \in \mathbf{C}\,\mathbf{S}(G)$ and define for any fixed $\chi \in G^\wedge$ the complex-valued function φ_χ by

$$\varphi_\chi(t) := \hat{\mu}_t(\chi)$$

whenever $t > 0$. By Remark 5.2.2 φ_χ is continuous and satisfies

$$\varphi_\chi(t+u) = \varphi_\chi(t)\varphi_\chi(u)$$

for $t, u \in \mathbf{R}_+^\times$ as well as

$$\lim_{t \to 0} \varphi_\chi(t) = 1\,.$$

Consequently φ_χ is of the form

$$\varphi_\chi = \exp(-t\psi(\chi))$$

for some unique $\psi(\chi) \in \mathbf{C}$ $(t > 0)$. Clearly the function

$$\chi \mapsto \psi(\chi)$$

satisfies $\psi(0) \geq 0$, and the function

$$\chi \mapsto \exp(-t\psi(\chi)) = \hat{\mu}_t(\chi)$$

belongs to $CPD(G^\wedge)$ for all $t > 0$. From Schoenberg's duality theorem 5.1.6 we now infer that $\psi \in ND(G^\wedge)$. In order to show that ψ is indeed continuous we look at the equalities

$$\begin{aligned} \int_0^\infty e^{-t}\hat{\mu}_t(\chi)dt &= \int_0^\infty \exp(-t(1+\psi(\chi)))dt \\ &= \frac{1}{1+\psi(\chi)} \end{aligned}$$

valid for all $\chi \in G^\wedge$ and note that the left hand integral of the first equality is the Fourier transform of the Radon measure

$$f \mapsto \int_0^\infty e^{-t}\left(\int f d\mu_t\right) dt$$

in $M_+^{(1)}(G)$, and as such it is continuous.

2. Conversely, let $\psi \in CND(G^\wedge)$. Then for every $t > 0$ the function $\exp(-t\psi)$ belongs to $CPD(G^\wedge)$ by Theorem 5.1.6. From Bochner's theorem 4.3.5 we conclude that there exists a measure $\mu_t \in M_+^b(G)$ satisfying

$$\hat{\mu}_t = \exp(-t\psi)$$

for each $t > 0$. We shall show that $(\mu_t)_{t\geq 0} \in \mathbf{C}\,\mathbf{S}(G)$. First of all $\psi(0) \geq 0$ implies that $\|\mu_t\| \leq 1$ for all $t \geq 0$. Next we conclude from the equalities

$$\begin{aligned}\hat{\mu}_t(\chi)\hat{\mu}_s(\chi) &= \exp(-t\psi(\chi))\exp(-s\psi(\chi)) \\ &= \exp(-(t+s)\psi(\chi)) \\ &= \hat{\mu}_{t+s}(\chi)\end{aligned}$$

valid for all $\chi \in G^\wedge$ that $t \mapsto \mu_t$ is a homomorphism from $\mathbf{R}_+^\times$ into $M_+^{(1)}(G)$, as follows from the injectivity of the Fourier transform stated as Property 4.2.16.3. Since ψ is continuous, hence bounded on compact sets, we obtain that

$$\lim_{t\to 0} \hat{\mu}_t = \lim_{t\to 0} \exp(-t\psi) = 1$$

with respect to the topology τ_{co}. But Theorem 4.3.7 yields

$$\tau_w - \lim_{t\to 0} \mu_t = \varepsilon_0$$

which implies the assertion. ■

Corollary 5.2.4 *(Schoenberg duality, continuous case). For any complex-valued function ψ on $G^\wedge$ the following statements are equivalent:*

(i) $\psi \in CND(G^\wedge)$.

(ii) *(a)* $\psi(0) \geq 0$ *and*

(b) $\exp(-t\psi) \in CPD(G^\wedge)$ *for all* $t > 0$.

Corollary 5.2.5 *Let* $(\mu_t)_{t\geq 0} \in \mathbf{C}\,\mathbf{S}(G)$ *and* $\psi \in CND(G^\wedge)$ *with*

$$(\mu_t)_{t\geq 0} \longleftrightarrow \psi\,.$$

Then

(i) $\|\mu_t\| = \exp(-t\psi(0))$ *for all* $t > 0$.

In particular,

(ii) $\mu_t \in M^1(G)$ *for all* $t \geq 0$ *if and only if* $\psi(0) = 0$.

The **proofs** of these corollaries are clear.

Application 5.2.6 *(Generation of convolution semigroups)*

5.2.6.1 *(Symmetry) A semigroup* $(\mu_t)_{t>0} \in \mathbf{S}(G)$ *is said to be* ***symmetric*** *if* $\mu_t = \mu_t^\sim$ *for all* $t > 0$. *Let* $(\mu_t)_{t\geq 0} \in \mathbf{C}\,\mathbf{S}(G)$ *and* $\psi \in CND(G^\wedge)$ *with*

$$(\mu_t)_{t\geq 0} \longleftrightarrow \psi\,.$$

Then

$$(\mu_t^\sim)_{t\geq 0} \longleftrightarrow \overline{\psi}\,.$$

Clearly $(\mu_t)_{t\geq 0} \in \mathbf{C}\,\mathbf{S}(G)$ *is symmetric if and only if* ψ *is real. For any* $(\mu_t)_{t\geq 0} \in \mathbf{CS}(G)$ *the convolution semigroup* $(\mu_t * \mu_t^\sim)_{t\geq 0}$ *is symmetric.*

If $(\mu_t)_{t\geq 0}$ *is symmetric, then* $\mu_t \in M_{p,+}(G)$, *since*

$$\mu_t = \mu_{\frac{t}{2}} * \mu_{\frac{t}{2}}^\sim$$

for all $t \geq 0$.

5.2.6.2 *(Convolutions) For* $i = 1, 2$ *let* $(\mu_t^{(i)})_{t\geq 0} \in \mathbf{C}\,\mathbf{S}(G), \psi_i \in CND(G^\wedge)$ *with*

$$(\mu_t^{(i)})_{t\geq 0} \longleftrightarrow \psi_i$$

and let $\alpha_i \in \mathbf{R}_+^\times$. *Then*

$$\alpha_1\psi_1 + \alpha_2\psi_2 \longleftrightarrow (\mu_{\alpha_1 t}^{(1)} * \mu_{\alpha_2 t}^{(2)})_{t\geq 0}\,.$$

5.2.6.3 *(Products) For* $i = 1, 2$ *let* G_i *be a locally compact Abelian group with dual* $G_i{}^\wedge$,

$(\mu_t^{(i)})_{t\geq 0} \in \mathbf{C}\,\mathbf{S}(G_i), \psi_i \in CND(G_i{}^\wedge)$ *such that*

$$\psi_i \longleftrightarrow (\mu_t^{(i)})_{t>0}\,.$$

Then the function ψ *on* $G_1{}^\wedge \times G_2{}^\wedge$ *defined by*

$$\psi(\chi_1, \chi_2) := \psi_1(\chi_1) + \psi_2(\chi_2)$$

for all $(\chi_1, \chi_2) \in G_1{}^\wedge \times G_2{}^\wedge$ *belongs to* $CND(G_1{}^\wedge \times G_2{}^\wedge)$, *and*

$$\psi \quad \longleftrightarrow \quad \mu_t^{(1)} \otimes \mu_t^{(2)}\ .$$

5.2.7 First Examples *of convolution semigroups*

5.2.7.1 *A semigroup in* $M_+^{(1)}(G)$ *is called a* **translation semigroup** *in* $M_+^{(1)}(G)$ *if there exists a continuous semigroup homomorphism* $x : \mathbf{R}_+ \to G$ *such that*

$$\mu_t = \varepsilon_{x(t)}$$

for all $t \geq 0$.

One observes that $(\mu_t)_{t\geq 0} \in \mathbf{C}\,\mathbf{S}(G)$.

Extending x *to a continuous group homomorphism* $\varphi : \mathbf{R} \to G$ *by*

$$\varphi(s) := \begin{cases} x(s) & \text{if} \quad s \geq 0 \\ -x(s) & \text{if} \quad s < 0 \end{cases}$$

and looking at the dual homomorphism $h := \varphi^\wedge : G^\wedge \to \mathbf{R}$ *one obtains that*

$$\hat{\varepsilon}_{x(t)} = \exp(-t\,i\,h)$$

for all $t \geq 0$.

Conversely, given a continuous homomorphism $h : G^\wedge \to \mathbf{R}$ *and introducing*

$$x := \mathrm{Res}\,_{\mathbf{R}_+} h^\wedge : \mathbf{R}_+ \to G^{\wedge\wedge} \cong G$$

the translation semigroup $(\varepsilon_{x(t)})_{t\geq 0} \in \mathbf{C}\,\mathbf{S}(G)$ *satisfies the above Fourier representation. Here Pontryagin's theorem 4.2.13 has been applied.*

In summing up we have established a one-to-one correspondence between the sets of translation semigroups $(\varepsilon_{x(t)})_{t\geq 0}$ in $M_+^{(1)}(G)$ and of continuous homomorphisms $h : G^\wedge \to \mathbf{R}$ such that

$$(\varepsilon_{x(t)})_{t\geq 0} \longleftrightarrow \psi,$$

where $\psi := ih$.

5.2.7.2 *For $\mu \in M_+^b(G)$ with $\|\mu\| \leq \alpha$ one introduces the convolution semigroup $(\mu_t)_{t\geq 0}$ in $M_+^{(1)}(G)$ by*

$$\mu_t := e^{-t\alpha} \exp(t\mu)$$

with

$$\exp(t\mu) := \sum_{n\geq 0} \frac{(t\mu)^n}{n!}$$

for all $t \geq 0$, where $\mu^0 := \varepsilon_0$. Obviously

$$\hat{\mu}_t = \exp(-t(\alpha - \hat{\mu}))$$

with $\alpha - \hat{\mu} \in CND(G^\wedge)$, hence

$$(\mu_t)_{t\geq 0} \longleftrightarrow \alpha - \hat{\mu}\,.$$

For $\alpha := 1$ $(\mu_t)_{t\geq 0}$ is said to be the ***Poisson semigroup determined by*** ψ.

In order to stress the determining measure μ we shall write $(e(\mu)_t)_{t\geq 0}$ instead of $(\mu_t)_{t\geq 0}$. But then $(e(\varepsilon_{x_0})_t)_{t\geq 0}$ appears to be the (classical) ***Poisson semigroup with parameter*** $x_0 \in G$.

Let $(\mu_t)_{t>0}$ be a semigroup in $\mathbf{S}(G)$. For every $t > 0$ we shall examine the set

$$H_t := \{\chi \in G^\wedge : \hat{\mu}_t(\chi) \neq 0\}\,.$$

Properties 5.2.8

5.2.8.1 H_t *is independent of* $t > 0$ *and will therefore be denoted by* H.

5.2.8.2 H *is an open subgroup of* $G^\wedge$.

5.2.8.3 $K := H^\perp = \operatorname{Inv}(\mu_t)$ *for all* $t > 0$, *and* K *will be called the* ***invariance group*** *of the semigroup* $(\mu_t)_{t>0}$.

5.2.8.4 *If* $(\mu_t)_{t>0}$ *is symmetric then*

$$\tau_w - \lim_{t\to 0} \mu_t = \omega_K \,,$$

hence $(\mu_t)_{t\geq 0} \in \mathbf{C}\,\mathbf{S}(G)$.

More generally,

5.2.8.5 $(\mu_t)_{t>0} \in \mathbf{S}(G)$ *is* τ_w*-continuous if and only if the limit relationship of Property 5.2.8.4 holds.*

5.2.8.6 $(\mu_t)_{t>0} \in \mathbf{S}(G)$ *is a convolution semigroup* ($\in \mathbf{C}\,\mathbf{S}(G)$) *if and only if* $(\mu_t)_{t>0}$ *is* τ_w*-continuous and has invariance group* $K = \{0\}$.

It suffices to provide arguments for Properties 5.2.8.1 to 5.2.8.4. Concerning 5.2.8.1 we note that from the homomorphism property of $t \mapsto \mu_t$ we obtain that

$$H_s \subset H_t = H_{nt} \subset H_s$$

whenever $0 < t < s$ and $n \in \mathbf{N}$ is chosen such that $s < nt$. The assertion follows.

For 5.2.8.2 we observe that H is a subgroup of G if and only if $1_H \in PD(G)$. In order to show this characterizing property we assume without loss of generality that $(\mu_t)_{t>0}$ is symmetric; the general case follows via symmetrization (See 5.2.6.1) by just looking at the equality

$$H_t = \{\chi \in G^\wedge : (\mu_t * \mu_t^\sim)^\wedge(\chi) \neq 0\}$$

for $t > 0$. For symmetric $(\mu_t)_{t>0}$, however, we have $0 \le \hat{\mu}_t \le 1$, hence

$$\tau_p - \lim_{n\to\infty} \hat{\mu}_{\frac{1}{n}} = \lim_{n\to\infty} \hat{\mu}_1^n = 1_H \, ,$$

and from Property 4.3.3.4 we conclude that $1_H \in PD(G^\wedge)$. Since $\hat{\mu}_1 > 0$ in a neighborhood of 0, H is open (hence closed), thus $1_H \in CPD(G^\wedge)$.

For a proof of 5.2.8.3 one just refers to the fact that supp $(\hat{\mu}_t) = H$ and applies Property 4.4.18.1.

Finally we provide an argument for 5.2.8.4. Since the function $t \mapsto \hat{\mu}_\psi(\chi)$ is decreasing for each $\chi \in G^\wedge$ and consequently

$$\tau_p - \lim_{t\to 0} \hat{\mu}_t = 1_H$$

holds (by the above limit relations), the continuity theorem 4.3.8 applies, and together with 4.4.22 it implies that

$$\tau_w - \lim_{t\to 0} \mu_t = \omega_K \, .$$

In the subsequent discussion we shall employ the symbol $\mathbf{S}(G, K)$ for the set of all τ_w-continuous semigroups in $\mathbf{S}(G)$ admitting K as invariance group. Clearly, $\mathbf{S}(G, \{0\}) = \mathbf{C}\,\mathbf{S}(G)$.

Theorem 5.2.9 *(Generalized Schoenberg correspondence). There is a one-to-one correspondence*

$$(\mu_t)_{t>0} \longleftrightarrow \psi$$

between the set $\mathbf{S}(G, K)$ and the set $CND(H)$ of continuous negative definite functions ψ on the open subgroup $H := K^\perp$ of $G^\wedge$ given by

$$\hat{\mu}_t(\chi) = \begin{cases} \exp(-t\psi(\chi)) & \text{if } \chi \in H \\ 0 & \text{if } \chi \notin H. \end{cases}$$

Remark 5.2.10 *First of all we realize that by 4.2.17.5 $K^\perp$ is indeed an open subgroup of $G^\wedge$.*

Moreover, given any open subgroup H of $G^\wedge$ and $\psi \in CND(H)$ the τ_w-continuous semigroup $(\mu_t)_{t>0} \in \mathbf{S}(G)$ with

$$(\mu_t)_{t>0} \longleftrightarrow \psi$$

has $K := H^\perp$ as its invariance group, hence belongs to $\mathbf{S}(G, K)$.

The proof of the theorem is performed similar to that of Theorem 5.2.3. One just has to observe that in constructing the semigroup $(\mu_t)_{t>0} \in \mathbf{S}(G)$ the functions appearing in the Fourier representation of $(\mu_t)_{t>0}$ are belonging to $CND(G^\wedge)$. But this follows from Property 4.3.2.5, since H is an open subgroup of $G^\wedge$.

Remark 5.2.11 *For the special choice $K = \{0\}$, i.e. $H = G^\wedge$, the correspondence theorems 5.2.3 and 5.2.9 coincide.*

Until now we studied convolution semigroups corresponding to continuous negative definite functions. A further object to associate with them is the resolvent family originating from the potential theory of semigroups of operators.

Definition 5.2.12 *A (one-parameter) family $(\varrho_\lambda)_{\lambda>0}$ of measures in $M_+^b(G)$ with $\|\lambda\varrho_\lambda\| \leq 1$ for all $\lambda > 0$ is called a* **resolvent** *(of measures) in $M_+^b(G)$ if it satisfies the* **resolvent equation**

$$\varrho_\lambda - \varrho_\mu = (\mu - \lambda)\varrho_\lambda * \varrho_\mu$$

valid for all $\lambda, \mu > 0$.

In analogy to the invariance group of a semigroup of measures we introduce for a resolvent of measures $(\varrho_\lambda)_{\lambda>0}$ in $M_+^b(G)$ the set

$$L_\lambda := \{\chi \in G^\wedge : \hat{\varrho}_\lambda(\chi) \neq 0\}$$

for $\lambda > 0$ and show the following

Properties 5.2.13

5.2.13.1 *L_λ is independent of $\lambda > 0$ and will therefore be denoted by L.*

5.2.13.2 *L is an open subgroup of $G^\wedge$.*

5.2.13.3 $M := L^\perp = \mathrm{Inv}\,(\varrho_\lambda)$ *for all $\lambda > 0$, and M will be called the* ***invariance group*** *of the resolvent $(\varrho_\lambda)_{\lambda>0}$.*

5.2.13.4

$$\tau_w - \lim_{\lambda\to\infty} \lambda\varrho_\lambda = \omega_M \, .$$

For the proofs of these properties we simulate the arguments provided for the Properties 5.2.8. It suffices to verify that $1_L \in CPD(G^\wedge)$.

In fact

$$\tau_p - \lim_{\lambda\to\infty} \lambda\hat{\varrho}_\lambda = 1_L,$$

since

$$\hat{\varrho}_\lambda = \frac{\hat{\varrho}_1}{1 + (\lambda - 1)\hat{\varrho}_1}$$

for all $\lambda > 0$ by the resolvent equation, and this implies

$$\lim_{n\to\infty} \lambda\hat{\varrho}_\lambda(\chi) = 1$$

whenever $\chi \in L$. Property 4.3.3.4 together with the openness of L yields the assertion.

In what follows we shall use the symbol $\mathbf{R}(G, M)$ for the set of all resolvents in $M_+^b(G)$ admitting M as their invariance group. An analogue of the generalized Schoenberg correspondence theorem reads as follows

Theorem 5.2.14 *(Resolvent correspondence). There is a one-to-one correspondence*

$$(\mu_t)_{t>0} \longleftrightarrow (\varrho_\lambda)_{\lambda>0}$$

between the sets $\mathbf{S}(G, K)$ and $\mathbf{R}(G, K)$ given by

$$\varrho_\lambda = \int_0^\infty e^{-\lambda t}\mu_t \, dt,$$

where this equality is understood as an equality of Radon measures on G.

*In the case of this correspondence $(\mu_t)_{t>0}$ and $(\varrho_\lambda)_{\lambda>0}$ are said to be **associated**.*

Proof 1. Starting with a semigroup $(\mu_t)_{t>0} \in \mathbf{S}(G,K)$ we define Radon measures $\varrho_\lambda \in M_+(G)$ by

$$\int f d\varrho_\lambda := \int_0^\infty e^{-\lambda t} \left(\int f d\mu_t \right) dt$$

for all $f \in C^c(G)$. Applying the generalized Schoenberg correspondence theorem 5.2.9 in the sense of

$$(\mu_t)_{t>0} \longleftrightarrow \psi$$

with $\psi \in CND(H)$, where $H := K^\perp$, we obtain that $\|\lambda \varrho_\lambda\| \leq 1$, hence that $\varrho_\lambda \in M_+^b(G)$, and that

$$\hat{\varrho}_\lambda = \begin{cases} \frac{1}{\lambda + \psi(\chi)} & \text{if} \quad \chi \in H \\ 0 & \text{if} \quad \chi \notin H. \end{cases}$$

From this Fourier representation it follows that $(\varrho_\lambda)_{\lambda>0}$ is a resolvent in $M_+^b(G)$, and that it determines the function ψ as well as the τ_w-continuous semigroup $(\mu_t)_{t>0}$ uniquely. It is clear that $(\varrho_\lambda)_{\lambda>0}$ has the same invariance group as $(\mu_t)_{t>0}$, hence $\in \mathbf{R}(G,K)$.

2. Now, let $\chi \in H := K^\perp$. From the resolvent equation we deduce that the number

$$\psi_\lambda(\chi) := \frac{1 - \lambda \hat{\varrho}_\lambda(\chi)}{\hat{\varrho}_\lambda(\chi)}$$

is independent of $\lambda > 0$. Consequently $\psi := \psi_\lambda$ is a well-defined complex-valued function on H. In the proof of Properties 5.2.13 we showed that

$$1_H = \tau_p - \lim_{\lambda \to \infty} \lambda \varrho_\lambda .$$

On the other hand

$$\lambda \hat{\varrho}_\lambda(\chi) \psi(\chi) = \lambda(1 - \lambda \hat{\varrho}_\lambda(\chi))$$

for all $\chi \in H, \lambda > 0$, hence

$$\psi(\chi) = 1_H(\chi)\psi(\chi) = \lim_{\lambda\to\infty} \lambda\hat{\varrho}_\lambda(\chi)\psi(\chi) = \lim_{\lambda\to\infty} \lambda(1 - \lambda\hat{\varrho}_\lambda(\chi))$$

for all $\chi \in H$, and consequently $\psi \in CND(H)$ by Property 4.3.3.4. But now Theorem 5.2.9 implies the assertion, since

$$\left(\int_0^\infty e^{-\lambda t}\mu_t dt\right)^\wedge = \frac{1}{\lambda+\psi}1_H = \hat{\varrho}_\lambda,$$

whenever $\lambda > 0$. ■

Corollary 5.2.15 *A resolvent $(\varrho_\lambda)_{\lambda>0}$ in $M_+^b(G)$ determines a convolution semigroup $(\mu_t)_{t\geq 0}$ in $M_+^{(1)}$ if and only if the invariance group of $(\varrho_\lambda)_{\lambda>0}$ equals $\{0\}$.*

This is immediate from the theorem together with Property 5.2.8.5.

Theorem 5.2.16 *(Support correspondence) Let $(\mu_t)_{t\geq 0} \in \mathbf{C\,S}(G)$, and let $(\varrho_\lambda)_{\lambda>0}$ be a resolvent in $M_+^b(G)$ with*

$$(\mu_t)_{t\geq 0} \longleftrightarrow (\varrho_\lambda)_{\lambda>0}\,.$$

Then for each $\lambda > 0$

(i) $\quad \operatorname{supp}(\varrho_\lambda) = \left(\bigcup_{t>0} \operatorname{supp}(\mu_t)\right)^-,$

(ii) $\quad \operatorname{supp}(\varrho_\lambda)$ *is a $\sigma-$compact semigroup in G with* $0 \in \operatorname{supp}(\varrho_\lambda)$,

and

(iii) $\quad S := [\bigcup_{t>0} \operatorname{supp}(\mu_t)]^-$ *is $\sigma-$compact.*

Proof. We fix $\lambda > 0$ and $f \in C_+^c(G)$. Then

$$\int f d\varrho_\lambda = 0 \text{ if and only if } \int f d\mu_t = 0 \text{ for all } t > 0\,.$$

Since by assumption on $(\mu_t)_{t\geq 0}$

$$\tau_v - \lim_{t\to 0} \mu_t = \varepsilon_0,$$

$f(0) > 0$ implies $\int f d\varrho_\lambda = 0$, hence $0 \in \operatorname{supp}(\varrho_\lambda)$. Similarly, given an open subset U of G we have

$$\operatorname{supp}(\varrho_\lambda) \subset U^c \text{ if and only if } \operatorname{supp}(\mu_t) \subset U^c$$

for all $t > 0$, and

$$\operatorname{supp}(\varrho_\lambda) = \left(\bigcup_{t>0} \operatorname{supp}(\mu_t)\right)^-$$

as asserted in *(i)*.

From the support formula quoted in Theorem 4.1.10 we conclude that for $s, t > 0$

$$\operatorname{supp}(\mu_s) + \operatorname{supp}(\mu_t) \subset \operatorname{supp}(\mu_{t+s}),$$

hence that

$$\left(\bigcup_{t>0} \operatorname{supp}(\mu_t)\right)^-$$

is a semigroup which we have shown to contain 0. Since every measure in $M_+^b(G)$ has σ-compact support, *(i)* implies *(ii)*.

Finally,

$$S = [\bigcup_{t>0} \operatorname{supp}(\mu_t)]^- = (\operatorname{supp}(\varrho_\lambda) - \operatorname{supp}(\varrho_\lambda))^- .$$

But the set $\operatorname{supp}(\varrho_\lambda) - \operatorname{supp}(\varrho_\lambda)$ being σ-compact, also its closure and hence S is σ-compact. This proves *(iii)*. ∎

Theorem 5.2.17 *Let* $(\mu_t)_{t\geq 0} \in \mathbf{C}\,\mathbf{S}(G)$ *and* $\psi \in CND(G^\wedge)$ *with*

$$(\mu_t)_{t\geq 0} \longleftrightarrow \psi,$$

and let $S := [\bigcup_{t>0} \operatorname{supp}(\mu_t)]^-$.

Then

(i) $T := \operatorname{Inv}(\psi) = \{\chi \in G^\wedge : \psi(\chi) = \psi(0)\}$.

and

(ii) $S = T^\perp$, *hence* $T = S^\perp$.

Proof. For each $t > 0$ we define the set

$$A_t := \{\chi \in G^\wedge : \exp(-t\psi(\chi)) = \exp(-t\psi(0))\}.$$

Clearly,

$$\chi \in \operatorname{Inv}(\psi) \text{ if and only if } \chi \in \operatorname{Inv}(\exp(-t\psi))$$

for all $t > 0$. By Property 4.4.18.2 this implies that

$$\operatorname{Inv}(\psi) = \bigcap_{t>0} A_t = \{\chi \in G^\wedge : \psi(\chi) = \psi(0)\},$$

hence *(i)*.

Moreover, Properties 4.4.18.1 and 4.4.18.2 lead to

$$A_t = \operatorname{Inv}(\hat{\mu}_t) = (\operatorname{supp}(\mu_t))^\perp$$

for all $t > 0$, and therefore to

$$\begin{aligned}(\operatorname{Inv}(\psi))^\perp &= \left(\bigcap_{t>0} (\operatorname{supp}(\mu_t))^\perp\right)^\perp \\ &= \left(\bigcap_{t>0} \operatorname{supp}(\mu_t)\right)^{\perp\perp}\end{aligned}$$

which by Property 4.2.17.1 yields the assertion in *(ii)*. ■

By passing to quotient groups and applying the Pontryagin duality the discussion starting with Properties 4.4.19 can be extended in an obvious way.

Properties 5.2.18

5.2.18.1 *If* $(\mu_t)_{t>0}$ *is a semigroup in* $M_+^{(1)}(G)$ *with invariance group* K*, then* $(\dot{\mu}_t)_{t>0}$ *is a semigroup in* $M_+^{(1)}(G/K)$ *with invariance group* $\{0\}$ *(in* (G/K)*).*

If, moreover, $(\mu_t)_{t>0} \in \mathbf{S}(G, K)$*, then* $(\dot{\mu}_t)_{t>0} \in \mathbf{S}(G/K, \{0\}) = \mathbf{C}\,\mathbf{S}(G/K)$.

5.2.18.2 *Let* $(\varrho_\lambda)_{\lambda>0} \in \mathbf{R}(G, K)$ *and let* $(\mu_t)_{t>0} \in \mathbf{S}(G, K)$ *with*

$$(\mu_t)_{t>0} \longleftrightarrow (\varrho_\lambda)_{\lambda>0}\,.$$

Then $(\dot{\varrho}_\lambda)_{\lambda>0}$ *is a resolvent in* $M_+^b(G/K)$ *with*

$$(\dot{\varrho}_\lambda)_{\lambda>0} \longleftrightarrow (\dot{\mu}_t)_{t>0} \in \mathbf{C}\,\mathbf{S}(G/K)\,.$$

5.2.18.3 *Let* $(\mu_t)_{t\geq 0} \in \mathbf{C}\,\mathbf{S}(G)$ *and* $\psi \in CND(G^\wedge)$ *with*

$$(\mu_t)_{t\geq 0} \longleftrightarrow \psi\,.$$

Then the quotient function $\dot{\psi} \in CND(G^\wedge/S)$ *corresponds to* $(\mu_t)_{t\geq 0}$ *considered as an element of* $\mathbf{C}\,\mathbf{S}(T)$*, where* $S := \mathrm{Inv}\,(\psi)$ *and* $T := S^\perp$.

5.3 Lévy functions

In the previous section it was shown that the constant functions, homomorphisms and nonnegative quadratic forms on the dual $G^\wedge$ of a locally compact Abelian group G are negative definite. The question arises of whether these types of negative definite functions can be viewed as building blocks of general negative definite functions on $G^\wedge$. An answer to this question can be given within different frame works. In probabilistic terms it covers the canonical decomposition due to P. Lévy and A.I. Khintchine of infinitely divisible or embeddable probability measures on Euclidean space (See Theorem 3.4.20). In the present exposition we shall present a canonical representation of negative definite functions on the dual $G^\wedge$ of a locally compact Abelian group G which by the Schoenberg correspondence provides also a canonical representation of the associated convolution semigroups on G.

We start with the discussion of a fundamental centering method developed in terms of a centering function (or local inner product) for G.

Definition 5.3.1 *A function $g \in C(G \times G^\wedge)$ is called a* ***Lévy function*** *for G if the following conditions are satisfied*

($LF1$) *For every compact $C \subset G^\wedge$*

$$\sup_{x \in G} \sup_{\chi \in C} |g(x, \chi)| < \infty .$$

$$(LF2) \qquad g(x, \chi + \varrho) = g(x, \chi) + g(x, \varrho)$$

and

$$g(-x, \chi) = -g(x, \chi)$$

for all $x \in G, \chi, \varrho \in G^\wedge$.

($LF3$) *For every compact $C \subset G^\wedge$ there exists a $U := U_C \in \mathfrak{V}_G(0)$ such that*

$$\chi(x) = \exp i\, g(x, \chi)$$

whenever $x \in U, \chi \in C$.

(*LF*4) *For every compact* $C \subset G^\wedge$

$$\lim_{x\to 0} \sup_{\chi\in C} g(x,\chi) = 0\,.$$

Examples 5.3.2 *of Lévy functions*

5.3.2.1 *If* $G := \mathbf{R}^d$ *for* $d \geq 1$ *and if for every* $i = 1,\ldots,d$ *a function* $\zeta_i \in C^b(\mathbf{R})$ *is given with the properties that*

$$\zeta_i(t) = t$$

for all t *from some* $U \in \mathfrak{V}_G(0)$ *and that*

$$\zeta_i(-t) = -\zeta_i(t)$$

for all $t \in \mathbf{R}$, *then the function* $g : \mathbf{R}^d \times \mathbf{R}^d \to \mathbf{R}$ *defined by*

$$g(x,y) := \sum_{i=1}^{d} \zeta_i(x_i) y_i$$

for all $x = (x_1,\ldots,x_d)$ *and* $y := (y_1,\ldots,y_d)$ *in* $\mathbf{R}^d$ *is a Lévy function for* G.

5.3.2.2 *Let* $G := \mathbf{T}^d$ *for* $d \geq 1$. *Then* G *can be viewed as the group*

$$\{(x_1,\ldots,x_d) \in \mathbf{R}^d : x_i \in]-1,1] \text{ for all } i = 1,\ldots,d\}$$

with addition modulo 2. Employing for every $i = 1,\ldots,d$ *the function* ζ_i *introduced in the previous example one obtains a Lévy function* g *for* G *by putting*

$$g(x,m) := \sum_{i=1}^{d} \zeta_i(x_i) m_i$$

for all $x = (x_1,\ldots,x_d) \in \mathbf{T}^d$ *and* $m := (m_1,\ldots,m_d) \in \mathbf{Z}^d$.

5.3.2.3 *Let* $G := \mathbf{Q}_d{}^\wedge$ *(the dual of the discretely topologized rationals),* $\zeta \in C^b(G)$, *and let* $\chi_0 \in \mathbf{Q}_d$ *with* $\chi_0 \neq 0$ *be such that*

$\exp i\zeta(x) = \chi_0(x)$ *for all* x *in some* $U \in \mathfrak{V}_G(1)$. *Then the real-valued function* g *on* $G \times G^\wedge$ *defined by*

$$g(x, \chi) := \zeta(x) \frac{\chi}{\chi_0}$$

for all $x \in G, \chi \in G^\wedge$ *is a Lévy function for* G.

5.3.2.4 *If* G *is totally disconnected, then the zero function on* $G \times G^\wedge$ *is a Lévy function for* G, *since every homomorphism from* $G^\wedge$ *into* $\mathbf{R}$ *is trivial.*

Properties 5.3.3 *of Lévy functions*

5.3.3.1 *(Products) Let* $G = G_1 \times G_2$ *be the product of two locally compact Abelian groups* G_1 *and* G_2 *admitting Lévy functions* g_1 *and* g_2 *respectively. Then* G *admits the Lévy function* g *given by*

$$g(x, \chi) := g_1(x_1, \chi_1) + g_2(x_2, \chi_2)$$

whenever $x := (x_1, x_2) \in G_1 \times G_2$ *and* $\chi := (\chi_1, \chi_2) \in {G_1}^\wedge \times {G_2}^\wedge$.

5.3.3.2 *(Extensions) Let* H *be an open subgroup of a locally compact Abelian group* G *admitting a Lévy function* g_0. *Let* $\pi^\wedge$ *denote the canonical homomorphism* $G^\wedge \to H^\wedge \cong G^\wedge / H^\perp$. *Then the function* g *on* $G \times G^\wedge$ *defined by*

$$g(x, \chi) := \begin{cases} g_0(x, \pi^\wedge(\chi)) & \text{if } x \in H \\ 0 & \text{if } x \notin H \end{cases}$$

and all $\chi \in G^\wedge$ *is a Lévy function for* G.

In fact, $g \in C(G \times G^\wedge)$, since H is an open, hence a closed subgroup of G. The remaining properties of g follow from those of g_0.

Theorem 5.3.4 *On any locally compact Abelian group* G *there exists a Lévy function.*

Proof 1. From Property 5.3.3.2 we infer that it is sufficient to prove the theorem for an open subgroup H of G. Choosing H as the subgroup generated by an open compact neighborhood of $0 \in G$ we can apply the structure theorem 4.2.19 and obtain the direct product decomposition

$$H \cong \mathbf{R}^d \times \mathbf{Z}^e \times K$$

where $d, e \geq 0$ and K is a compact Abelian group. But now it is obvious by Property 5.3.3.1 that we may without loss of generality assume G to be a compact group, since for $\mathbf{R}^d$ and $\mathbf{Z}^d$ Lévy functions exist by Examples 5.3.2.1 and 5.3.2.2 respectively.

2. Let G be a compact Abelian group. For the connected component G_0 of the identity $0 \in G$ we have

$$G_0{}^\wedge \cong G^\wedge / G_0{}^\perp \quad \text{and} \quad (G/G_0)^\wedge \cong G_0{}^\perp$$

by the functorial properties 4.2.17 of the duality. Since G_0 is connected and compact, $G_0{}^\wedge$ is a discrete group in which each element is of infinite order. An application of Zorn's lemma yields a maximal family $\{d_\alpha : \alpha \in A\}$ in $G_0{}^\wedge$ with the following linear independence property: If

$$\sum_{i=1}^{l} n_{\alpha_i} d_{\alpha_i} = 0$$

for some $l \in \mathbf{N}, n_{\alpha_1}, \ldots, n_{\alpha_l} \in \mathbf{Z}$ and $d_{\alpha_1}, \ldots, d_{\alpha_l} \in \{d_\alpha : \alpha \in A\}$, then $n_{\alpha_i} = 0$ for all $i = 1, \ldots, l$. But then for any $d \in G_0{}^\wedge$ there exist $d_{\alpha_1}, \ldots, d_{\alpha_k} \in \{d_\alpha : \alpha \in A\}$ and $n, n_1, \ldots, n_k \in \mathbf{Z}, n > 0$ such that

$$nd = \sum_{i=1}^{k} n_i d_{\alpha_i}, \tag{1}$$

where the representation is unique within multiplication by integers. Each element of $G_0{}^\wedge$ being a coset of $G_0{}^\perp$ in $G^\wedge$ we take the coset d_α and choose an element $\chi_\alpha \in G^\wedge$ from this coset. χ_α will be fixed for the moment. One observes that there exists a function $h_{\chi_\alpha} \in C(G)$ having the following properties:

$$|h_{\chi_\alpha}| \leq \pi,$$

$$h_{\chi_\alpha}(-x) = -h_{\chi_\alpha}(x)$$

for all $x \in G$, and

$$\chi_\alpha(x) = \exp ih\chi_\alpha(x)$$

for all $x \in G$ such that $|\chi_\alpha(x) - 1| \leq \frac{1}{2}$. Next we define

$$g(x, \chi_\alpha) := h\chi_\alpha(x)$$

for all $x \in G$ and $\alpha \in A$.

Now, let $\chi \in G^\wedge$ be arbitrary. Then χ belongs to some coset of ${G_0}^\perp$ being an element of $G^\wedge/{G_0}^\perp$. This element d admits a representation of the form (1) with $n, n_1, \ldots, n_k \in \mathbf{Z}, n > 0$ and $d_{\alpha_1}, \ldots, d_{\alpha_k} \in \{d_\alpha : \alpha \in A\}$ $(k \geq 1)$. The function g defined by

$$g(x, \chi) := \sum_{j=1}^{k} \frac{n_j}{n} g(x, \chi_{\alpha_j})$$

for all $x \in G, \chi \in G^\wedge$ will serve as a Lévy function for G. We shall show the required properties in

3. First of all $g \in C(G \times G^\wedge)$, since $x \mapsto g(x, \chi)$ is continuous on G for each $\chi \in G^\wedge$ and $G^\wedge$ is discrete. Properties $(LF1)$ and $(LF2)$ follow directly from the construction.

Since compact sets in $G^\wedge$ are finite it suffices to prove Property $(LF3)$ for each $\chi \in G^\wedge$. For any $\chi \in G^\wedge$ we denote the coset of ${G_0}^\perp$ to which χ belongs, by $\dot{\chi}$. Then (1) can be rewritten in the form

$$n\dot{\chi} = \sum_{j=1}^{k} n_j \dot{\chi}_{\alpha_j}. \tag{2}$$

Clearly, $\chi_1 - \chi_2 \in {G_0}^\perp$ whenever $\chi_1, \chi_2 \in \dot{\chi}$. Since ${G_0}^\perp = (G/G_0)^\wedge$ is totally disconnected, every element of ${G_0}^\perp$ is of finite order. Hence for any $\chi \in {G_0}^\perp$ there exists a neighborhood $V \in \mathfrak{V}_G(0)$ such that $\chi(x) = 1$ for all $x \in V$, and consequently, for $\chi_1, \chi_2 \in \dot{\chi}$ we have $\chi_1(x) = \chi_2(x)$ for all x in some $V_1 \in \mathfrak{V}_G(0)$.

But now we refer to the construction of the function h_{χ_α} above. There exists a neighborhood $V_2 \in \mathfrak{V}_G(0)$ such that

$$\chi_\alpha(x) = \exp ig(x, \chi_\alpha)$$

for all $x \in V_2$ $(\alpha \in A)$. For arbitrary $\chi \in G^\wedge$ we employ (2) in order to obtain the representation

$$n\chi = \sum_{j=1}^{k}(\chi_{\alpha_j 1} + \ldots + \chi_{\alpha_j n_j}), \tag{3}$$

where $\chi_{\alpha_j 1}, \ldots, \chi_{\alpha_j n_j} \in \dot{\chi}_{\alpha_j}$ for $j = 1, \ldots, k$. For the above discussion it follows that there exists a neighborhood $V_3 \in \mathfrak{V}_G(0)$ (depending on the characters $\chi_{\alpha_j r}$ and χ_{α_j}) such that

$$\chi_{\alpha_j r}(x) = \chi_{\alpha_j}(x)$$

for all $x \in V_3$ $(r = 1, \ldots, n_j, j = 1, \ldots, k)$. Let U_1 denote the intersection of all neighborhood of type V_3 arising from the choices of the $\chi_{\alpha_j r}(r = 1, \ldots, n_j, j = 1, \ldots, k)$. Then

$$\chi_{\alpha_j r}(x) = \chi_{\alpha_j}(x) \tag{4}$$

for all $x \in U_1, r = 1, \ldots, n_j, j = 1, \ldots, k$. But (3) and (4) imply that

$$\chi(x)^n = \prod_{j=1}^{k} \chi_{\alpha_j}(x)^{n_j}$$

for all $x \in U_1$. Since there are neighborhoods $\in \mathfrak{V}_G(0)$, on which

$$\chi_{\alpha_j} = \exp ig(\cdot, \chi_{\alpha_j})$$

holds, there is also a neighborhood $\in \mathfrak{V}_G(0)$ where

$$\chi^n = \exp(ing(\cdot, \chi))$$

is valid. χ and $\exp ig(\cdot, \chi)$ being continuous and $\neq 0$ at the identity 0 of G we obtain the desired neighborhood $U \in \mathfrak{V}_G(0)$ such that

$$\chi(x) = \exp ig(x, \chi)$$

for all $x \in U, \chi \in G^\wedge$.

Finally we note that Property $(LF4)$ follows from the facts that $g \in C(G \times G^\wedge)$ and that $g(x,\chi) = 0$ whenever either x or χ are the identities of G or $G^\wedge$ respectively. The proof is complete. ■

Applying a Lévy function g for G we can provide a new type of continuous negative definite functions on $G^\wedge$ which will be crucial for the canonical representation we are aiming at

Theorem 5.3.5 *Let g be a Lévy function for G, and let $\mu \in M_+(G^\times)$ with $G^\times := G\backslash\{0\}$ be such that*

$$\int_{G^\times} (1 - \text{Re } \chi) d\mu < \infty$$

for all $\chi \in G^\wedge$. Then the function ψ_μ on G defined by

$$\psi_\mu(\chi) := \int_{G^\times} (1 - \overline{\chi(x)} + ig(x,\chi))\mu(dx)$$

for all $\chi \in G^\wedge$ belongs to $CND(G^\wedge)$.

Proof. For a fixed $\chi_0 \in G^\wedge$ and a compact neighborhood $W \in \mathfrak{V}_{G^\wedge}(0)$ there exists by Property $(LF3)$ of g a neighborhood $U = U(W) \in \mathfrak{V}_G(0)$ such that

$$\overline{\chi(x)} = \exp\ ig(x,\chi)$$

for all $x \in U$ and $\chi \in \chi_0 + W$. Then Property $(LF4)$ implies the existence of a neighborhood $U' \in \mathfrak{V}_G(0)$ satisfying the inequality

$$|\sin g(x,\chi) - g(x,\chi)| \leq 1 - \cos g(x,\chi)$$

valid for all $x \in U'$ and $\chi \in \chi_0 + W$. It follows that for all $x \in U'$ and $\chi \in \chi_0 + W$

$$\begin{aligned} |1 - \overline{\chi(x)} + i\, g(x,\chi)| &= |1 - \cos g(x,\chi) - i \sin g(x,\chi) + i\, g(x,\chi)| \\ &\leq 2|1 - \text{Re } \chi(x)| \end{aligned} \tag{5}$$

Since $\mu(G^\times \backslash U') < \infty$ by assumption, Property $(LF1)$ implies that $\psi_\mu(\chi_0)$ is a well-defined number $\in \mathbf{C}$ and moreover that there exists a constant $c := c(\chi_0) \geq 0$ such that

$$|\psi_\mu(\chi)| \leq 2\,\mathrm{Re}\ \psi_\mu(\chi) + c(\chi_0) \tag{6}$$

whenever $\chi \in \chi_0 + W$. Since for every $x \in G$ the functions

$$\chi \mapsto 1 - \overline{\chi(x)} + ig(x, \chi)$$

and

$$\chi \mapsto 1 - \mathrm{Re}\ \chi(x)$$

belong to $ND(G^\wedge)$, also ψ_μ and $\mathrm{Re}\ \psi_\mu$ are elements of $ND(G^\wedge)$. Here Example 5.1.8.2 and Properties 5.1.3 are applied.

It remains to show that ψ_μ is continuous. Once we know that $\mathrm{Re}\ \psi_\mu$ is locally bounded, inequality (6) justifies the local boundedness of ψ_μ. Let therefore ψ_μ be of the form

$$\chi \mapsto \int_{G^\times} (1 - \mathrm{Re}\ \chi) d\mu\,.$$

Let $\mathcal{K}$ denote the family of the compact subsets K of G with $0 \notin K$, and set for each $K \in \mathcal{K}$

$$\mu_K := \mathrm{Res}\ _K\mu\,.$$

Clearly $\psi_{\mu_K} \in CND(G^\wedge)$ for all $K \in \mathcal{K}$, hence

$$\psi_\mu := \sup_{K \in \mathcal{K}}\ \psi_{\mu_K} \in ND(G^\wedge)$$

is lower semicontinuous. For every $n \geq 1$ the set $A_n := \{\chi \in G^\wedge : \psi_\mu(\chi) \leq n\}$ is closed. Since ψ_μ is finite everywhere we have $G^\wedge = \bigcup\limits_{n \geq 1} A_n$ and by Baire's theorem there exists an $n_0 \geq 1$ such that $\overset{\circ}{A}_{n_0} \neq \emptyset$. Let V be an open neighborhood $\in \mathfrak{V}_G(0)$ and let $\chi_1 \in \overset{\circ}{A}_{n_0}$

be such that $\chi_1 + V \subset A_{n_0}$. Since $\psi_\mu \in ND(G^\wedge)$ we obtain from Property 5.1.2.3 for all $\chi \in \chi_0 + V$ $(\chi_0 \in G)$ and $\chi_2 := \chi_0 - \chi_1$ that

$$\begin{aligned}\sqrt{\psi_\mu(\chi)} &= \sqrt{\psi_\mu(\chi - \chi_2 + \chi_2)}\\ &\leq \sqrt{\psi_\mu(\chi - \chi_2)} + \sqrt{\psi_\mu(\chi_2)}\\ &\leq \sqrt{\psi_\mu(\chi_2)} + \sqrt{n_0},\end{aligned}$$

hence that ψ_μ is locally bounded. Since ψ_μ is a complex linear combination of linear semicontinuous functions, it is locally $\omega_{G^\wedge}$-integrable, and given a function $f \in C_+^c(G^\wedge)$ satisfying $f(\chi) = f(-\chi)$ for all $\chi \in G^\wedge$ and $\int f d\omega_{G^\wedge} = 1$ we conclude that

$$\begin{aligned}\psi_\mu * f(\chi) &= \int\limits_{G^\wedge} f(\chi - \varrho)\Big(\int\limits_{G^\times} (1 - \overline{\varrho(x)} + i\, g(x,\varrho))\mu(dx)\Big)\omega_{G^\wedge}(d\varrho)\\ &= \int\limits_{G^\times} (1 - \overline{\chi(x)}\mathcal{F}_{G^\wedge} f(x) + i\, g(x,\chi))\mu(dx)\\ &= \psi_\mu(\chi) + \int\limits_{G^\times} \overline{\chi(x)}(1 - \mathcal{F}_{G^\wedge} f(x))\mu(dx)\end{aligned}$$

whenever $\chi \in G^\wedge$. But this implies the continuity of ψ_μ, since the last integral in the above chain of equalities is the Fourier transform of the measure $(1 - \mathcal{F}_{G^\wedge} f) \cdot \mu \in M_+^b(G^\times)$ considered as a measure on G. ■

5.4 The Lévy-Khintchine representation

For any locally compact space E we denote by $M^c(E)$ the set of Radon measures with compact support. In the case of a locally compact Abelian group G with dual $G^\wedge$ the set

$$S := \{\sigma \in M^1(G^\wedge) \cap M^c(G^\wedge) : \sigma = \sigma^\sim\}$$

of symmetric probability measures on $G^\wedge$ with compact support will be efficiently applied.

Theorem 5.4.1 *Let $\psi \in CND(G^\wedge)$. Then*

(i) for any $\sigma \in S$ the functions $\psi * \sigma - \psi$ belongs to $CPD(G^\wedge)$ and hence admits a Bochner measure $\mu_\sigma \in M_+^b(G)$ in the sense that

$$\mathcal{F}_G \mu_\sigma = \psi * \sigma - \psi\,.$$

(ii) There exists a measure $\mu \in M_+(G^\times)$ satisfying $(1 - \mathcal{F}_{G^\wedge}\sigma) \cdot \mu =$ $\mathrm{Res}\,_{G^\times} \mu_\sigma$ whenever $\sigma \in S$.

(iii) If ψ is associated with a convolution semigroup $(\mu_t)_{t \geq 0}$ in $M_+^{(1)}(G)$ according to the Schoenberg correspondence, then

$$\mu = \tau_\psi - \lim_{t \to 0} \frac{1}{t} \mathrm{Res}\,_{G^\times} \mu_t\,.$$

Remark 5.4.2 *The relevant references to statements (i) and (iii) of the theorem are Theorems 4.3.5 and 5.2.3 respectively.*

Definition 5.4.3 *The measure μ established in Theorem 5.4.1 is called the* ***Lévy measure*** *for the continuous negative definite function ψ or for the convolution semigroup $(\mu_t)_{t \geq 0}$ associated with ψ.*

Proof of Theorem 5.4.1. Let $\psi \leftrightarrow (\mu_t)_{t \geq 0}$ (via Theorem 5.2.3) and let $\sigma \in S$. Then for any $t > 0$

$$(1 - \mathcal{F}_{G^\wedge}\sigma) \cdot \frac{1}{t} \mu_t \in M_+^b(G)$$

and

$$\begin{aligned}\mathcal{F}_G[(1-\mathcal{F}_{G^\wedge}\sigma)\cdot\frac{1}{t}\mu_t](\chi) &= \frac{1}{t}\int\overline{\chi(x)}(1-\mathcal{F}_{G^\wedge}\sigma(x))\mu_t(dx)\\ &= \frac{1}{t}\int\overline{\chi(x)}(1-\int\varrho(x)\sigma(d\varrho))\mu_t(dx)\\ &= \frac{1}{t}\Big(\hat{\mu}_t(\chi)-\int\int\overline{(\chi-\varrho)(x)}\sigma(d\varrho)\mu(dt)\Big)\\ &= \frac{1}{t}(\hat{\mu}_t(\chi)-\hat{\mu}*\sigma(\chi))\\ &= \frac{1}{t}(1-\exp(-t\psi)*(\sigma-\varepsilon_0))(\chi)\end{aligned}$$

for all $\chi \in G^\wedge$. Since

$$\tau_{co}-\lim_{t\to 0}\frac{1}{t}(1-\exp(-t\varphi)) = \psi,$$

we obtain that

$$\lim_{t\to 0}\mathcal{F}_G[(1-\mathcal{F}_{G^\wedge}\sigma)\cdot\frac{1}{t}\mu_t] = \psi*\sigma-\psi$$

in either of the topologies τ_p and τ_{co}. As a conclusion we obtain that $\psi*\sigma-\psi \in CPD(G^\wedge)$, and by an application of Theorem 4.3.7 that

$$\tau_w-\lim_{t\to 0}(1-\mathcal{F}_{G^\wedge}\sigma)\cdot\frac{1}{t}\mu_t = \mu_\sigma,$$

where $\mu_\sigma \in M_+^b(G)$ such that

$$\hat{\mu}_\sigma = \psi*\sigma-\psi\,.$$

Given $\psi \in C_+^c(G)$ with $\operatorname{supp}(\varphi) \subset G^\times$ we choose $\sigma \in S$ such that $\mathcal{F}_{G^\wedge}\sigma(x) \le \frac{1}{2}$ for all x in a neighborhood of $\operatorname{supp}(\varphi)$. The function φ' on G defined by

$$\varphi'(x) := \begin{cases}\frac{\varphi(x)}{1-\mathcal{F}_{G^\wedge}\sigma(x)} & \text{if } x\in\operatorname{supp}(\varphi)\\[1ex] 0 & \text{otherwise}\end{cases}$$

belongs to $C_+^c(G)$, and

$$\lim_{t\to 0}\int \varphi\, d\Big(\frac{1}{t}\mu_t\Big) = \lim_{t\to 0}\int \varphi' d\Big((1-\mathcal{F}_{G^\wedge}\sigma)\frac{1}{t}\varphi_t)\Big)$$
$$= \int \varphi' d\mu_\sigma\,.$$

This implies that there exists a measure $\mu \in M_+(G^\times)$ such that

$$\mu = \tau_v - \lim_{t\to 0}\frac{1}{t}\text{Res}\,_{G^\times}\mu_t$$

and that

$$(1-\mathcal{F}_{G^\wedge}\sigma)\cdot\mu = \text{Res}\,_{G^\times}\mu_\sigma$$

holds for all $\sigma \in S$. ■

5.4.4 Properties *of Lévy measures*

5.4.4.1 *Any Lévy measure ψ of a function $\psi \in CND(G^\wedge)$ (or of its associated convolution semigroup in $M_+^{(1)}(G)$) has the following properties:*

(*a*)
$$\int_{G^\times}(1-\text{Re}\,\chi)d\mu < \infty$$

for all $\chi \in G^\wedge$.

(b) For every compact neighborhood $V \in \mathfrak{V}_G(0)$

$$\text{Res}\,_{V^c}\mu \in M_+^b(G)\,.$$

While the statement in (a) follows directly from Theorem 5.4.1 by specializing σ to the measure $\frac{1}{2}(\varepsilon_\chi+\varepsilon_{-\chi})$ for $\chi\in G^\wedge$, the statement in (b) makes use of the existence of $\sigma \in S$ satisfying

$$1-\mathcal{F}_{G^\wedge}\sigma(x) \geq \frac{1}{2}$$

for all x in the complement V^c of a compact neighborhood $V \in \mathfrak{V}_G(0)$; one just applies (ii) of Theorem 5.4.1 and deduces the inequalities

$$\begin{aligned}\int \varphi \, d\mu &\leq 2 \int \varphi \, d[(1 - \mathcal{F}_{G^\wedge}\sigma) \cdot \mu] \\ &= 2 \int \varphi \, d\mu_\sigma\end{aligned}$$

valid for each $\varphi \in C_+^c(G)$ with $\operatorname{supp}(\varphi) \subset V^c$.

5.4.4.2 *Let $\psi_\mu \in CND(G^\wedge)$ be the function introduced for a given measure $\mu \in M_+(G^\times)$ in Theorem 5.3.5. The Lévy measure of ψ_μ is just μ, hence μ is uniquely determined by ψ_μ.*

In fact, for $\sigma \in S$ and $\chi \in G^\wedge$ we obtain the identity

$$\psi_\mu * \sigma(\chi) - \psi_\mu(\chi) = \int_{G^\times} \overline{\chi(x)}(1 - \mathcal{F}_{G^\wedge}\sigma)(x)\mu(dx) \,,$$

and it follows that the measure $(1 - \mathcal{F}_{G^\wedge}\sigma) \cdot \mu$ considered as an element of $M_+^b(G)$ has $\psi_\mu * \sigma - \psi_\mu$ as its Fourier transform.

5.4.4.3 *Let $\psi_1, \psi_2 \in CND(G^\wedge)$ (associated with convolution semigroups $(\mu_t^{(1)})_{t\geq 0}$ and $(\mu_t^{(2)})_{t\geq 0}$ respectively) admit μ_1 and μ_2 as their Lévy measures. Then the function $\psi := \psi_1 + \psi_2 \in CND(G^\wedge)$ (associated with the convolution semigroups $(\mu_t^{(1)} * \mu_t^{(2)})_{t\geq 0}$) admits $\mu := \mu_1 + \mu_2$ as its Lévy measure.*

Example 5.4.5 *Let $\mu \in M_+^b(G)$ with $\|\mu\| \leq 1$. Then by Example 5.2.7.2 $\psi := 1 - \hat{\mu}$ is the continuous negative definite function associated with the convolution semigroup determined by μ. It turns out that the Lévy measure for ψ is* $\operatorname{Res}_{G^\times} \mu$.

By a simple computation of Fourier transforms and an application of their sequential bicontinuity (Theorem 4.3.8) one shows that any convolution semigroup $(\mu_t)_{t\geq 0}$ in $M_+^{(1)}(G)$ with associated resolvent $(\varrho_\lambda)_{\lambda>0}$ can be obtained as a limit of convolution semigroups $(\mu_t^\lambda)_{t\geq 0}$ determined by $\lambda^2 \varrho_\lambda$ in the sense that

$$\mu_t = \tau_w - \lim_{\lambda\to\infty} \mu_t^\lambda$$

for all $t > 0$. It is easy to see that for the Lévy measure μ of $(\mu_t)_{t\geq 0}$ one obtains

$$\mu = \tau_v - \lim_{\lambda\to\infty} \text{Res}_{\,G^\times} \lambda^2 \varrho_\lambda .$$

We are now prepared to *characterize the two types of continuous negative definite functions* introduced in Examples 5.1.8.2 and 5.1.8.3 in terms of measures in S.

5.4.6.1 *A real-valued function $h \in C(G^\wedge)$ with $h(0) = 0$ is a (group) homomorphism if and only if $h * \sigma - h = 0$ for all $\sigma \in S$.*

In fact, for any homomorphism $h : G^\wedge \to \mathbf{R}$ and each $\sigma \in S$ we have

$$\begin{aligned}\int h(\varrho)\sigma(d\varrho) &= \int h(-\varrho)\sigma^\sim(d\varrho) \\ &= -\int h(\varrho)\sigma(d\varrho),\end{aligned}$$

hence $\int h d\sigma = 0$ and consequently

$$\begin{aligned}h * \sigma(\chi) &= \int h(\chi - \varrho)\sigma(d\varrho) \\ &= h(\chi) - \int h(\varrho)\sigma(d\varrho) \\ &= h(\chi)\end{aligned}$$

whenever $\chi \in G^\wedge$. If conversely this latter condition holds for $\sigma \in S$, then we obtain for $\chi, \varrho \in G^\wedge$ and $\sigma := \frac{1}{2}(\varepsilon_\chi + \varepsilon_{-\chi})$ that

$$0 = h * \sigma(\varrho) - h(\varrho) = \frac{1}{2}[h(\varrho - \chi) + h(\varrho + \chi)] - h(\varrho),$$

in particular for the choice $\varrho = 0$ that

$$\frac{1}{2}[h(-\chi) + h(\chi)] = 0 .$$

Analogously we obtain

$$0 = \frac{1}{2}[h(\chi - \varrho) + h(\chi + \varrho)] - h(\chi),$$

hence adding the two equalities yields

$$h(\chi + \varrho) = h(\chi) + h(\varrho)\,.$$

As an immediate consequence of this property we note that

5.4.6.2 *the Lévy measure for a function* $\psi \in CND(G^\wedge)$ *of the form* $\psi := ih$ *with a continuous homomorphism* $h : G^\wedge \to \mathbf{R}$ *is the zero measure.*

With similar arguments one proves that

5.4.6.3 *for a function* $\psi \in CND(G^\wedge)$ *with Lévy measure* μ *the function* $i\,\mathrm{Im}\,\psi \in CND(G^\wedge)$ *if and only if* μ *is symmetric.*

5.4.6.4 *A symmetric real-valued function* $q \in C(G^\wedge)$ *with* $q(0) = 0$ *is a quadratic form on* $G^\wedge$ *if and only if there exists a constant* $c \in \mathbf{R}$ *such that*

$$q * \sigma - q = c$$

whenever $\sigma \in S$. *In the affirmative case* $q \geq 0$ *if and only if*

$$q * \sigma - q \geq 0$$

for all $\sigma \in S$.

For the proof let q be a quadratic form on $G^\wedge$. Then for $\sigma \in S$ and $\chi \in G^\wedge$ we have

$$\begin{aligned} q * \sigma(\chi) &= \int q(\chi - \varrho)\sigma(d\varrho) \\ &= \int q(\chi + \varrho)\sigma(d\varrho) \\ &= \int \frac{1}{2}[q(\chi - \varrho) + q(\chi + \varrho)]\sigma(d\varrho) \\ &= \int [q(\chi) + q(\varrho)]\sigma(d\varrho) \\ &= q(\chi) + \int q(\varrho)\sigma(d\varrho). \end{aligned}$$

If conversely $q * \sigma - q$ equals to a constant $c \in \mathbf{R}$ for all $\sigma \in S$, then for $\chi, \varrho \in G^\wedge$ and $\sigma := \frac{1}{2}(\varepsilon_\varrho + \varepsilon_{-\varrho})$ we obtain

$$q * \sigma(\chi) - q(\chi) = q * \sigma(0) - q(0)$$

or

$$\frac{1}{2}[q(\chi + \varrho) + q(\chi - \varrho)] - q(\chi) = \frac{1}{2}[q(\varrho) + q(-\varrho)] = q(\varrho)\,.$$

These equalities also yield the remaining part of the assertion.

As an immediate consequence of 5.4.6.4 we arrive at the assertion that

5.4.6.5 *a real-valued function $\psi \in CND(G^\wedge)$ with $\psi(0) = 0$ having Lévy measure μ is a quadratic form ≥ 0 if and only if μ is the zero measure.*

In fact, from 5.4.6.4 follows that ψ is a quadratic form if and only if the measure μ_σ introduced in *(i)* of Theorem 5.4.1 is supported by $\{0\}$ for all $\sigma \in S$. But this statement is equivalent to μ being the zero measure.

Theorem 5.4.7 *(Canonical representation of continuous negative definite functions)*

Let g be a Lévy function for G. For any complex-valued function ψ on G the following statements are equivalent:

(i) $\psi \in CND(G^\wedge)$.

(ii) *There exist a constant $c \geq 0$, a continuous homomorphism $h : G^\wedge \to \mathbf{R}$, a nonnegative continuous quadratic form on $G^\wedge$ and a measure $\mu \in M_+(G^\times)$ satisfying*

$$\int_{G^\times} (1 - \mathrm{Re}\ \chi) d\mu < \infty$$

for all $\chi \in G^\wedge$ such that

$$\psi(\chi) = c + ih(\chi) + q(\chi) + \int_{G^\times} (1 - \overline{\chi(x)} + ig(x, \chi))\mu(dx) \quad (7)$$

whenever $\chi \in G^\wedge$.

Moreover, the **canonical quadruple** (c, h, q, μ) *with* $c := \psi(0)$, q *given by*

$$q(\chi) = \lim_{n\to\infty} \frac{1}{n^2} \operatorname{Re} \psi(n\chi) \tag{8}$$

for all $\chi \in G^\wedge$, *and* μ *being the Lévy measure for* ψ *is uniquely determined by* ψ, *the function* ih, *however, depending on the choice of the Lévy function* g.

Proof. *(i)* $\Rightarrow$ *(ii)*. Let $\psi \in CND(G^\wedge)$. The Lévy measure μ of ψ coincides with the Lévy measure $\psi' := \psi - c$. Introducing the function

$$\psi'' := \psi' - \psi_\mu,$$

where ψ_μ has been defined in Theorem 5.3.5, we obtain that

$$\psi'' * \sigma - \psi'' = \mu_\sigma(\{0\}), \tag{9}$$

where $\mu_\sigma \in M_+^b(G)$ for $\sigma \in S$ has Fourier transform

$$\mathcal{F}_G(\mu_\sigma) = \psi * \sigma - \psi$$

(See the proof of Property 5.4.4.2). In fact,

$$\begin{aligned} \psi'' * \sigma - \psi'' &= \psi' * \sigma - \psi' - (\psi_\mu * \sigma - \psi_\mu) \\ &= \mathcal{F}_G(\mu_\sigma(\{0\})\varepsilon_0 + \operatorname{Res}_{G^\times} \mu_\sigma) - \mathcal{F}_G((1 - \mathcal{F}_{G^\wedge}\sigma) \cdot \mu) \\ &= \mu_\sigma(\{0\}) \,. \end{aligned}$$

Now taking real and imaginary parts in the identity (9) we arrive at the equalities

$$(\operatorname{Re} \psi'') * \sigma - \operatorname{Re} \psi'' = \mu_\sigma(\{0\})$$

and

$$(\operatorname{Im} \psi'') * \sigma - \operatorname{Im} \psi'' = 0$$

for all $\sigma \in S$. Since ψ'' is a difference between negative definite functions we have that

$$\operatorname{Im} \psi''(0) = \operatorname{Re} \psi''(0) = 0$$

and

$$\text{Re } \psi''(\chi) = \text{Re } \psi''(-\chi)$$

for all $\chi \in G^\wedge$. Application of the characterizations 5.4.6.1 and 5.4.6.5 provides the desired homomorphism $h = \text{Im } \psi''$ and the non-negative quadratic form $q := \text{Res } \psi''$. The representation (7) has been established.

(ii) $\Rightarrow$ *(i)*. We are now given the function ψ represented as in (7) via the quadruple (c, l, q, μ). Then Example 5.1.8.4 and Theorem 5.3.5 imply that $\psi \in CND(G^\wedge)$. But by Property 5.4.4.2 together with characterizations 5.4.6.1 and 5.4.6.5 μ is the Lévy measure for ψ and $c = \psi(0)$. It remains to be shown that q satisfies the limit relation (8). Since h is a homomorphism and q a quadratic form (on $G^\wedge$), we have

$$\begin{aligned} q(\chi) &= \frac{1}{n^2} q(n\chi) \\ &= \frac{1}{n^2}\psi(n\chi) - \frac{1}{n} i\, l(\chi) - \frac{c}{n^2} - \frac{1}{n^2}\psi_\mu(n\chi) \end{aligned}$$

for all $\chi \in G^\wedge$ and $n \geq 1$. Hence it suffices to prove that

$$\lim_{n\to\infty} \frac{1}{n^2}\psi_\mu(n\chi) = 0$$

or by the dominated convergence theorem that

$$\lim_{n\to\infty} \frac{1}{n^2}(1 - \overline{n\chi(x)} + i\, g(x, n\chi)) = 0$$

as well as

$$\frac{1}{n^2}|1 - \overline{n\chi(x)} + i\, g(x, n\chi)| \leq C|1 - \text{Re } \chi(x)|$$

holds for all $x \in G, n \geq 1$ and some constant $C > 0$. While the above limiting relationship is evident, the inequality requires a look back to the estimate (5) in the proof of Theorem 5.3.5 and an argument in support of the inequality

$$\frac{1}{n^2}(1 - \text{Re } n\chi(x)) \leq C(1 - \text{Re } \chi(x))$$

valid for all $x \in G, n \geq 1$. The latter can be established as follows: For fixed $x \in G$ and $\chi \in G^\wedge$ there exists $\vartheta \in [-\pi, \pi]$ such that $\chi(x) = e^{i\vartheta}$. It follows that

$$\begin{aligned}\frac{1}{n^2}(1 - \text{Re } n\chi(x)) &= \frac{1}{n^2}(1 - \cos n\vartheta) \\ &= \frac{2\sin^2 \frac{n\vartheta}{2}}{n^2} \; \frac{2\sin^2 \frac{\vartheta}{2}}{2\sin^2 \frac{\vartheta}{2}} \; \frac{(\frac{\vartheta}{2})^2}{(\frac{\vartheta}{2})^2} \\ &= \left(\frac{\sin \frac{n\vartheta}{2}}{\frac{n\vartheta}{2}}\right)^2 \left(\frac{\frac{\vartheta}{2}}{\sin \frac{\vartheta}{2}}\right)^2 (1 - \cos \vartheta) \\ &\leq C(1 - \text{Re } \chi(x)),\end{aligned}$$

since the functions $y \mapsto \frac{\sin y}{y}$ is bounded away from 0 on $[-\frac{\pi}{2}, \frac{\pi}{2}]$.

Now, by Property 5.4.4.2 the quadruple (c, h, q, μ) is uniquely determined by ψ and deserves to be called canonical. ■

Remark 5.4.8 *In the special cases of bounded or symmetric Lévy measures μ for ψ the canonical representation given in Theorem 5.4.7 does not require the use of the Lévy function g which had been introduced in order to achieve the absolute convergence of the integral defining ψ_μ.*

Discussion 5.4.9 *of the* ***classical Lévy-Khintchine formula*** *(See Remark 3.4.21).*

In the case $G = G^\wedge = \mathbf{R}^d$ $(d \geq 1)$ various Lévy functions (apart from the one introduced in Example 5.3.2.1) can be given. Unfortunately the classical function g_0 defined by

$$g_0(x, y) := \frac{\langle x, y \rangle}{1 + \|x\|^2}$$

for all $(x, y) \in \mathbf{R}^d \times \mathbf{R}^d$ is not a Lévy function in the sense of Definition 5.3.1. A slight modification of the above approach, however, makes it possible to obtain the classical analogue of Theorem 5.4.7.

In fact, for a measure $\mu \in M_+((\mathbf{R}^d)^\times)$ the condition

$$\int_{(\mathbf{R}^d)^\times} (1 - \mathrm{Re}\ e^{-i\langle x,y\rangle})\mu(dx) < \infty$$

valid for all $y \in \mathbf{R}^d$ is equivalent to the condition

$$\int_{U_1\setminus\{0\}} \|x\|^2 \mu(dx) + \int_{U_1^c} \mu(dx) < \infty,$$

where U_1 denotes the unit ball with center 0 and radius 1. Moreover, it can be shown that for any compact subset K of $\mathbf{R}^d$ there exist a neighborhood $V \in \mathfrak{V}_{\mathbf{R}^d}(0)$ and a constant $c > 0$ such that

$$|1 - e^{i\langle x,y\rangle} + ig_0(x,y)| \leq c\|x\|^2$$

for all $x \in V, y \in K$. This inequality together with the above finiteness condition is the proper replacement for the canonical estimate (5) in the proof of Theorem 5.3.5.

With these tools at hand the previous arguments yield the equivalence of the following statements valid for any complex-valued function ψ on $\mathbf{R}^d$

(i') $\psi \in CND(\mathbf{R}^d)$.

(ii') There is a canonical quadruple (c, h, q, μ') as in (ii) of Theorem 5.4.7 with a bounded Lévy measure $\mu' \in M_+((\mathbf{R}^d)^\times)$ such that

$$\psi(y) = c + ih(y) + q(y) + \int_{(\mathbf{R}^d)^\times} \left(1 - e^{i\langle x,y\rangle} + \frac{i\langle x,y\rangle}{1+\|x\|}\right) \frac{1+\|x\|^2}{\|x\|} \mu'(dx)$$

for all $y \in \mathbf{R}^d$.

In order to relate Theorem 5.4.7 for locally compact Abelian groups to the Lévy-Khintchine decomposition theorem 3.4.20 for separable Banach spaces we proceed as follows. As a direct consequence of Theorem 5.4.7 we obtain

Theorem 5.4.10 *(Canonical representation of convolution semigroups)*

Let g be a Lévy function for the underlying locally compact Abelian group G. Then any convolution semigroup $(\mu_t)_{t \geq 0} \in \mathbf{C\,S}(G)$ *has a Fourier representation*

$$\hat{\mu}_t = \exp(-t\psi)$$

valid for all $t > 0$*, where the negative definite function* $\psi \in CND(G^\wedge)$ *associated with* $(\mu_t)_{t \geq 0}$ *admits a canonical quadruple* (c, h, q, μ) *of the form (7) in Theorem 5.4.7.*

The **proof** of this statement follows from the Schoenberg correspondence theorem 5.2.3 (together with Theorem 5.4.7).

As in Section 2.3 for separable Banach space we may also for locally compact Abelian group G introduce the notions of infinite divisibility and embeddability of probability measures on G.

Definition 5.4.11 *A measure* $\mu \in M^1(G)$ *is called* ***infinitely divisible*** *if for every* $n \in \mathbf{N}$ *there exists an n-th* **root** *of* μ*, i.e. a measure* $\mu_n \in M^1(G)$ *such that*

$$\mu_n^n = \mu\,.$$

$\mu \in M^1(G)$ *is said to be* ***embeddable*** *if there exists a convolution semigroup* $(\mu_t)_{t \geq 0}$ *in* $M^1(G)$ *such that* $\mu_1 = \mu$.

The sets of infinitely divisible and embeddable probability measures on G will be denoted by $I(G)$ *and* $EM(G)$ *respectively.*

Clearly,

$$EM(G) \subset I(G),$$

but the inverse inclusion fails to be true in general. In the special case $G := \mathbf{R}^d$, however, the sets $I(G)$ and $EM(G)$ coincide (See the embedding theorem 2.3.9 for arbitrary separable Banach spaces).

With this notation we add

Corollary 5.4.12 *(to Theorem 5.4.10). Any measure* $\mu \in EM(G)$ *with embedding convolution semigroups* $(\mu_t)_{t\geq 0} \in \mathbf{C}\,\mathbf{S}(G)$ *has a Fourier representation*

$$\hat{\mu} = \exp(-\psi),$$

where $\psi \in CND(G^\wedge)$ *associated with* $(\mu_t)_{t\geq 0}$ *admits a canonical quadruple* (c, h, q, μ) *of the form (7) in Theorem 5.4.7.*

6

Probabilistic Properties of Convolution Semigroups

6.1 Transient convolution semigroups

The probabilistic notions of transience and recurrence for stationary independent increment processes can be discussed in terms of convolution semigroups. This analytic treatment yields further interesting results of potential-theoretic nature.

Let G denote a locally compact Abelian group. For a convolution semigroup $(\mu_t)_{t\geq 0}$ in $\mathbf{C}\,\mathbf{S}(G)$ with associated resolvent $(\varrho_\lambda)_{\lambda>0}$ and any function $f \in C^c_+(G)$ we observe that the mapping

$$\lambda \mapsto \int f d\varrho_\lambda = \int_0^\infty e^{-\lambda t} \mu_t(f) dt$$

for $\mathbf{R}^\times_+$ into $\mathbf{R}$ is decreasing. Hence the monotone convergence theorem implies that

$$\lim_{\lambda \to 0} \int f d\varrho_\lambda = \int_0^\infty \mu_t(f) dt \leq \infty\,.$$

Definition 6.1.1 *A convolution semigroup $(\mu_t)_{t\geq 0}$ in $\mathbf{C\,S}(G)$ is called* ***transient*** *if*

$$\lim_{\lambda\to 0}\int f d\varrho_\lambda = \int_0^\infty \mu_t(f)dt < \infty$$

for all $f \in C_+^c(G)$, and in this case

$$\kappa := \tau_v - \lim_{\lambda\to 0} \varrho_\lambda \in M_+(G)$$

is said to be the ***potential measure*** *of $(\mu_t)_{t\geq 0}$.*

$(\mu_t)_{t\geq 0}$ is called ***recurrent*** *if $(\mu_t)_{t\geq 0}$ is not transient.*

Discussion 6.1.2

6.1.2.1 *Theorem 5.2.16 and its proof imply that the potential measure κ of a transient convolution semigroup $(\mu_t)_{t\geq 0} \in \mathbf{C\,S}(G)$ with associated resolvent $(\varrho_\lambda)_{\lambda>0}$ has the property that*

$$\operatorname{supp}(\kappa) = \operatorname{supp}(\varrho_\lambda) = \Big(\bigcup_{t>0} \operatorname{supp}(\mu_t)\Big)^-$$

(for all $\lambda > 0$) and that $\operatorname{supp}(\kappa)$ is a semigroup in G.

6.1.2.2 *Let $(\mu_t)_{t\geq 0} \in \mathbf{C\,S}(G)$ and $\psi \in CND(G^\wedge)$ with*

$$(\mu_t)_{t>0} \longleftrightarrow \psi\,.$$

If $\psi(0) > 0$ then $(\mu_t)_{t\geq 0}$ is transient, $\kappa \in M_+^b(G)$ with

$$\|\kappa\| = \frac{1}{\psi(0)},$$

and

$$\hat{\kappa} = \frac{1}{\psi}\,.$$

On the other hand we shall see later, that there exist transient as well as recurrent convolution semigroups $(\mu_t)_{t\geq 0}$ in $M^1(G)$ with

$$(\mu_t)_{t\geq 0} \longleftrightarrow \psi$$

such that $\psi(0) = 0$. *In fact, for convolution semigroups* $(\mu_t)_{t\geq 0}$ *in* $M^1(G)$ *the potential measure* κ *is always unbounded.*

6.1.2.3 *If* G *is compact, a convolution semigroup* $(\mu_t)_{t\geq 0} \in \mathbf{C}\,\mathbf{S}(G)$ *with*

$$(\mu_t)_{t\geq 0} \longleftrightarrow \psi,$$

where $\psi \in CND(G^\wedge)$, *is transient if and only if* $\psi(0) > 0$. *This equivalence holds true for an arbitrary locally compact Abelian group* G *provided there exists a compact subset* K *of* G *such that*

$$\operatorname{supp}(\mu_t) \subset K$$

for all $t \geq 0$.

We add an important

Example 6.1.3 *Let* $(\mu_t)_{t\geq 0}$ *be the Poisson semigroup determined by a measure* $\mu \in M_+^{(1)}(G)$ *introduced in Example 5.2.7.2. We recall that*

$$\mu_t = e^{-t} \exp(t\mu)$$

for all $t \geq 0$.

(μ_t) *is transient if and only if the series*

$$\sum_{n\geq 0} \mu^n \quad \text{is } \tau_v - \text{convergent}\,.$$

In this case $\sum_{n\geq 0} \mu^n \in M_+(G)$, *and as such it coincides with the potential measure* κ *of* $(\mu_t)_{t\geq 0}$.

In fact, for the resolvent $(\varrho_\lambda)_{\lambda\geq 0}$ with

$$(\varrho_\lambda)_{\lambda>0} \longleftrightarrow (\mu_t)_{t\geq 0}$$

we obtain the following equalities for Radon measures

$$\begin{aligned}\varrho_\lambda &= \int_0^\infty e^{-\lambda_t} e^{-t} \exp(t\mu) dt \\ &= \int_0^\infty e^{-(\lambda+1)t} \left(\sum_{n\geq 0} \frac{t^n}{n\,!} \mu^n \right) dt \\ &= \sum_{n\geq 0} \mu^n \frac{1}{n\,!} \int_0^\infty t^n e^{-(\lambda+1)t} dt \\ &= \sum_{n\geq 0} \left(\frac{1}{\lambda+1} \right)^{n+1} \mu^n,\end{aligned}$$

hence for $\lambda \to 0$ the assertion.

We note that measures $\kappa \in M_+(G)$ of the form

$$\kappa := \sum_{n\geq 0} \mu^n$$

with $\mu \in M_+^{(1)}(G)$ are called ***elementary kernels*** determined by μ.

Properties 6.1.4 *of potential measures of transient convolution semigroups* $(\mu_t)_{t\geq 0} \in \mathbf{C\,S}(G)$ *with associated* $\psi \in CND(G^\wedge)$ *and resolvent* $(\varrho_\lambda)_{\lambda>0}$.

6.1.4.1 *For every* $\lambda > 0$ *we have that*

$$\frac{1}{2}(\varrho_\lambda + \varrho_\lambda{}^\sim) \in M_p(G),$$

hence that

$$\frac{1}{2}(\kappa + \kappa^\sim) \in M_p,$$

and $\kappa \in M^{sb}(G)$.

6.1.4.2 $\mathrm{Inv}\,(\kappa) = \{0\}$.

6.1.4.3 *For each* $\lambda > 0$ *the measure*

$$\lambda\kappa + \varepsilon_0$$

is the elementary kernel determined by the measure $\lambda\varrho_\lambda$.

Both Properties 6.1.4.1 and 6.1.4.3 are proved by taking Fourier transforms. For Property 6.1.4.1 we just note that $\frac{1}{2}(\varrho_\lambda + \varrho_\lambda^{\sim}) \in M_p(G)$, since

$$\begin{aligned}\mathcal{F}_G[\frac{1}{2}(\varrho_\lambda + \varrho_\lambda{}^{\sim})] &= \operatorname{Re} \frac{1}{\lambda + \psi} \\ &= \frac{\lambda + \operatorname{Re} \psi}{|\lambda + \psi|^2} \geq 0\end{aligned}$$

(See Theorem 4.4.14). The τ_v-closeness of $M_p(G)$ in $M(G)$ implies that $\frac{1}{2}(\kappa + \kappa^{\sim}) \in M_p(G)$, and the assertion follows from Proposition 4.4.9 *(ii)*.

In the proof of Property 6.1.4.3 one verifies the equality

$$\varrho_{\lambda'} = \sum_{n \geq 0} (\lambda - \lambda')^n \varrho_\lambda{}^{n+1}$$

for $\lambda' \in]0, \lambda[$ by applying the Fourier transform and then takes the limit for $\lambda' \downarrow \lambda$.

Terminology 6.1.5 *Let* $\psi \in CND(G^\wedge)$.

(a) The function $\frac{1}{\psi}$ *is said to be* ***integrable over some*** *measurable* ***subset*** W *of* $G^\wedge$ *provided*

$$\psi \neq 0 \ [\omega_{G^\wedge}] \text{ on } W$$

and $\frac{1}{\psi}$ *(defined* $[\omega_{G^\wedge}]$ *on* W*) is (*$\omega_{G^\wedge}$*-)integrable over* W.

(b) The real function $\operatorname{Re} \frac{1}{\psi}$ *is said to be* ***locally integrable*** *(on* $G^\wedge$*) if there exists a neighborhood* $U \in \mathfrak{V}_{G^\wedge}(0)$ *such that*

$$\psi \neq 0 \ [\omega_{G^\wedge}] \text{ on } U$$

and $\operatorname{Re} \frac{1}{\psi}$ *is* $\operatorname{Res}_U \omega_{G^\wedge}$*-integrable.*

We note that $\operatorname{Re} \frac{1}{\psi}$ *is locally integrable if and only if it is integrable over every open, relatively compact subset of* $G^\wedge$.

With this slightly sophisticated terminology in mind we are able to establish a generalization of Property 5.1.7.1

Proposition 6.1.6 *Let* $\psi \in CND(G^\wedge)$ *and assume that* $\frac{1}{\psi}$ *is integrable over some open, relatively compact neighborhood* $\in \mathfrak{V}_{G^\wedge}(0)$. *Then*

(i) $\frac{1}{\psi}$ *is locally integrable*

and

(ii) $\frac{1}{\psi} \cdot \omega_{G^\wedge} \in M_p(G)$.

Proof. From the discussion 6.1.2.3 we know that it suffices to consider the case that $\psi(0) = 0$. Let K be a compact subset of $G^\wedge$. If $\psi(\chi) \neq 0$ for all $\chi \in K$, we have that

$$\int_K \frac{1}{|\psi(\chi)|} \omega_{G^\wedge}(d\chi) < \infty .$$

Now let $\psi(\chi) = 0$ for some $\chi \in K$. Then the set

$$K_1 := \{\chi \in K : \psi(\chi) = 0\}$$

is compact, hence

$$K_1 \subset \bigcup_{i=1}^{n} (\chi_i + V),$$

where $\chi_1, \ldots, \chi_n \in K_1$ and V is an open, relatively compact neighborhood $\in \mathfrak{V}_{G^\wedge}(0)$ such that $\frac{1}{\psi}$ is integrable over V. The set

$$F := K \backslash \bigcap_{i=1}^{n} (\chi_i + V)$$

being compact with $\psi(\chi) \neq 0$ for all $\chi \in F$ we see that $\psi \neq 0 \quad [\omega_{K^\wedge}]$ on K, and that

$$\int_K \frac{1}{|\psi(\chi)|} \omega_{K^\wedge}(d\chi)$$
$$\leq \int_F \frac{1}{|\psi(\chi)|} \omega_{G^\wedge}(d\chi) + \int_{\bigcup_{i=1}^n (\chi_i + V)} \frac{1}{|\psi(\chi)|} \omega_{G^\wedge}(d\chi)$$
$$\leq \int_F \frac{1}{|\psi(\chi)|} \omega_{G^\wedge}(d\chi) + n \int_V \frac{1}{|\psi(\chi)|} \omega_{G^\wedge}(d\chi) < \infty.$$

Here we employed the fact that $K_1 \subset \operatorname{Inv}(\psi)$ which is part of Theorem 5.2.17.

We turn to the proof of statement *(ii)* of the theorem. For a given $f \in C^c(G^\wedge)$ and $n \geq 1$ we have

$$\int f * f^\sim \frac{1}{\psi + \frac{1}{n}} d\, \omega_{G^\wedge} \geq 0,$$

since $\frac{1}{\psi + \frac{1}{n}} \in CPD(G^\wedge)$ by Property 5.1.7.1. The function $f * f^\sim \frac{1}{\psi}$ is defined $[\omega_{G^\wedge}]$, and

$$\lim_{n \to \infty} f * f^\sim \frac{1}{\psi + \frac{1}{n}} = f * f^\sim \frac{1}{\psi} \quad [\omega_{G^\wedge}] \text{ (on } G^\wedge).$$

But Re $\psi \geq 0$ implies that

$$|\psi| \leq \left| \psi + \frac{1}{n} \right|$$

for all $n \geq 1$, hence Lebesgue's dominated convergence theorem yields the assertion *(ii)*. ■

Remark 6.1.7 *Proposition 6.1.6 remains valid if one replaces* $\frac{1}{\psi}$ *by* Re $\frac{1}{\psi}$ *with the modification that (ii) reads as* Re $\frac{1}{\psi} \cdot \omega_{G^\wedge} \in M_+(G^\wedge)$.

We are now aiming at a characterization of transience of convolution semigroups in terms of integrability of their associated negative definite functions. The following statement takes care of the less sophisticated part of that equivalence.

Theorem 6.1.8 *Let* $(\mu_t)_{t\geq 0} \in \mathbf{C\,S}(G)$ *and* $\psi \in CND(G^\wedge)$ *with*

$$(\mu_t)_{t\geq 0} \longleftrightarrow \psi\,.$$

If $(\mu_t)_{t\geq 0}$ *is transient, then* $(\mu_t)_{t\geq 0}$ *is* ***locally integrable*** *in the sense that* $\mathrm{Re}\,\frac{1}{\psi}$ *is locally integrable.*

Proof. Without loss of generality we assume that $\psi(0) = 0$. We know that the set

$$T := \{\chi \in G^\wedge : \psi(\chi) = 0\}$$

is a closed subgroup of $G^\wedge$ and by Theorem 5.2.17 *(ii)* that

$$\mathrm{supp}\,(\mu_t) \subset T^\perp$$

for all $t > 0$. Since $(\mu_t)_{t\geq 0}$ is transient by assumption, $T^\perp$ is not compact, hence T is not open. But a closed subgroup of $G^\wedge$ is either open or a local $\omega_{G^\wedge}$-null set. This implies that $\psi \neq 0$ locally $[\omega_{G^\wedge}]$ on $G^\wedge$. Now, let K be a compact subset of $G^\wedge$. As usual we choose a function $f \in C_+^c(G)$ with

$$|\overline{\mathcal{F}}_G f| \geq 1 \text{ on } K\,.$$

But then the monotone convergence theorem together with formula

(4) in the proof of Theorem 4.4.14 yields

$$\begin{aligned}
\int_K \mathrm{Re}\,\frac{1}{\psi}\,d\omega_{G^\wedge} &= \int_K \lim_{\lambda\to 0} \mathrm{Re}\,\frac{1}{\lambda+\psi} d\omega_{G^\wedge} \\
&\le \int_{G^\wedge} \liminf_{\lambda\to 0} |\overline{\mathcal{F}}_G f|^2\,\mathrm{Re}\,\frac{1}{\lambda+\psi} d\omega_{G^\wedge} \\
&\le \liminf_{\lambda\to 0} \int_{G^\wedge} |\overline{\mathcal{F}}_G f|^2\,\mathrm{Re}\,\frac{1}{\lambda+\psi} d\omega_{G^\wedge} \\
&= \liminf_{\lambda\to 0} \int_{G^\wedge} |\overline{\mathcal{F}}_G f|^2 \mathcal{F}_G[\frac{1}{2}(\varrho_\lambda + \varrho_\lambda{}^\sim)] d\omega_{G^\wedge} \\
&= \liminf_{\lambda\to 0} \frac{1}{2} \int_G f * f^\sim d(\varrho_\lambda + \varrho_\lambda^\sim) \\
&= \frac{1}{2} \int_G f * f^\sim d(\kappa + \kappa^\sim) < \infty,
\end{aligned}$$

where, as in the proof of Property 6.1.4.1, the identity

$$\mathcal{F}_G[\frac{1}{2}(\varrho_\lambda + \varrho_\lambda^\sim)] = \mathrm{Re}\,\frac{1}{\lambda+\psi}$$

has been applied (for $\lambda > 0$), $(\varrho_\lambda)_{\lambda>0}$ and κ being as usual associated with $(\mu_t)_{t\ge 0}$. ∎

Corollary 6.1.9 *Let* $\mu \in M_+^{(1)}(G)$ *be such that*

$$\sum_{n\ge 0} \mu^n \quad \text{is } \tau_v - \text{convergent}.$$

Then $\mathrm{Re}\,\frac{1}{1-\hat{\mu}}$ *is locally integrable (on* $G^\wedge$*).*

Proof. We look at the convolution semigroup $(\mu_t)_{t\ge 0}$ determined by the measure μ and its associated function $\psi = 1 - \hat{\mu} \in CND(G^\wedge)$. By Example 6.1.3 $(\mu_t)_{t\ge 0}$ is transient, hence the theorem implies the assertion. ∎

The discussion in the remaining part of this section will be devoted to the implication (inverse to that of Theorem 6.1.8)

(TC) $(\mu_t)_{t\geq 0}$ locally integrable $\Rightarrow (\mu_t)_{t\geq 0}$ transient.

Remark 6.1.10 *Until now the implication (TC) is only available in the special case of a compact group G. See Discussion 6.1.2.3.*

In other words, (TC) holds for convolution semigroups $(\mu_t)_{t\geq 0}$ in $M_+^b(G)$ which are not in $M^1(G)$.

If, in fact, $\|\mu_t\| < 1$ for some $t > 0$, then $(\mu_t)_{t\geq 0}$ is transient, and for the associated $\psi \in CND(G^\wedge)$ we have Re $\psi > 0$ by Corollary 5.2.5 *(ii)*.

For the somewhat lengthy proof of the implication (TC) which in its course will contain useful side results on transient convolution semigroups in $M_+^{(1)}(G)$ we shall prepare several steps of reducing the statement to be shown and two basic facts from the potential theory of convolution semigroups in $M_+^{(1)}(G)$.

6.1.11 Properties *(Reducing the proof of (TC)).*

6.1.11.1 *Let H be a compact subgroup of G and let $(\dot{\mu}_t)_{t\geq 0} \in$ $\mathbf{C}\,\mathbf{S}(G/H)$ be the canonical image of $(\mu_t)_{t\geq 0} \in \mathbf{C}\,\mathbf{S}(G)$. Then (TC) holds for $(\mu_t)_{t\geq 0}$ provided it holds for $(\dot{\mu}_t)_{t\geq 0}$.*

In fact, given a recurrent convolution semigroup $(\mu_t)_{t\geq 0} \in \mathbf{C}\,\mathbf{S}(G)$, $(\dot{\mu}_t)_{t\geq 0}$ is a recurrent convolution semigroup $\in \mathbf{C}\,\mathbf{S}(G/H)$. Since $(G/H)^\wedge$ is isomorphic to the open subgroup $H^\perp$ of $G^\wedge$ by the functorial property 4.2.17.3 and since $\psi \in CND((G/H)^\wedge)$ with

$$\dot{\psi} \longleftrightarrow (\dot{\mu}_t)_{t\geq 0}$$

coincides with Res $_{H^\perp}\psi$, where

$$\psi \longleftrightarrow (\mu_t)_{t\geq 0},$$

Re $\frac{1}{\psi}$ is not locally integrable.

6.1.11.2 *Without loss of generality the given $(\mu_t)_{t\geq 0} \in \mathbf{C}\,\mathbf{S}(G)$ can be assumed to be* ***adapted*** *in the sense that*

$$S := \left[\bigcup_{t>0} \operatorname{supp}(\mu_t)\right]^- = G\,.$$

In fact, if $(\mu_t)_{t\geq 0}$ is not adapted, then we replace G by S. Considering the canonical surjection $\pi : G^\wedge \Rightarrow G^\wedge/S^\perp \cong S^\wedge$ (Functorial property 4.2.17.4) we obtain that

$$\psi = \dot{\psi} \circ \pi$$

holds for $\psi \in CND(G^\wedge)$ with $\psi \longleftrightarrow (\mu_t)_{t\geq 0}$ and $\dot{\psi} \in CND(S^\wedge)$ with

$$\dot{\psi} \longleftrightarrow (\dot{\mu}_t)_{t\geq 0}\,.$$

But then $\operatorname{Re}\frac{1}{\psi}$ is locally integrable on $G^\wedge$ if and only if $\operatorname{Re}\frac{1}{\dot{\psi}}$ is locally integrable on $S^\wedge$.

6.1.11.3 *It is sufficient to prove (TC) for adapted convolution semigroups in* $\mathbf{C}\,\mathbf{S}(G)$*, where G is of the form*

$$G = \mathbf{R}^d \times \mathbf{Z}^e$$

for $d, e \geq 0$.

This follows from the structure theorem 4.2.20 which implies that

$$G/H \cong \mathbf{R}^d \times \mathbf{Z}^e$$

for some compact subgroup H of G and integers $d, e \geq 0$. In fact, since every closed subgroup of a group of the form $\mathbf{R}^d \times \mathbf{Z}^e$ is of the same type (being a closed subgroup of $\mathbf{R}^{d+e}$), the previous properties yield the assertion.

Finally

6.1.11.4 *it is sufficient to establish (TC) for (adapted) Poisson semigroups on $G = \mathbf{R}^d \times \mathbf{R}^e$ for $d, e \geq 0$.*

In order to see this one employs the resolvent $(\varrho_\lambda)_{\lambda>0}$ associated with $(\mu_t)_{t\geq 0}$ and considers the Poisson semigroup $(e_t(\varrho_1))_{t\geq 0}$ determined by $\varrho_1 \in M_+^{(1)}(G)$. It is easily shown that (TC) for $(\mu_t)_{t\geq 0}$ is a consequence of the implication

$$(e_t(\varrho_1))_{t\geq 0} \quad \text{locally integrable} \Rightarrow \sum_{n\geq 0} \varrho_1^n \quad \text{is } \tau_v\text{-convergent}\,.$$

Moreover,

$$[\operatorname{supp}(\varrho_1)]^- = G$$

by Theorem 5.2.16, once $(\mu_t)_{t\geq 0}$ is adapted.

Theorem 6.1.12 *(The Choquet-Deny* ***convolution equation****) Let G be a locally compact Abelian group, and let $\sigma \in M^1(G)$ be a fixed measure with* ***support subgroup***

$$G(\sigma) := [\operatorname{supp}(\sigma)]^- .$$

Then, for every measure $\mu \in M_+(G)$ the following statements are equivalent:

(i) $\mu \in M^{sb}(G)$ *and*

$$\mu * \sigma = \mu.$$

(ii) $\operatorname{Inv}(\mu) \supset G(\sigma)$.

Proof. It suffices to verify the implication *(i)* $\Rightarrow$ *(ii)*. Let $\mu \in M_+(G)$ be a measure satisfying *(i)* of the theorem. For $\varphi \in C^c(G)$ we introduce the function $f := f_\varphi$ on G defined by

$$f(x) := \int \varphi(x+y)\mu(dy)$$

whenever $x \in G$. Clearly, f is uniformly continuous (with respect to the uniform structure of G), and it is bounded since $\mu \in M^{sb}(G)$. Moreover, f satisfies the integral equations

$$f(x) = \int f(x+z)\sigma(dz)$$

for all $x \in G$. Then for $a \in \operatorname{supp}(\sigma)$ the functions $T_{-a}f$ and $g := T_{-a}f - f$ are uniformly continuous, bounded and satisfy the above integral equation.

Now let $c \in \mathbf{R}_+$ be such that $|g| \leq c$ and let

$$\gamma := \frac{1}{2} \sup_{x\in G} g(x) .$$

There is a sequence $(x_n)_{n\geq 1}$ in G satisfying

$$\lim_{n\to\infty} g(x_n) = 2\gamma\,.$$

Then the set $\{g_n : n \in \mathbf{N}\}$ of functions $g_n := T_{-x_n}g$ $(n \geq 1)$ is equicontinuous, since g is uniformly continuous. Hence, by the Arzelà-Ascoli theorem there exists a subnet $(g_{n_\alpha})_{\alpha\in A}$ of $(g_n)_{n\geq 1}$ which converges with respect to the topology τ_{co} towards a function $h \in C(G)$ with $|h| \leq c$. Furthermore h satisfies the above integral equation. But then

$$\begin{aligned} 2\gamma &= \lim_{\alpha\in A} g(x_{n_\alpha}) \\ &= \lim_{\alpha\in A} g_{n_\alpha}(e) \\ &= h(e) \\ &= \int h d\sigma. \end{aligned}$$

From $h \leq 2\gamma$ we conclude that $h(x) = 2\gamma$ for all $x \in \operatorname{supp}(\sigma)$. Given $x \in \operatorname{supp}(\sigma)$ we further obtain

$$2\gamma = h(x) = \int h(x+z)\sigma(dz),$$

hence

$$h(x+z) = 2\gamma$$

for all $z \in \operatorname{supp}(\sigma)$. An iteration procedure yields

$$h(x) = 2\gamma$$

for all x belonging to the closed semigroup generated by $\operatorname{supp}(\sigma)$ and therefore

$$h(ka) = 2\gamma$$

for all $a \in \operatorname{supp}(\sigma), k \geq 1$. Now, for every $l \geq 1$ there exists an $\alpha \in A$ with

$$g_{n_\alpha}(ka) > \gamma$$

whenever $k = 1, \ldots, l$, provided $\gamma > 0$. But then by looking at the equalities

$$\begin{aligned} g_{n_\alpha}(ka) &= f(x_{n_\alpha} + ka + a) - f(x_{n_\alpha} + ka) \\ &= f(x_{n_\alpha} + (k+1)a) - f(x_{n_\alpha} + ka), \end{aligned}$$

and after summing for $k = 1, \ldots, l$ we obtain

$$f(x_{n_\alpha} + (l+1)a) - f(x_{n_\alpha} + a) > l\gamma .$$

Since f is bounded, this yields a contradiction. Therefore $\gamma \leq 0$, hence $g \leq 0$. Replacing g by $-g$ one gets $-g \leq 0$, thus altogether $g = 0$ or

$$f = T_{-a} f$$

for all $a \in \operatorname{supp}(\sigma)$. From this follows

$$\int \varphi(y+a)\mu(dy) = \int \varphi(y)\mu(dy),$$

hence

$$\mu * \varepsilon_a = \mu$$

for all $a \in \operatorname{supp}(\sigma)$ and finally for all $a \in G(\sigma)$. ■

Corollary 6.1.13 *(Characterizing* **idempotent measures***) For any $\mu \in M_+^b(G)\backslash\{0\}$ the following statement are equivalent:*

(i) $\mu * \mu = \mu$.

(ii) $\mu = \omega_H$ *for some compact subgroup* H *of* G.

Proof. Again it suffices to show the implication *(i)* ⇒ *(ii)*. Without loss of generality we may assume that $\mu \in M^1(G)$, since for $\mu \in M_+^b(G)\backslash\{0\}$ one has

$$\|\mu * \mu\| = \|\mu\|\|\mu\| = \|\mu\|,$$

hence $\|\mu\| = 1$.

From the boundedness of μ follows that $\mathrm{Inv}(\mu)$ is a compact subset of G (See the discussion following Definition 4.4.17). Now, Theorem 6.1.12 says that

$$G(\mu) \subset \mathrm{Inv}(\mu),$$

hence that $H := G(\mu)$ is a compact subgroup of G and μ is H-invariant on H. Thus $\mu = \omega_H$ (the normed Haar measure of H). ■

The following result from potential theory of convolution semigroups goes back to Deny and has been reproved in the book [6] of Berg and Forst.

Theorem 6.1.14 *(Existence of an equilibrium measure). Let G be a locally compact Abelian group, $(\mu_t)_{t\geq 0}$ a convolution semigroup $\in$* $\mathbf{C\,S}(G)$ *with potential measure $\kappa \in M_+(G)$, and let W be an open, relatively compact subset of G. Then there exists an* ***equilibrium measure*** *$\gamma \in M_+^b(G)$ for $(\mu_t)_{t\geq 0}$ which by definition has the following properties:*

(a) $\mathrm{supp}\,(\gamma) \subset \overline{W}$

(b) $\kappa * \gamma \leq \omega_G$

(c) $\kappa * \gamma = \omega_G$ *on* W.

Corollary 6.1.15 *For every $\sigma \in M_+^{(1)}(G)$ and $f \in C_+^c(G)$ with* $\mathrm{supp}\,(f^\sim) \subset W$ *we have that*

$$\kappa * \gamma * (\varepsilon_0 - \sigma) * f(0) \geq 0\,.$$

The proof follows immediately from the inequalities

$$\begin{aligned}\kappa * \gamma * (\varepsilon_0 - \sigma) * (f(0) &= (\kappa * \gamma * f)(0) - (\kappa * \gamma + \sigma * f)(0)\\ &= \int f^\sim d\omega_G - \int f^\sim d(\kappa * \gamma * \sigma) \geq 0,\end{aligned}$$

since

$$\kappa * \gamma * \sigma \leq \omega_G * \sigma \leq \omega_G. \qquad ■$$

6.2 The transience criterion

This section will be devoted entirely to the announced measure - theoretic proof of the famous transience criterion for convolution semigroups which for second countable locally compact Abelian groups and with probabilistic methods has been established by Port and Stone in [34].

Theorem 6.2.1 *(The Port-Stone criterion) Let G be a compactly generated locally compact Abelian group. For any $(\mu_t)_{t\geq 0} \in \mathbf{C}\,\mathbf{S}(G)$ the following statements are equivalent:*

(i) $(\mu_t)_{t\geq 0}$ is transient.

(ii) $(\mu_t)_{t\geq 0}$ is locally integrable.

In view of the proof of Theorem 6.2.1 we now restrict the discussion to *adapted* convolution semigroups $(\mu_t)_{t\geq 0} \in \mathbf{C}\,\mathbf{S}(G)$ with

$$(\mu_t)_{t\geq 0} \longleftrightarrow \psi \longleftrightarrow (\varrho_\lambda)_{\lambda>0}\,.$$

Lemma 6.2.2 *(The Chung-Fuchs criterion) Suppose there exists a relatively compact neighborhood $U \in \mathfrak{V}_{G^\wedge}(0)$ such that*

$$\limsup_{\lambda\to 0} \int_U \mathrm{Re}\,\frac{1}{\lambda+\psi}\,d\,\omega_{G^\wedge} < \infty\,.$$

Then $(\mu_t)_{t\geq 0}$ is transient.

Proof. Since $(\mu_t)_{t\geq 0}$ is adapted, we infer from Theorem 5.2.17 that $\psi(\chi) \neq 0$ for $\chi \neq 0$. Taking into account that

$$\mathrm{Re}\,\frac{1}{\lambda+\psi} \leq \frac{\lambda + \mathrm{Re}\,\psi}{|\psi|^2}$$

the hypothesis of the lemma yields

$$\limsup_{\lambda\to 0} \int_C \mathrm{Re}\,\frac{1}{\lambda+\psi}\,d\omega_{G^\wedge} < \infty$$

for all compact subsets C of $G^\wedge$. It follows that

$$\limsup_{\lambda\to 0} \int_{G^\wedge} \operatorname{Re} \frac{1}{\lambda+\psi} f * f^\sim d\omega_{G^\wedge} < \infty$$

for all $f \in C^c(G^\wedge)$. But given $g \in C_+(G)$ there exists an $f \in C^c(G^\wedge)$ with $|\check{f}|^2 \geq g$. This implies that

$$\begin{aligned} \limsup_{\lambda\to 0} \int_G g\, d\,[\frac{1}{2}(\varrho_\lambda + \varrho_\lambda^\sim)] &\leq \limsup_{\lambda\to 0} \int_G |\check{f}|^2 d\,[\frac{1}{2}(\varrho_\lambda + \varrho_\lambda^\sim)] \\ &= \limsup_{\lambda\to 0} \int_G f * f^\sim \operatorname{Re} \frac{1}{\lambda+\psi}\, d\omega_{G^\wedge} < \infty, \end{aligned}$$

where the last equality follows from formula (4) in the proof of Theorem 4.4.14 together with Theorem 4.4.16. So we have

$$\limsup_{\lambda\to 0} \int_G g d\varrho_\lambda < \infty,$$

hence by monotonicity that the limit

$$\lim_{\lambda\to 0} \int_G g d\varrho_\lambda$$

exists and is finite for all $g \in C_+^c(G)$. ∎

Remark 6.2.3 *Applying Theorem 6.1.8 one obtains the converse of the statement in Lemma 6.2.2.*

Lemma 6.2.4 *If $(\mu_t)_{t\geq 0}$ is locally integrable and recurrent, then for every $f \in C_+^c(G)$ with $f \not\equiv 0$ one has*

(a) $$\limsup_{\lambda\to 0} \lambda\varrho_\lambda * \varrho_\lambda^\sim * f * f^\sim(0) = \infty$$

and

(b) $$\lambda\varrho_\lambda * \varrho_\lambda^\sim * f * f^\sim(0) > 0$$

whenever $\lambda > 0$.

Proof. Since $(\mu_t)_{t\geq 0}$ is assumed to be locally integrable, we have for all relatively compact neighborhoods $U \in \mathfrak{V}_{G^\wedge}(0)$ that

$$\begin{aligned} \int_U \frac{\mathrm{Re}\,\psi}{|\lambda+\psi|^2} d\omega_{G^\wedge} &\leq \int\limits_U \frac{\mathrm{Re}\,\psi}{|\psi|^2} d\omega_{G^\wedge} \\ &= \int\limits_U \mathrm{Re}\,\frac{1}{\psi} d\omega_{G^\wedge} < \infty. \end{aligned} \tag{1}$$

But we also assumed $(\mu_t)_{t\geq 0}$ to be recurrent. This implies by Lemma 6.2.2 that

$$\limsup_{\lambda\to 0} \int\limits_U \frac{\lambda + \mathrm{Re}\,\psi}{|\lambda+\psi|^2}\, d\omega_{G^\wedge} = \limsup_{\lambda\to 0} \int\limits_U \mathrm{Re}\,\frac{1}{\lambda+\psi}\, d\omega_{G^\wedge} = \infty. \tag{2}$$

Moreover, for $f \in C_+^c(G)$ with $f \not\equiv 0$, and all $\lambda > 0$

$$\begin{aligned} \lambda\varrho_\lambda * f * (\varrho_\lambda * f)^\sim(0) &= \lambda \int\limits_G |\varrho_\lambda * f|^2 d\omega_G \\ &= \lambda \int\limits_{G^\wedge} |\widehat{\varrho_\lambda}|^2 |\hat{f}|^2 d\omega_{G^\wedge} \\ &= \int\limits_{G^\wedge} |\hat{f}|^2 \frac{\lambda}{|\lambda+\psi|^2} d\omega_{G^\wedge} > 0. \end{aligned}$$

We note that in fact

$$\int\limits_{G^\wedge} |\hat{f}|^2 \frac{\lambda}{|\lambda+\psi|^2} d\omega_{G^\wedge} \geq \alpha^2 \int\limits_U \frac{\lambda}{|\lambda+\psi|^2} d\omega_{G^\wedge},$$

where α is chosen such that $\hat{f}(0) > \alpha > 0$ and U is a relatively compact neighborhood in $\mathfrak{V}_{G^\wedge}(0)$ within the set $\{\chi \in G^\wedge : |\hat{f}(\chi)| \geq \alpha\}$. Together with (1) and (2) this yields

$$\limsup_{\lambda\to 0} \lambda\varrho_\lambda * f * (\varrho_\lambda * f)^\sim(0) = \infty\,.$$

■

Principal Lemma 6.2.5 *Let G be a non-compact second countable locally compact Abelian group, and let $(\mu_t)_{t\geq 0}$ be an adapted continuous convolution semigroup in $M^1(G)$. Suppose that*

(α) $(\mu_t)_{t\geq 0}$ *is locally integrable*

and that

(β) *there exists an $f \in C_+^c(G)$ with $f \not\equiv 0$ such that*

$$\sup_{\lambda\in]0,1]} \sigma_\lambda * f * f^\sim(0) < \infty,$$

where

$$\sigma_\lambda := \frac{1}{2}(\varrho_\lambda + \varrho_\lambda^\sim) - \lambda\varrho_\lambda * \varrho_\lambda^\sim \in M_+^b(G).$$

Then $(\mu_t)_{t\geq 0}$ is transient.

Proof. 1. Assuming that $(\mu_t)_{t\geq 0}$ is recurrent we infer from Lemma 6.2.4 that with the notation

$$a_\lambda := \lambda\varrho_\lambda * \varrho_\lambda^\sim * f * f^\sim(0)^{-1}$$

the statements

$$\liminf_{\lambda\to 0} a_\lambda^{-1} = \infty$$

and

$$a_\lambda^{-1} > 0$$

for all $\lambda > 0$ are available. One notes that the numbers a_λ $(\lambda > 0)$ depend on the choice of a function $f \in C_+^c(G)$ with $f \not\equiv 0$. Let $(a_\lambda)_{\lambda\in]0,1]}$ be a bounded subfamily of $(a_\lambda)_{\lambda\in\mathbf{R}_+^\times}$. Then $(a_\lambda\varrho_\lambda)_{\lambda\in]0,1]}$ is a τ_v-relatively compact subset of $M_+(G)$.

In order to see this it suffices to prove that

$$\sup_{\lambda\in]0.1]} a_\lambda(\varrho_\lambda + \varrho_\lambda^\sim)(K) < \infty$$

for each compact subset K of G. But this will be clear from the following reasoning: We choose a compact subset K of G and a number $c > 0$ such that

$$L := \{x \in G : f * f^{\sim}(x) > c\} \neq \emptyset\,.$$

Then, for a suitable sequence $\{x_1, \ldots, x_n\}$ in G $(n \geq 1)$ we get

$$K \subset \bigcup_{i=1}^{n} (x_i - L)\,.$$

It follows that

$$\begin{aligned}(\varrho_\lambda + \varrho_\lambda^{\sim})(K) &\leq \sum_{i=1}^{n} (\varrho_\lambda + \varrho_\lambda^{\sim})(x_i - L) \\ &\leq \frac{1}{c} \sum_{i=1}^{n} (\varrho_\lambda + \varrho_\lambda^{\sim}) * f * f^{\sim}(x_i).\end{aligned}$$

Since $\varrho_\lambda + \varrho_\lambda^{\sim} \in M_p(G)$, Proposition 4.4.9 implies that $(\varrho_\lambda + \varrho_\lambda^{\sim}) * f * f^{\sim} \in CPD(G)$, and the above sequence of inequalities can be extended to

$$\begin{aligned}&\leq \frac{1}{c} \sum_{i=1}^{n} (\varrho_\lambda + \varrho_\lambda^{\sim}) * f * f^{\sim}(0) \\ &= \frac{2n}{c} (\sigma_\lambda * f * f^{\sim}(0) + a_\lambda^{-1})\end{aligned}$$

for all $\lambda \in]0, 1]$.

2. The assumptions on G imply that $M_+(G)$ is τ_v-metrizable. Therefore there exists a sequence $(\lambda_k)_{k \geq 1}$ in $]0, 1]$ with $\lambda_k \downarrow 0$ such that

$$\lim_{k \to \infty} a_{\lambda_k} = 0$$

and

$$(\tau_v-) \lim_{k \to \infty} a_{\lambda_k} \varrho_{\lambda_k} =: \zeta \in M_+(G)\,.$$

From

$$\begin{aligned}(\frac{1}{2}(\zeta+\zeta^{\sim})) * f * f^{\sim}(0) &= \lim_{k\to\infty} a_{\lambda_k}(\frac{1}{2}(\varrho_{\lambda_k}+\varrho_{\lambda_k}^{\sim})) * f * f^{\sim}(0)\\ &= \lim_{k\to\infty} a_{\lambda_k}\sigma_{\lambda_k} * f * f^{\sim}(0) + 1 = 1\end{aligned}$$

we conclude that $\zeta \neq 0$.

3. The next aim will be to show that

(a) $\zeta \in M^{sb}(G)$

and that

(b) $\zeta = \lambda\zeta * \varrho_\lambda$ for all $\lambda > 0$.

This being done Theorem 6.1.12 together with Theorem 5.2.16 (and the adaptation of $(\mu_t)_{t\geq 0}$) will imply that

$$\zeta = c\omega_G$$

for some $c > 0$.

Now concerning (a) we just note that $\zeta + \zeta^{\sim} \in M_p(G)$, since $\varrho_\lambda + \varrho_\lambda^{\sim} \in M_p(G)$, hence $\zeta + \zeta^{\sim} \in M^{sb}$ and so also $\zeta \in M^{sb}$.

As for the proof of (b) we deduce from the resolvent equation that

$$(\lambda - \lambda_k)\varrho_\lambda * a_{\lambda_k}\varrho_{\lambda_k} \leq a_{\lambda_k}\varrho_{\lambda_k}$$

for every $\lambda > 0$ and $k \geq 1$. Given $\lambda > 0$ we choose $\varepsilon \in]0, \lambda[$ and suppose that $\lambda_k \in]0, \varepsilon[$. Then

$$(\lambda - \varepsilon)\varrho_\lambda * a_{\lambda_k}\varrho_{\lambda_k} \leq a_{\lambda_k}\varrho_{\lambda_k},$$

hence

$$(\lambda - \varepsilon)\varrho_\lambda * \zeta \leq \zeta,$$

and for $\varepsilon \to 0$ we arrive at

$$\lambda\zeta * \varrho_\lambda \leq \zeta$$

valid for all $\lambda > 0$.

Suppose now that $\lambda\zeta * \varrho_\lambda \neq \zeta$ for some $\lambda > 0$. Since

$$\sum_{k=0}^{n} (\lambda\varrho_\lambda)^k * (\zeta - \lambda\varrho_\lambda * \zeta) = \zeta - (\lambda\varrho_\lambda)^{n+1} * \zeta \leq \zeta,$$

the series $\sum_{n\geq 1} (\lambda\varrho_\lambda)^n$ norm-converges for $\lambda > 0$, and

$$\begin{aligned}\sum_{k=0}^{n} \|(\lambda\varrho_\lambda)^k\| &= \|\sum_{k=0}^{n}(\lambda\varrho_\lambda)^k\| \\ &\leq \frac{\|\zeta\|}{\|\zeta - \lambda\varrho_\lambda * \zeta\|}\,.\end{aligned}$$

But as in the proof of Property 6.1.4.3 we apply the formula

$$\varrho_{\lambda'} = \sum_{n\geq 0} (\lambda - \lambda')\varrho_\lambda^{n+1}$$

valid for $\lambda' \in]0, \lambda[$ and obtain

$$\lim_{\lambda' \to 0} \varrho_{\lambda'} = \frac{1}{\lambda} \sum_{n \geq 1} (\lambda\varrho_\lambda)^n$$

which implies the transience of $(\mu_t)_{t\geq 0}$, hence a contradiction.

4. Now let W be an open, relatively compact subset of G such that $\operatorname{supp}(f * f^{\sim}) \subset W$. For each $k \geq 1$ we consider the convolution semigroup $(e^{-\lambda_k t}\mu_t^{\sim})_{t\geq 0} \in \mathbf{C\,S}(G)$ with potential measure $\varrho_{\lambda_k}^{\sim}$ and equilibrium measure γ_{λ_k} depending on W, the latter existing by Theorem 6.1.14. For every $k \geq 1$ we introduce the measure

$$\nu_{\lambda_k} := a_{\lambda_k}^{-1}\gamma_{\lambda_k},$$

and we obtain

$$\begin{aligned}\nu_{\lambda_k}(\overline{W})a_{\lambda_k}\varrho_{\lambda_k}(\overline{W}) &\leq \int_G \int_G 1_{\overline{W}-\overline{W}}(x+y)\gamma_{\lambda_k}(dx)\varrho_{\lambda_k}^{\sim}(dy) \\ &= (\gamma_{\lambda_k} * \varrho_{\lambda_k}^{\sim})(\overline{W} - \overline{W}) \\ &\leq \omega_G(\overline{W} - \overline{W}) < \infty.\end{aligned}$$

Since

$$\liminf_{k\to\infty} a_{\lambda_k}\varrho_{\lambda_k}(\overline{W}) > 0,$$

the sequence $(\nu_{\lambda_k}(\overline{W}))_{k\geq 1}$ is bounded, and since $\operatorname{supp}(\gamma_{\lambda_k}) \subset \overline{W}$ for all $k \geq 1, (\nu_k)_{k\geq 1}$ is norm-bounded. Without loss of generality we may assume that $(\nu_{\lambda_k})_{k\geq 1}$ τ_v-converges to a measure $\nu \in M_+(G)$. ν is bounded, since $\operatorname{supp}(\nu) \subset \overline{W}$, and $\nu \neq 0$. The latter statement requires a detailed argument. First of all we have that

$$\begin{aligned}
0 < \omega_G(W) &= \nu_{\lambda_k} * a_{\lambda_k}\varrho^{\sim}_{\lambda_k}(W) \\
&= a_{\lambda_k}\int\int 1_W(x+y)\nu_{\lambda_k}(dx)\varrho^{\sim}_{\lambda_k}(dy) \\
&\leq a_{\lambda_k}\int\limits_{W-W}\left(\int\limits_{W} d\nu_{\lambda_k}\right) d\varrho^{\sim}_{\lambda_k} \\
&= a_{\lambda_k}\nu_{\lambda_k}(W)\varrho^{\sim}_{\lambda_k}(W-W).
\end{aligned}$$

Since W is relatively compact, there exists a function $g \in C^c_+(G)$ such that

$$1_{W\cup(W-W)} \leq g$$

which implies that

$$\limsup_{k\to\infty} \nu_{\lambda_k}(W) \leq \lim_{k\to\infty} \nu_{\lambda_k}(g) = \nu(g)\,.$$

Under the assumption that $\nu \equiv 0$ we obtain

$$\lim_{k\to\infty} \nu_{\lambda_k}(W) = 0\,.$$

From

$$\begin{aligned}
\limsup_{k\to\infty} a_{\lambda_k}\varrho^{\sim}_{\lambda_k}(W-W) &\leq \lim_{k\to\infty} a_{\lambda_k}\varrho^{\sim}_{\lambda_k}(g) \\
&= \zeta^{\sim}(g) < \infty
\end{aligned}$$

we conclude that the sequence $(a_{\lambda_k}\varrho^{\sim}_{\lambda_k}(W-W))_{k\geq 1}$ is bounded, so the product sequence $(a_{\lambda_k}\varrho^{\sim}_{\lambda_k}(W-W)\nu_{\lambda_k}(W))_{k\geq 1}$ tends to 0 and its members are not $\geq \omega_G(W) > 0$ which provides a contradiction. Consequently $\nu \in M^b_+(G)\backslash\{0\}$.

5. By the resolvent equation

$$\lim_{k\to\infty} \lambda_k \varrho_{\lambda_k} * \varrho_{\lambda_k}^{\sim} * (\varepsilon_e - (\lambda - \lambda_k)\varrho_\lambda) * \nu_{\lambda_k} \tag{3}$$

$$= \lim_{k\to\infty} \lambda_k \varrho_{\lambda_k}^{\sim} * \nu_{k_\lambda} * \varrho_\lambda,$$

and this limit equals 0. For the latter assertion we need to realize that

$$\begin{aligned}\lambda_k \varrho_{\lambda_k}^{\sim} * \nu_{\lambda_k} * \varrho_\lambda &\le a_{\lambda_k}^{-1} \lambda_k (\varrho_\lambda * \omega_G) \\ &= a_{\lambda_k}^{-1} \lambda_k \|\varrho_\lambda\| \omega_G\end{aligned}$$

and to prove that

$$\lim_{k\to\infty} a_{\lambda_k}^{-1} \lambda_k = 0\,.$$

This limit relationship can be shown as follows: From the proof of Lemma 6.2.4 we infer that

$$a_{\lambda_k}^{-1} \lambda_k = \int_{G^\wedge} |\hat{f}|^2 \left(\frac{\lambda_k}{|\lambda_k + \psi|}\right)^2 d\omega_{G^\wedge}$$

for all $k \ge 1$. The integrands are bounded by $|\hat{f}|^2$ and converge (for $k \to \infty$) pointwise to 0 on $G^\wedge \backslash \{0\}$ (since $\psi(\chi) \ne 0$ for $\chi \ne 0$). An application of the Plancherel theorem 4.2.11 (iii) yields that it is also $\omega_{G^\wedge}$-integrable. But then the dominated convergence theorem implies the assertion.

6. We are ready to finish the proof by establishing a contradiction to the boundedness of $(a_{\lambda_k})_{k\ge 1}$.

By the choice of F and W according to 4. of this proof we note that

$$\operatorname{supp}(f * f^{\sim})^{\sim} = \operatorname{supp}(f * f^{\sim}) \subset W\,.$$

Moreover

$$\|(\lambda - \lambda_k)\varrho_k\| \le 1$$

for all $\lambda \ge \lambda_k \quad (k \ge 1)$. Corollary 6.1.15 now yields

$$\varrho_{\lambda_k}^{\sim} * (\varepsilon_0 - (\lambda - \lambda_k)\varrho_k) * \nu_{\lambda_k} * f * f^{\sim}(0) \ge 0 \tag{4}$$

for $\lambda \geq \lambda_k$. But $\sigma_\lambda \in M_p(G)$, since

$$\hat{\sigma}_\lambda = \frac{\operatorname{Re} \psi}{|\lambda + \psi|^2} \geq 0\,.$$

Hence

$$\sigma_{\lambda_k} * f * f^\sim \leq \sigma_{\lambda_k} * f * f^\sim(0)$$

and consequently,

$$\begin{aligned}
&\sigma_{\lambda_k} * (\varepsilon_0 - (\lambda - \lambda_k)\varrho_\lambda) * \nu_{\lambda_k} * f * f^\sim(0) \\
&\quad = \int_G (\sigma_{\lambda_k} * f * f^\sim)(-x)[(\varepsilon_0 - (\lambda - \lambda_k)\varrho_k) * \nu_{\lambda_k}](dx) \qquad (5) \\
&\quad \leq \sigma_{\lambda_k} * f * f^\sim(0)\|\nu_{\lambda_k}\|\|\varepsilon_0 - (\lambda - \lambda_k)_{\varrho k}\|.
\end{aligned}$$

But this expression admits a bound M (independent of λ) for all $\lambda_k \leq \lambda$, since $(\sigma_{\lambda_k} * f * f^\sim(0))_{k\geq 1}$ is bounded by assumption and

$$\begin{aligned}
\|\varepsilon_0 - (\lambda - \lambda_k)\varrho_k\| &\leq 1 + \|(\lambda - \lambda_k)\varrho_k\| \\
&\leq 1 + \|\lambda\varrho_\lambda\| \leq 2.
\end{aligned}$$

Now we deduce from (3) that

$$\lim_{k\to\infty} \lambda_k * \varrho_{\lambda_k} * \varrho_{\lambda_k}^\sim * (\varepsilon_0 - (\lambda - \lambda_k)\varrho_k) * \nu_{\lambda_k} * f * f^\sim(0) = 0,$$

whence by (5)

$$\limsup_{k\to\infty} (\varrho_{\lambda_k} + \varrho_{\lambda_k}^\sim) * (\varepsilon_0 - (\lambda - \lambda_k)\varrho_\lambda) * \nu_{\lambda_k} * f * f^\sim(0) \leq 2M$$

and thus by (4)

$$\limsup_{k\to\infty} \varrho_{k_k} * (\varepsilon_0 - (\lambda - \lambda_k)\varrho_k) * \nu_{\lambda_k} * f * f^\sim(0) \leq 2M$$

which together with the resolvent equation implies that

$$\limsup_{k\to\infty} \varrho_\lambda * \nu_{\lambda_k} * f * f^\sim(0) \leq 2M\,.$$

From $\tau_v - \lim_{k\to\infty} \nu_{\lambda_k} = \nu$ (Part 4. of the proof) and the norm-boundedness of $(\nu_{\lambda_k})_{k\geq 1}$ we derive

$$\varrho_\lambda * \nu * f * f^\sim(0) \leq 2M,$$

i.e.

$$\int \nu^\sim * f * f^\sim d\varrho_\lambda \leq 2M$$

for all $\lambda > 0$, in particular for $\lambda := \lambda_k$ whenever $k \geq 1$. Consequently the sequence

$$(a_{\lambda_k}^{-1} \int_G \nu^\sim * f * f^\sim d(a_{\lambda_k}\varrho_{\lambda_k}))_{k\geq 1}$$

is bounded. On the other hand

$$\lim_{k\to\infty} \int_G \nu^\sim * f * f^\sim d(a_{\lambda_k}\varrho_{\lambda_k}) = \int_G \nu^\sim * f * f^\sim d\zeta > 0,$$

the strict positivity following from the facts that $\zeta = c\omega_G$ for $c > 0$ and that

$$\nu^\sim * f * f^\sim \not\equiv 0\,.$$

But then $(a_{\lambda_k}^{-1})_{k\geq 1}$ must be bounded, and this contradicts the assumption made in Part 1. of the proof. The demonstration of the lemma is finally complete. ∎

Proof of the Port-Stone criterion 6.2.1 By Property 6.1.11.4 it suffices to establish the implication (TC) for an adapted Poisson semigroup $(\mu_t)_{t\geq 0}$ determined by a measure $\mu \in M^1(G)$, where G is of the form $\mathbf{R}^d \times \mathbf{Z}^e$ for $a, e \geq 1$. For the associated continuous negative definite function of the form $\psi = 1 - \hat{\mu}$ we assume Re $\frac{1}{\psi}$ to be locally integrable. It will be shown that $(\mu_t)_{t\geq 0}$ is transient.

1. The case $d + e \geq 3$.

The dual $G^\wedge = \mathbf{R}^d \times \mathbf{T}^e$ of G will be interpreted as the subset $\mathbf{R}^d \times]-\pi, \pi]^e$ of $\mathbf{R}^{d+e}$ equipped with the Euclidean norm $\|\cdot\|$.

At first we show that there exists a constant $c > 0$ such that

$$\text{Re } \psi(y) \geq c\|y\|^2 \tag{6}$$

for all y in a relatively compact neighborhood $U \in \mathfrak{V}_{G^\wedge}(0)$.

In fact, since $G = [\text{supp}\,(\mu)]^-$, one finds $m := d + e$ linearly independent elements $x_1, \ldots, x_m \in \text{supp}\,(\mu)$. Let $V := \{x \in G : \|x\| \leq \gamma\}$, where $\gamma := \max\,\{\|x_i\| : i = 1, \ldots, m\}$. The quadratic form Q defined by

$$Q(y) := \int_V y(x)^2 \mu(dx)$$

for all $y \in G^\wedge$ is positive definite, hence its smallest eigenvalue τ is strictly positive and

$$Q(y) \geq \tau\|y\|^2$$

for all $y \in G^\wedge$. On the other hand

$$\begin{aligned} \text{Re } \psi(y) &= \int_G (1 - \cos y(x))\mu(dx) \\ &\leq 2 \int_V \sin^2(\frac{1}{2} y(x))\mu(dx). \end{aligned}$$

for all $y \in G^\wedge$. But the modulus of the latter integrand being $\geq \frac{1}{\pi}|y(x)|$ if $|y(x)| \leq \pi$ we obtain that

$$\text{Re } \psi(y) \geq \frac{2}{\pi^2} \int_W |y(x)|^2 \mu(dx),$$

for all $y \in G^\wedge$, where

$$W := \{x \in G : \|x\| \leq \gamma, |y(x)| \leq \pi\}.$$

If $\|y\| \leq \frac{\pi}{\gamma}$ and $\|x\| \leq \gamma$ we obtain $|y(x)| \leq \pi$. But then

$$\text{Re } \psi(y) \geq \frac{2}{\pi^2} \int_V |y(x)|^2 \mu(dx) \geq \frac{2}{\pi^2} \tau\|y\|^2$$

for all y in the relatively compact neighborhood $U := \{u \in G^\wedge : \|u\| \le \frac{\pi}{\gamma}\} \in \mathfrak{V}_{G^\wedge}(0)$. With $c := \frac{2}{\pi^2}\tau$ we arrive at the assertion.

Applying the inequality (6) we now obtain that

$$\int\limits_U \operatorname{Re} \frac{1}{\lambda + \psi} d\lambda^{d+e} \le \int\limits_U \frac{1}{\lambda + \operatorname{Re} \psi} d\lambda^{d+e} \le \frac{1}{c} \int\limits_U \frac{1}{\|y\|^2} \lambda^{d+e}(dy),$$

and the Chung-Fuchs criterion 6.2.2 yields the transience of $(\mu_t)_{t \ge 0}$, provided $d + e \ge 3$.

In

2. the *case* $d + e = 2$ we note that for all $\lambda > 0$

$$\begin{aligned}\int\limits_U \frac{\lambda}{|\lambda + \psi|^2} d\lambda^2 &\le \int\limits_U \frac{\lambda}{|\lambda + \operatorname{Re} \psi|^2} d\lambda^2 \\ &\le \int\limits_U \frac{\lambda}{(\lambda + c\|y\|^2)^2} \lambda^2(dy) \\ &\le \int\limits_{\mathbf{R}^2} \frac{1}{(1 + c\|y\|^2)^2} \lambda^2(dy) < \infty.\end{aligned}$$

On the other hand we obtain from the hypothesis that

$$\sup_{\lambda \in]0,1]} \int\limits_U \operatorname{Re} \frac{1}{\lambda + \psi} d\lambda^2 = \sup_{\lambda \in]0,1]} \int\limits_U \frac{\lambda + \operatorname{Re} \psi}{|\lambda + \psi|^2} d\lambda^2 < \infty$$

which, again by Lemma 6.2.2, implies the assertion.

It remains to treat

3. the *case* $d + e = 1$, i.e. the groups $G = \mathbf{R}$ or $G = \mathbf{Z}$.

From Principal Lemma 6.2.5 we infer that it suffices to establish the existence of a function $f \in C_+^c(G)$ with $f \not\equiv 0$ satisfying

$$\sup_{\lambda \in]0,1]} \sigma_\lambda * f * f^\sim(0) < \infty\,.$$

At first we observe that for all $f \in C_+^c(G)$

$$\begin{aligned}
\sigma_\lambda * f * f^\sim(0) &= \int_G (\sigma_\lambda * f)\overline{f} d\omega_G = \int_{G^\wedge} \widehat{\sigma_\lambda * f}\,\overline{\hat{f}} d\omega_{G^\wedge} \\
&= \int_{G^\wedge} \hat{\sigma}_\lambda |\hat{f}|^2 d\omega_{G^\wedge} = \int_{G^\wedge} \frac{\text{Re}\ \psi}{|\lambda + \psi|^2} |\hat{f}|^2 d\omega_{G^\wedge} \\
&\le \int_{G^\wedge} \frac{\text{Re}\ \psi}{|\psi|^2} |\hat{f}| d\omega_{G^\wedge} = \int_{G^\wedge} \text{Re}\ \frac{1}{\psi} |\hat{f}|^2 d\omega_{G^\wedge}.
\end{aligned}$$

If $G = \mathbf{Z}$, the last integral is finite, since $G^\wedge = \mathbf{T}$ is compact.

Now let $G = \mathbf{R}$. For each relatively compact neighborhood $U \in \mathfrak{V}_{G^\wedge}(0)$ we have that

$$\begin{aligned}
\sigma_\lambda * f * f^\sim(0) &\le \int_{G^\wedge} \text{Re}\ \frac{1}{\psi} |\hat{f}|^2 d\omega_{G^\wedge} \\
&= \int_U \text{Re}\ \frac{1}{\psi} |\hat{f}|^2 d\omega_{G^\wedge} + \int_{U^c} \text{Re}\ \frac{1}{\psi} |\hat{f}|^2 d\omega_{G^\wedge}.
\end{aligned}$$

Obviously the integral over U exists. Given a symmetric measure $\nu \in M_+^b(\mathbf{R})$ with compact support such that

$$\nu \le \frac{1}{2}(\mu * \mu^\sim)$$

holds we introduce the desired function f on $\mathbf{R}$ by

$$f(x) := \int_{\mathbf{R}} |x - y| \nu(dy) - \|\nu\| |x|$$

whenever $x \in \mathbf{R}$. Since $\nu \neq \varepsilon_0$, we may assume that ν is not a multiple of ε_0, hence $f(0) \neq 0$. It is easily shown that $f \in C_+^c(\mathbf{R})$ and that

$$\hat{f}(x) = \frac{2}{x^2}(\|\nu\| - \hat{\nu}(x))$$

for all $x \in \mathbf{R}^{\times}$. From the inequalities

$$\begin{aligned} 0 &\le \|\nu\| - \hat{\nu}(x) = \int (1 - \cos xy)\nu(dy) \\ &\le \int (1 - \cos xy)\mu(dy) = 1 - \text{Re } \hat{\mu}(x) \\ &= \text{Re } \psi(x) \le 2 \end{aligned}$$

valid for all $x \in \mathbf{R}$ we conclude that

$$\begin{aligned} \int_{U^c} \text{Re } \frac{1}{\psi(x)} |\hat{f}(x)|^2 \lambda(dx) &\le \int_{U^c} \frac{4(\text{Re } \psi(x))^2}{x^4 \text{ Re } \psi(x)} \lambda(dx) \\ &\le \int_{U^c} \frac{8}{x^4} \lambda(dx) < \infty. \end{aligned}$$

The proof of the Port-Stone criterion 6.2.1 now is complete also for $G = \mathbf{R}$. ■

Examples 6.2.6

We are considering continuous convolution semigroups $(\mu_t)_{t \ge 0}$ *in* $M^1(G)$ *with associated continuous negative definite function* ψ *on* $G^\wedge$ *admitting a canonical quadruple (in the sense of Theorem 5.4.7) of the form* $(0, h, q, 0)$. *These semigroups determined by*

$$\psi = ih + q$$

with a homomorphism $h : G^\wedge \to \mathbf{R}$ *and a positive quadratic form* q *on* $G^\wedge$ *are called (non symmetric)* ***Gaussian***. *The subsequent discussion is designed to study the transience of Gaussian semigroups.*

6.2.6.1 *Let* $(\varepsilon_{x(t)})_{t \ge 0}$ *be a translation semigroup in* $M^1(G)$ *as introduced in Example 5.2.7.1. Recall that the mapping* $x : \mathbf{R}_+ \to G$ *can be extended to a continuous homomorphism* $\varphi : \mathbf{R} \to G$. *It was shown in Example 5.2.7.1 that* $(\varepsilon_{x(t)})_{t \ge 0}$ *admits a canonical quadruple of the form* $(0, h, 0, 0)$.

Now, the semigroup $(\varepsilon_{x(t)})_{t\geq 0}$ *is transient if and only if* $\overline{\varphi(\mathbf{R})}$ *is not compact (in* G*), and in the affirmative case its potential measure can be described as the image* $\varphi(\kappa)$ *under* φ *of the potential measure*

$$\kappa = 1_{\mathbf{R}_+^\times} \cdot \lambda \in M_+(\mathbf{R})$$

of the translation semigroup $(\varepsilon_t)_{t\geq 0}$ *in* $M^1(\mathbf{R})$*. In other words,*

$$\varphi(\kappa) = \operatorname{Res}_{\varphi(\mathbf{R}_+^\times)} \omega_{\varphi(\mathbf{R})}\,.$$

In fact, by Appendix A 3.6 a continuous homomorphism $\varphi : \mathbf{R} \to G$ admits the following alternative: either (1) $\overline{\varphi(\mathbf{R})}$ is compact or (2) φ is a homeomorphism onto $\varphi(\mathbf{R})$, and $\varphi(\mathbf{R})$ is a closed subgroup of G. In the case (1) $(\varepsilon_{x(t)})_{t\geq 0}$ is obviously recurrent (See Discussion 6.1.2.3), in the case (2) it is transient, since for every $f \in C_+^c(G)$ the set $\operatorname{supp}(f) \cap \varphi(\mathbf{R})$ is compact, hence for some $t_0 > 0$

$$\int f d\varepsilon_{x(t)} = f \circ \varphi(t) = 0$$

whenever $t \geq t_0$.

6.2.6.2 *Here we discuss the transience of symmetric Gaussian semigroups* $(\nu_t)_{t\geq 0}$ *admitting a canonical quadruple of the form* $(0, 0, q, 0)$*. More specifically, we restrict ourselves to the case that* $G = \mathbf{R}^d$ $(d \geq 1)$ *and to the symmetric* ***Brownian semigroup*** $(\nu_t)_{t\geq 0}$ *given by* $\nu_t = n_t \cdot \lambda^d$ *with*

$$n_t(x) = \frac{1}{(4\pi t)^{\frac{d}{2}}} \exp\left(-\frac{\|x\|^2}{4t}\right)$$

for all $x \in \mathbf{R}^d$ $(t > 0)$*.*

The following computation shows (by an application of the Schoenberg correspondence theorem 5.2.3) that the function ψ *on* $\mathbf{R}^d$ *defined by*

$$\psi(y) := \|y\|^2$$

for all $y \in \mathbf{R}^d$ is the continuous negative definite function associated with $(\nu_t)_{t\geq 0}$.

Indeed, for all $y \in \mathbf{R}^d, t > 0$ we have

$$\begin{aligned}\widehat{n_t}(y) &= \int\limits_{\mathbf{R}^d} \exp(-i\langle x, y\rangle)\frac{1}{(4\pi t)^{\frac{d}{2}}}\exp\left(-\frac{\|x\|^2}{4t}\right)dx \\ &= \prod_{k=1}^{d}\int\limits_{\mathbf{R}} \exp(-ix_k y_k)\frac{1}{(4\pi t)^{\frac{d}{2}}}\exp\left(-\frac{x_k^2}{4t}\right)dx_k \\ &= \prod_{k=1}^{d}\exp(-ty_k^2) = \exp(-t\|y\|^2)\end{aligned}$$

(with the obvious notation x_k and y_k for the k-th component of the vectors x and y respectively).

Now the Port-Stone criterion 6.2.1 implies that $(\nu_t)_{t\geq 0}$ is transient if and only if $d \geq 3$.

This result can also be obtained by an explicit computation of the potential measure. In fact, one easily sees that

$$\int_0^\infty n_t(x)dt = \begin{cases} \infty & \text{if } d = 1,2 \text{ for all } x \in \mathbf{R}^d, \\ & \text{and if } d \geq 3 \ \text{ for } x = 0 \\ \dfrac{\Gamma(\frac{d}{2})}{2\pi^{\frac{d}{2}}(d-2)}\|x\|^{2-d} & \text{if } d \geq 3 \text{ and } x \in \mathbf{R}^d, x \neq 0. \end{cases}$$

Putting for $d \geq 3$

$$c_d := \frac{\Gamma(\frac{d}{2})}{2\pi^{\frac{d}{2}}(d-2)}$$

and

$$N_d(x) := \begin{cases} \|x\|^{2-d} & \text{if } x \in \mathbf{R}^d, x \neq 0 \\ \infty & \text{if } x = 0 \end{cases}$$

we see that the potential measure

$$\kappa_d := c_d N_d \cdot \lambda^d$$

of $(\nu_t)_{t\geq 0}$ is a multiple of the classical Newton kernel $N_d \cdot \lambda^d$. We also note that $\kappa_d \in M^\infty(\mathbf{R}^d)$.

6.2.6.3 *Symmetric stable semigroups of order* $\alpha \in]0,2]$ *are given as families* $(\mu_t^\alpha)_{t\geq 0}$ *of measures in* $M^1(\mathbf{R}^d)$ $(d \geq 1)$ *admitting a Fourier representation*

$$\widehat{\mu_t^\alpha}(y) = \exp(-t\|y\|^\alpha)$$

for all $y \in \mathbf{R}^d, t \geq 0$. *It can be shown that a symmetric stable semigroup* $(\mu_t^\alpha)_{t\geq 0}$ *of order* α *is a convolution semigroup subordinated to the symmetric Brownian semigroup* $(\nu_t)_{t\geq 0}$ *defined in Example 6.2.6.2 by means of the one-sided stable semigroup of order* $\frac{\alpha}{2}$.

Clearly, the symmetric stable semigroup of order $\alpha = 2$ coincides with the symmetric Brownian semigroup in $M^1(\mathbf{R}^d)$.

Since $\widehat{\mu_t^\alpha} \in L^1(\mathbf{R}^d, \lambda^d)$, an application of the inverse Fourier transform (introduced next to Properties 4.2.16) yields an integral representation of the λ^d-density m_t^α of μ_t^α as a function in $C_+^0(\mathbf{R}^d)$. In the special cases $\alpha = 1$ and $\alpha = 2$ one may compute m_t^α explicitly. It turns out that

$$m_t^1(x) = \Gamma\left(\frac{d+1}{2}\right) \frac{t}{[\pi(\|x\|^2 + t^2)]^{\frac{d+1}{2}}}$$

whenever $x \in \mathbf{R}^d, t \geq 0$. If, in addition, $d = 1$, then $(\mu_t^1)_{t\geq 0}$ reduces to the **Cauchy semigroup**.

Since

$$(\mu_t^\alpha)_{t\geq 0} \longleftrightarrow \psi,$$

where $\psi(y) := \|y\|^\alpha$ for all $y \in \mathbf{R}^d$, the λ^d-local integrability of $\frac{1}{\psi}$ which occurs exactly in the cases

$$\begin{cases} d = 1 & \text{and } \alpha \in]0,1[, \\ d = 2 & \text{and } \alpha \in]0,2[, \\ d \geq 3 & \text{and } \alpha \in]0,2] \end{cases}$$

implies the transience of $(\mu_t^\alpha)_{t\geq 0}$ exactly for these choices of d and α.

For $d = 1$ and $\alpha \in [1,2]$ as well as for $d = 2$ and $\alpha = 2$ the convolution semigroup $(\mu_t^\alpha)_{t\geq 0}$ is recurrent.

In the cases of transience the potential measure $\kappa_{d,\alpha}$ of $(\mu_t^\alpha)_{t\geq 0}$ appears to be the **Riesz kernel of order** α having a λ^d-density

$$x \mapsto \frac{\Gamma\left(\frac{d-\alpha}{2}\right)}{2^\alpha \pi^{\frac{d}{2}} \Gamma(\frac{\alpha}{2})} \|x\|^{\alpha-d}.$$

For the computation of this density one employs the construction of $(\mu_t^\alpha)_{t\geq 0}$ by means of subordination.

6.2.6.4 *Let $(\mu_t)_{t\geq 0}$ denote the heat semigroup in $M^1(\mathbf{R}^d \times \mathbf{R})$ defined by*

$$\mu_t := \nu_t \otimes \varepsilon_t$$

for all $t \geq 0$, where $(\nu_t)_{t\geq 0}$ and $(\varepsilon_t)_{t\geq 0}$ denote the Brownian and the translation semigroups discussed in Examples 6.2.6.2 and 6.2.6.1 respectively. One has,

$$(\mu_t)_{t\geq 0} \longleftrightarrow \psi$$

with

$$\psi(y,s) := \|y\|^2 + is$$

whenever $(y,s) \in \mathbf{R}^d \times \mathbf{R}$. Clearly, $(\mu_t)_{t\geq 0}$ is locally integrable, and hence the Port-Stone criterion 6.2.1 implies the transience of $(\mu_t)_{t\geq 0}$.

The transience of $(\mu_t)_{t\geq 0}$ can also be obtained from an explicit computation of the potential measure.

In fact, let $(\mu_t)_{t\geq 0} \longleftrightarrow (\varrho_\lambda)_{\lambda>0}$. Then for each $g \in C^c(\mathbf{R}^d \times \mathbf{R})$ we have that

$$\begin{aligned}\varrho_\lambda(g) &= \int_0^\infty e^{-\lambda s} \mu_s(g) ds \\ &= \int_0^\infty e^{-\lambda s} \left(\int_{\mathbf{R}^d} g(x,s) n_s(x) dx \right) ds \qquad (\lambda > 0).\end{aligned}$$

But the function k on $\mathbf{R}^d \times \mathbf{R}$ given by

$$k(x,s) := \begin{cases} n_s(x) & \text{if } x \in \mathbf{R}^d, s \in \mathbf{R}_+^\times \\ 0 & \text{if } x \in \mathbf{R}^d, s \notin \mathbf{R}_+^\times \end{cases}$$

is locally (λ^{d+1}-) integrable, hence $(\mu_t)_{t\geq 0}$ is transient with potential measure

$$\kappa = k \cdot \lambda^{d+1}.$$

Moreover, $\kappa \in M^\infty(\mathbf{R}^d \times \mathbf{R})$.

6.2.6.5 *Let* $G := \mathbf{Z}$ *and let* $(\mu_t)_{t\geq 0}$ *denote the Poisson semigroup determined by the measure*

$$\mu := \sum_{n\geq 2} \frac{a}{n^2 \log n} \varepsilon_n \in M^1(G)$$

with a suitably chosen $a > 0$.

Clearly,

$$\kappa := \sum_{n\geq 0} \mu^n$$

exists, since $\operatorname{supp}(\mu^n) \subset \{2n, 2n+1, \ldots\}$ $(n \in \mathbf{N})$. *Observe that* $\kappa(n) = 0$ *for all* $n < 0$. *Thus* $(\mu_t)_{t\geq 0}$ *is transient. Moreover it follows from Spitzer's book [42] that* $\kappa \in M^\infty(G)$.

On the other hand, $(\mu_t)_{t\geq 0}$ *being determined by* μ *is associated with the negative definite function* ψ *on* $\mathbf{T}$ *given by*

$$\psi(\vartheta) := 1 - \hat{\mu}(\vartheta) = \sum_{n\geq 2} \frac{a}{n^2 \log n}(1 - \cos(n\vartheta)) + i \sum_{n\geq 2} \frac{a}{n^2 \log n} \sin(n\vartheta)$$

whenever $\vartheta \in \mathbf{R}$ *(after suitable identification of functions on* $\mathbf{Z}^\wedge = \mathbf{T}$ *and* $\mathbf{R}$*). A result in Zygmund's book [49] yields*

$$\psi(\vartheta) \sim ia\vartheta \log|\log \vartheta|$$

for $\vartheta \to 0+$, *hence* $\frac{1}{\psi}$ *is not integrable over any neighborhood of* 0.

Altogether we have seen that $(\mu_t)_{t\geq 0}$ *is transient, hence* $\operatorname{Re} \frac{1}{\psi}$ *is locally integrable by the transience criterion 6.1.8, but* $\frac{1}{\psi}$ *is not locally integrable.*

6.3 Recurrent random walks

Until now we have studied convolution semigroups on an arbitrary locally compact Abelian group G without particular reference to probability theory. In the case of a second countable group G this reference can be easily established. Given a convolution semigroup $(\mu_t)_{t\in\mathbf{R}_+}$ in $M^1(G)$ there exists by the Kolmogorov consistency theorem a translation invariant Markov process in G whose transition semigroup $(P_t)_{t\in\mathbf{R}_+}$ (of Markov kernels P_t on $K \times \mathfrak{B}(G)$) is related to $(\mu_t)_{t\in\mathbf{R}_+}$ by

$$P_t(x, B) = \mu_t * \varepsilon_x(B) \tag{1}$$

for all $(x, B) \in G \times \mathfrak{B}(G), t \in \mathbf{R}_+$. More precisely, there is a one-to-one correspondence between convolution semigroups $(\mu_t)_{t\in\mathbf{R}_+}$ in $M^1(G)$ and stationary independent increment processes with transition semigroup $(P_t)_{t\in\mathbf{R}_+}$ given by (1).

In the present section we shall study stationary independent increment processes in G with transition function $(P_n)_{n\in\mathbf{Z}_+}$, where $P_n := P^n$ for some transition kernel P of the form

$$P(x, B) := \mu * \varepsilon_x(B)$$

for all $(x, B) \in G\times\mathfrak{B}(G)$ with a measure μ in $M^1(G)$. Such processes are called **random walks** in G **with law** μ.

A sketch of the construction of such random walks now follows. Given P and a measure $\nu \in M^1(G)$ there exists a measurable space $(\Omega, \mathfrak{A})$ with $\Omega := G^{\mathbf{Z}_+}$ and $\mathfrak{A} := \mathfrak{B}(G)^{\otimes\mathbf{Z}_+}$ and a measure $\mathbf{P}^\nu$ on $(\Omega, \mathfrak{A})$ such that the sequence $(X_n)_{n\in\mathbf{Z}_+}$ of projections $X_n : G^{\mathbf{Z}_+} \to G$ (generating the σ-algebra $\mathfrak{A}$) forms a Markov chain in G with transition kernel P and starting measure ν. This statement can be made precise by the defining properties

$$\mathbf{E}^\nu(f \circ X_n \mid \mathfrak{A}_n) = P(f \circ X_n)\ [\mathbf{P}^\nu]$$

for all bounded measurable functions f on G, where $\mathfrak{A}_n := \sigma(\{X_0, X_1,$

$\ldots, X_n\})$ for every $n \in \mathbf{Z}_+$, and

$$\mathbf{P}^\nu([X_0 \in B]) = \nu(B)$$

for all $B \in \mathfrak{B}(G)$.

In the special case $\nu := \varepsilon_x$ for $x \in G$ we shall write $\mathbf{E}^x$ and $\mathbf{P}^x$ in place of $\mathbf{E}^\nu$ and $\mathbf{P}^\nu$ respectively.

Introducing the shift operator θ on Ω by

$$\theta((x_0, x_1, \ldots, x_n, \ldots)) := (x_1, x_2, \ldots, x_{n+1}, \ldots)$$

for all $(x_0, x_1, \ldots, x_n, \ldots) \in \Omega$ one obtains the *Markov property* of the chain $(X_n)_{n\in\mathbf{Z}_+}$ in the form

$$\mathbf{E}^\nu(Z \circ \theta^n \mid \mathfrak{A}_n) = \mathbf{E}^{X_n}(Z)[\mathbf{P}^\nu]$$

for every $\mathfrak{A}$-measurable function $Z \geq 0$ on $\Omega, n \in \mathbf{Z}_+$.

Moreover, it turns out that $(X_n)_{n\in\mathbf{Z}_+}$ enjoys the *strong Markov property* : Given any stopping time τ for $(X_n)_{n\in\mathbf{Z}_+}$ then for every $\mathfrak{A}$-measurable function $Z \geq 0$ on Ω

$$\mathbf{E}^\nu(Z \circ \theta_\tau \mid \mathcal{F}_\tau) = \mathbf{E}^{X_\tau}(Z)\ [\mathbf{P}^\nu],$$

where $\theta_\tau, \mathcal{F}_\tau$ and X_τ denote the τ-shift, τ-past and the τ-stopped chain respectively.

The random walk in G with law $\mu \in M^1(G)$ will be abbreviated by $X(\mu)$ if no other specification is needed.

It is easy to see that the random walk $X(\mu)$ constructed above has independent increments $Z_n := X_n - X_{n-1} (n \geq 1)$ (with respect to $\mathbf{P}^\nu$) and that these increments are identically distributed with

$$(\mathbf{P}^\nu)_{Z_n} = \mu$$

for all $n \geq 1$.

For a given random walk $X(\mu)$ with law $\mu \in M^1(G)$ we introduce the notation

$$S(\mu) := \langle \mathrm{supp}\,(\mu) \rangle^-$$

and

$$G(\mu) := [S(\mu)]^{-} .$$

Observation 6.3.1 *$x \in S(\mu)$ if and only if for each $V \in \mathfrak{V}_G(x)$ there exists an $n \geq 1$ such that*

$$\mu^n(V) = \mathbf{P}^0([X_n \in V]) > 0 .$$

Since G is assumed to be second countable, this implies that

$$\mathbf{P}^0\left(\bigcap_{n\geq 1} [X_n \in S(\mu)]\right) = 1$$

and, after translation in G,

$$\mathbf{P}^x\left(\bigcap_{n\geq 0} [X_n \in G(\mu)]\right) = 1$$

for all $x \in G$. This means that $G(\mu)$ is an **absorbing set** *of $X(\mu)$, hence we may assume without loss of generality that*

$$G(\mu) = G .$$

In terms of Fourier transforms this hypothesis can be rewritten as

$$\{\chi \in G^\wedge : \hat{\mu}(\chi) = 1\} = \{0\} .$$

Indeed, if μ is adapted in the sense of the hypothesis, then 1 being an extremal point of the unit disk, $\hat{\mu}(\chi) = 1$ implies that $\chi(x) = 1$ for μ - a.a. $x \in G$, hence for all $x \in G(\mu)$, and thus (by adaption of μ) for all $x \in G$ (without exception). Thus χ is the unit character 0 in $G^\wedge$.

In the following we shall look at sets of the form

$$R(B) := \limsup_{n\to\infty} [X_n \in B]$$

$$= [X_n \in B \text{ for infinitely many } n \geq 0],$$

where $B \in \mathfrak{B}(G)$.

Definition 6.3.2 *$x \in G$ is said to be **recurrent** if for every $V \in \mathfrak{V}_G(x)$ one has*

$$\mathbf{P}^0(R(V)) = 1\,.$$

*$x \in G$ is called **transient** if X is not recurrent.*

Theorem 6.3.3 *(Dichotomy) For each random walk $X(\mu)$ in G one has the following alternative: Either*

(1) every element of G is recurrent

or

(2) every element of G is transient.

Proof. Let R_μ denote the set of recurrent elements (for $X(\mu)$). Plainly, R_μ is a closed set, and hence it suffices to show that

$$R_\mu - S(\mu) \subset R_\mu\ .$$

In fact, if this implication holds true, then $R_\mu \subset S(\mu)$ and consequently

$$R_\mu - R_\mu \subset R_\mu\,.$$

So, if $R_\mu \neq \emptyset$, then R_μ is a closed subgroup of G. But $R_\mu \supset -S(\mu)$ implies that $R_\mu = G$, since $X(\mu)$ is adapted.

Now, let $x \in R_\mu$ and $y \in S(\mu)$. We shall show that $x - y \in R_\mu$. Let U be a neighborhood $\in \mathfrak{V}_g(x-y)$. Clearly, $U + (y-x) \in \mathfrak{V}_G(0)$, and there exists a neighborhood $V \in \mathfrak{V}_G(0)$ such that $V - V \subset U + (y-x)$. On the other hand, $V + y \in \mathfrak{V}_G(y)$, hence there exists a $k \in \mathbf{Z}_+$ satisfying

$$\mathbf{P}^0([X_k \in V + y]) > 0,$$

since $y \in S(\mu)$. Let

$$A := [X_k \in V + y]\,.$$

If $\omega \in A$, the relation $X_{n+k}(\omega) \in V + x$ implies that

$$(X_{n+k} - X_k)(\omega) \in V + x - (V + y) \subset U\,.$$

As x is recurrent, for $\mathbf{P}^0$ - a.a. $\omega \in A$ there exists an infinity of integers $n \in \mathbf{Z}_+$ such that $X_{n+k}(\omega) \in V + x$, therefore we obtain

$$\mathbf{P}^0(\limsup_{n\to\infty}[X_{n+k} - X_k \in U] \cap A) = \mathbf{P}^0(A)\,.$$

The members of the sequences $(X_{n+k} - X_k)_{n\in\mathbf{Z}_+}$ and $(X_n)_{n\in\mathbf{Z}_+}$ are independent G-valued random variables having the same distribution and are independent of A, therefore

$$\mathbf{P}^0(\limsup_{n\to\infty}[X_n \in U]) = 1$$

and $x - y \in R_\mu$. ∎

The dichotomy theorem enables us to make the following

Definition 6.3.4 *A random walk $X(\mu)$ with law $\mu \in M^1(G)$ is said to be* ***recurrent*** *if it satisfies (1) of the theorem. Otherwise $X(\mu)$ is said to be* ***transient.***

A locally compact Abelian group G is called ***recurrent*** *provided there exists an adapted recurrent random walk on it, and* ***transient*** *otherwise.*

We recall the notion of ***(first) return time*** of $X(\mu)$ into a set $B \in \mathfrak{B}(G)$ defined by

$$R_B := \begin{cases} \inf\{n \in \mathbf{N} : X_n \in B\} & \text{if } \{n \in \mathbf{N} : X_n \in B\} \neq \emptyset \\ \infty & \text{otherwise.} \end{cases}$$

It is known that R_B is a stopping time for $X(\mu)$ (with respect to the canonical filtration $(\mathfrak{A}_n)_{n\in\mathbf{Z}_+}$).

Theorem 6.3.5 *A (n adapted) random walk $X(\mu)$ with law $\mu \in M^1(G)$ is recurrent if and only if for each neighborhood $V \in \mathfrak{V}_G(0)$ one has*

$$\mathbf{P}^0([R_V < \infty]) = 1\,.$$

Proof. It suffices to prove that the condition is sufficient for $X(\mu)$ to be recurrent. Let V be a symmetric neighborhood $\in \mathfrak{V}_G(0)$ and let W

be a compact symmetric neighborhood $\in \mathfrak{V}_G(0)$ such that $W \subset V$. Clearly, there exists an open neighborhood $U \in \mathfrak{V}_G(0)$ satisfying

$$W + V^c = -W + V^c \subset U^c.$$

It follows that

$$[X_{m+n} \notin V] \cap [X_m \in W] \subset [X_{m+n} - X_m \notin U],$$

hence that

$$\begin{aligned}\mathbf{P}^0\left([X_m \in W] \cap \bigcap_{n\geq 1} [X_{m+n} \notin V]\right) \\ &\leq \mathbf{P}^0\left([X_m \in W] \cap \bigcap_{n\geq 1} [X_{m+n} - X_m \notin U]\right) \\ &= \mathbf{P}^0([X_m \in W])\mathbf{P}^0([R_U = \infty]),\end{aligned}$$

since $\mathbf{P}^0_{X_{m+n}-X_m} = \mathbf{P}^0_{X_n}$, and $X_{m+n} - X_m$ is independent of X_m $(m, n \in \mathbf{Z}_+)$. But by hypothesis

$$\mathbf{P}^0([X_m \in W] \cap \bigcap_{n\geq 1} [X_{m+n} \notin V]) = 0$$

for all $m \in \mathbf{Z}_+$. Let therefore $(W_r)_{r\geq 1}$ denote an increasing sequence of compact symmetric neighborhoods $W_r \in \mathfrak{V}_G(0)$ such that $\bigcup_{r\geq 1} W_r = V$.

Since

$$[X_m \in V] = \lim_{r\to\infty} [X_m \in W_r],$$

one has

$$\mathbf{P}^0\left([X_m \in V] \cap \bigcap_{n\geq 1} [X_{m+n} \notin V]\right) = 0$$

for all $m \in \mathbf{Z}_+$. Let now

$$T := \sup\{m \geq 0 : X_m \in V\}.$$

Then

$$\begin{aligned}\mathbf{P}^0([T<\infty]) &= \sum_{m\geq 0} \mathbf{P}^0([T=m]) \\ &= \sum_{m\geq 0} \mathbf{P}^0([X_m \in V] \cap \bigcap_{n\geq 1} [X_{m+n} \notin V]) = 0,\end{aligned}$$

hence 0 and therefore $X(\mu)$ is recurrent. ■

If $X(\mu)$ is a random walk in G with law $\mu \in M^1(G)$ admitting a transition kernel P, we introduce the ***potential kernel*** K of $X(\mu)$ by

$$Kf(x) := \sum_{n\geq 0} P^n f(x) = \mathbf{E}^x(\sum_{n\geq 0} f \circ X_n)$$

for all measurable functions ≥ 0 on G and all $x \in G$.

Properties 6.3.6 *of the potential kernel*

6.3.6.1 *K satisfies the* ***maximum principle*** *in the sense that*

$$\sup_{x\in G} Kf(x) = \sup\{Kf(y) : y \in \text{supp}\,(f)\}$$

for all measurable functions f on G.

6.3.6.2 *For each measurable function $f \geq 0$ on G one has*

$$f + PKf = f + KPf,$$

hence

6.3.6.3 *the* ***Poisson equation***

$$(I - P)Kf = f$$

is fulfilled whenever Kf is finite.

6.3.6.4 *For all $x \in G$ one has the commutation relationship*

$$T_x P = P T_x \,.$$

The proofs of these properties are straightforward except that of Property 6.3.6.1. Let $x \in G$ and let R^x denote the return time into $\mathrm{supp}\,(f)$ of the random walk $X(\mu)$ starting at x (Here we refer to the sequence $(S_n^x)_{n \geq 0}$ of shifted sums

$$S_n^x := \sum_{k=1}^{n} X_k + x$$

$(n \geq 1)$ and $S_0^x := x$). One has

$$\begin{aligned} Kf(x) &= \mathbf{E}^x\left(\sum_{n \geq 0} f \circ X_n\right) \\ &= \mathbf{E}^x\left(\sum_{n \geq R^x} f \circ X_n ; R^x < \infty\right) \\ &= \mathbf{E}^x\left(\sum_{n \geq 0} f \circ X_n \circ \theta_{R^x} ; R^x < \infty\right) \\ &= \mathbf{E}^x\left(\mathbf{E}^{X_{R^x}}\left(\sum_{n \geq 0} f \circ X_n\right) ; R_x < \infty\right) \end{aligned}$$

(by the strong Markov property of $X(\mu)$)

$$= \mathbf{E}^x(Kf \circ X_{R^x} ; R^x < \infty),$$

hence the assertion by taking the supremum over all $x \in G$.

For each pair $(x, B) \in G \times \mathfrak{B}(G)$ we set

$$\begin{aligned} K(x, B) :&= K 1_B(x) \\ &= \kappa * \varepsilon_x(B), \end{aligned}$$

where

$$\kappa := \sum_{n \geq 0} \mu^n$$

is the elementary kernel determined by $\mu \in M^1(G)$ in the sense of Example 6.1.3.

Theorem 6.3.7 *Let $X(\mu)$ be an adapted random walk in G with law $\mu \in M^1(G)$.*

Then

(i) $X(\mu)$ is recurrent if and only if $\kappa(U) = \infty$ for every non-empty open subset U of G.

(ii) $X(\mu)$ is transient if and only if $\kappa(U) < \infty$ for every relatively compact open subset U of G.

Proof. It suffices to show that $X(\mu)$ is recurrent provided $\kappa(U) = \infty$ for each relatively compact open neighborhood U from a neighborhood base of 0.

We are retaining the notation of the proof Theorem 6.3.5.

First of all we justify that without loss of generality we can choose U to be a symmetric relatively compact neighborhood $\in \mathfrak{V}_G(0)$. In fact, let

$$\kappa_N := \sum_{n=0}^{N} \mu^n \text{ for } N \geq 1.$$

We have

$$\begin{aligned}\kappa(U) &= \mathbf{E}^0\left(\sum_{n=0}^{N} 1_U \circ X_n\right)\\ &= 1 + \mathbf{E}^0\left(\sum_{n=R_U}^{N} 1_U \circ X_n\right),\end{aligned}$$

hence

$$\begin{aligned}\kappa_N(U) &\leq 1 + \mathbf{E}^0\left(\sum_{n=0}^{N} 1_U \circ X_n \circ \theta_{R_U}\right)\\ &= 1 + \mathbf{E}^0\left(\mathbf{E}^{X_{R_U}}\left(\sum_{n \geq 0} 1_U \circ X_n\right); R_U < \infty\right),\end{aligned}$$

which follows from the strong Markov property of $X(\mu)$, and by Property 6.3.6.4 we obtain that

$$\begin{aligned}\kappa_N(U) &\leq \mathbf{E}^0(\kappa_N(U - X_{R_U}); R_U < \infty) \\ &\leq \kappa_N(U - U),\end{aligned}$$

where $U - U$ is a symmetric relatively compact open neighborhood $\in \mathfrak{V}_G(0)$. Letting $N \to \infty$, one deduces that

$$\kappa(U - U) = \infty\,.$$

Now let U be a symmetric relatively compact open neighborhood in $\mathfrak{V}_G(0)$ such that

$$\kappa(U) = \infty\,.$$

One has

$$\begin{aligned}1 &\geq \mathbf{P}^0([T < \infty]) \\ &= \sum_{m\geq 0} \mathbf{P}^0\left([X_m \in U] \cap \bigcap_{n\geq 1} [X_{m+n} \notin U]\right)\end{aligned}$$

and, since $y - x \notin 2U$ implies $y \notin U$ (for $x \in U$),

$$1 \geq \sum_{m\geq 0} \mathbf{P}^0\left([X_m \in U] \cap \bigcap_{n\geq 1} [X_{m+n} - X_m \notin 2U]\right).$$

It follows that

$$1 \geq \kappa(U)\ \mathbf{P}^0\,[R_{2U} = \infty]\,,$$

and since $\kappa(U) = \infty$ by assumption, that

$$\mathbf{P}^0([R_{2U} = \infty]) = 0,$$

hence that

$$\mathbf{P}^0([R_{2U} < \infty]) = 1,$$

which by Theorem 6.3.5 completes the proof. ■

Corollary 6.3.8 *If $X(\mu)$ is transient, then*

(i) K is a proper kernel in the sense that $G = \bigcup_{n\geq 1} G_n$ for an increasing sequence $(G_n)_{n\geq 1}$ in G such that $K(\cdot, G_n)$ is a bounded function for all $n \geq 1$.

(ii) $K(x, \cdot) \in M_+(G)$ for all $x \in G$.

(iii) The set $\{K(x, \cdot) : x \in G\}$ is τ_v-relatively compact in $M_+(G)$.

Proof. From the theorem we infer that there exists an open $U \in \mathfrak{V}_G(0)$ such that $\kappa(U) < \infty$. Let V be an open neighborhood $\in \mathfrak{V}_G(0)$ satisfying $V - V \subset U$. Then we obtain for each $x \in V$ that

$$K(x, V) = \kappa(V - x) \leq \kappa(V - V) \leq \kappa(U)$$

and by the maximum principle 6.3.6.1 that

$$K(x, V) \leq \kappa(U)$$

for all $x \in G$. For every compact subset C of G there is a sequence $\{x_1, \ldots, x_n\}$ in C such that the sequence $\{V - x_i : i = 1, \ldots, n\}$ forms an open covering of C. But then

$$\begin{aligned} K(x, C) &\leq \sum_{i=1}^{n} K(x, V - x_i) \\ &= \sum_{i=1}^{n} K(x + x_i, V) \\ &\leq n\kappa(V) \end{aligned}$$

for all $x \in G$ which implies all the assertions *(i)* to *(iii)* at one time. ■

First **Examples 6.3.9** *of recurrent random walks*

6.3.9.1 *If G is compact, every (adapted) random walk $X(\mu)$ in G is recurrent.*

In fact, the potential kernel (measure) κ of $X(\mu)$ satisfies

$$\kappa(G) = \mathbf{E}^0(\sum_{n\geq 0} 1_G \circ X_n) = \infty$$

for the (relatively) compact open (sub)set G.

6.3.9.2 *For the* **Bernoulli walk** $X(\mu)$ *in* $\mathbf{Z}$ *with determining measure*

$$\mu := \frac{1}{2}\varepsilon_1 + \frac{1}{2}\varepsilon_{-1} \in M^1(\mathbf{Z})$$

(arising as the common law of independent $\mathbf{Z}$*-random variables* $Y_k (k \geq 1)$ *leading to partial sums* $X_n := \sum_{k=1}^n Y_k$ $(n \geq i)$*) one computes*

$$\kappa(\{0\}) = \sum_{n\geq 0} \mu^n(\{0\}) = \sum_{n\geq 0} \mathbf{P}([X_{2n} = 0]) = \sum_{n\geq 0} \binom{2n}{n} \frac{1}{2^{2n}},$$

where

$$\binom{2n}{n} \frac{1}{2^{2n}} = \frac{2n!}{n!n!} \cdot \frac{1}{2^{2n}} \sim \frac{1}{\sqrt{\pi n}}$$

(by Stirling's formula), and obtain that $\kappa(\{0\}) = \infty$ *which says that* $\mathbf{Z}$ *is a recurrent group.*

We note (and shall show later) that the random walk $X(\mu)$ *in* $\mathbf{Z}$ *with determining measure*

$$\mu := p\varepsilon_1 + q\varepsilon_{-1},$$

where $p, q \geq 0, p + q = 1$ *but* $p \neq q \neq \frac{1}{2}$*, is transient.*

6.3.9.3 *Considering the random walk* $X(\mu)$ *in* $\mathbf{Z}^2$ *with determining measure*

$$\mu := \frac{1}{4}(\varepsilon_{(1,0)} + \varepsilon_{(-1,0)} + \varepsilon_{(0,1)} + \varepsilon_{(0,-1)}) \in M^1(\mathbf{Z}^2)$$

one obtains that

$$\begin{aligned}\kappa(\{(0,0)\}) &= \sum_{n\geq 0} \mathbf{P}([X_{2n} = (0,0)]) \\ &= \sum_{n\geq 0} \sum_{i+j=n} \frac{2n!}{i!i!j!j!}\left(\frac{1}{4}\right)^{2n},\end{aligned}$$

where the inner sums

$$\begin{aligned}u_n &= \left(\frac{1}{4}\right)^{2n} \binom{2n}{n} \sum_{i+j=n} \binom{n}{i}\binom{n}{n-i} \\ &= \left(\frac{1}{4}\right)^{2n} \binom{2n}{n}^2\end{aligned}$$

are asymptotically equal to $\frac{1}{\pi n}$ *as* $n \to \infty$, *hence that* $\kappa(\{(0,0)\}) = \infty$. *Consequently* $\mathbf{Z}^2$ *is recurrent.*

We are now going to describe the class $\mathcal{R}$ of all recurrent (locally compact Abelian) groups. By a method similar to that applied in the proof of the Chung - Fuchs criterion 6.2.2 we shall at first characterize transient random walks in G.

Theorem 6.3.10 *For any random walk* $X(\mu)$ *with law* $\mu \in M^1(G)$ *the following statements are equivalent:*

(i) $X(\mu)$ *is transient.*

(ii) There exists a neighborhood $W \in \mathfrak{V}_{G^\wedge}(0)$ *such that*

$$\limsup_{t\uparrow 1} \int_W \operatorname{Re} \frac{1}{1-t\hat{\mu}} d\omega_{G^\wedge} < \infty .$$

Proof. *(ii)* $\Rightarrow$ *(i).* For any compact symmetric neighborhood $V \in$

$\mathfrak{V}_{G^\wedge}(0)$ we have

$$\Big(\int_V \overline{\chi(x)}\omega_{G^\wedge}(d\chi)\Big)^2$$
$$= \int_{G^\wedge} \overline{\chi(x)} \left(\int 1_V(\varrho)1_V(\chi-\varrho)\omega_{G^\wedge}(d\varrho)\right) \omega_{G^\wedge}(d\chi)$$
$$= \int_{G^\wedge} \overline{\chi(x)}\omega_{G^\wedge}(V\cap(V+\chi))\omega_{G^\wedge}(d\chi),$$

whenever $x \in G$. Integration with respect to μ yields

$$\int_{G^\wedge} \hat{\mu}(\chi)\omega_{G^\wedge}(V\cap(V+\chi))\omega_{G^\wedge}(d\chi)$$
$$= \int_G \left(\int_V \overline{\chi(x)}\omega_{G^\wedge}(d\chi)\right)^2 \mu(d\chi).$$

This identity holding true for the μ^n instead of μ, we may multiply by t^n for $t \in [0,1[$ and sum over n in order to obtain

$$\int_{G^\wedge} \frac{1}{1-t\hat{\mu}(\chi)}\omega_{G^\wedge}(V\cap(V+\chi))\omega_{G^\wedge}(d\chi)$$
$$= \sum_{n\geq 0} t^n \int_G \left(\int_V \overline{\chi(x)}\omega_{G^\wedge}(d\chi)\right)^2 \mu^n(d\chi)\,.$$

Now let W be a symmetric compact neighborhood $\in \mathfrak{V}_{G^\wedge}(0)$ and choose $V \in \mathfrak{V}_{G^\wedge}(0)$ such that $2V \subset W$. The function

$$\chi \mapsto \omega_{G^\wedge}(V\cap(V+\chi))$$

on $G^\wedge$ is zero on W^c, hence the left-hand side of the above equality is a real number

$$\leq \omega_{G^\wedge}(V)\int_W \operatorname{Re}\frac{1}{1-t\hat{\mu}}d\omega_{G^\wedge}\,.$$

Since the function

$$x \mapsto \Big(\int_V \overline{\chi}(x)\, d\omega_{G^\wedge}\Big)^2$$

on G is continuous and takes the value $\omega_{G^\wedge}(V)^2$ at $x = 0$, it is $\geq \frac{1}{2}(\omega_{G^\wedge}(V))^2$ on some neighborhood $U \in \mathfrak{V}_G(0)$. As a consequence we obtain

$$2\int_W \operatorname{Re} \frac{1}{1 - t\hat{\mu}} d\omega_{G^\wedge} \geq \omega_{G^\wedge}(V) \sum_{n\geq 0} t^n \mathbf{P}^0([X_n \in U])\,.$$

If, now, $X(\mu)$ is recurrent, then

$$\lim_{t\uparrow 1} \int_W \operatorname{Re} \frac{1}{1 - t\hat{\mu}} d\omega_{G^\wedge} = \infty\,,$$

and the implication *(i)* $\Rightarrow$ *(ii)* has been proved.

(i) $\Rightarrow$ *(ii)*. Let $X(\mu)$ be transient, and let V be a compact symmetric neighborhood $\in \mathfrak{V}_G(0)$. The function

$$y \longmapsto (1_V * 1_V * \mu)(y) = \int_G \omega_G((V + y) \cap (V + x))\mu(dx)$$

is an element of $CPD(G) \cap L^1(G, \omega_G)$, hence it is the inverse Fourier transform of the function

$$\chi \longmapsto \hat{\mu}(\chi)\Big(\int_V \overline{\chi}\, d\omega_G\Big)^2$$

and consequently

$$\begin{aligned}\int \hat{\mu}(\chi)\Big(\int_V \overline{\chi}\, d\omega_G\Big)^2 \omega_{G^\wedge}(d\chi) &= (1_V * 1_V * \mu)(0)\\ &= \int_G \omega_G(V \cap (V + x))\mu(dx) \leq \omega_G(V)\mu(2V).\end{aligned}$$

As in the first part of this proof we look at this inequality with μ^n instead of μ, multiply on both sides by t^n for $t \in [0,1[$ and sum over n. It follows that

$$\int_{G^\wedge} \left(\int_V \overline{\chi}\, d\omega_G \right)^2 \frac{1}{1 - t\hat{\mu}(\chi)} \omega_{G^\wedge}(d\chi) \leq \omega_G(V) \sum_{n \geq 0} t^n\, \mathbf{P}^e([X_n \in 2V]) .$$

Now we choose a compact symmetric neighborhood $W \in \mathfrak{V}_{G^\wedge}(0)$ and the neighborhood V such that $\kappa(2V) < \infty$ and Re $\chi(x) \geq \frac{1}{2}$ for all $(x, \chi) \in V \times W$. We obtain that

$$\omega_G(V) \int_W \operatorname{Re} \frac{1}{1 - t\hat{\mu}}\, d\omega_{G^\wedge} \leq 4 \sum_{n \geq 0} t^n\, \mathbf{P}^e([X_n \in 2V])$$

hence that

$$\limsup_{t \uparrow 1} \int_W \operatorname{Re} \frac{1}{1 - t\hat{\mu}} d\omega_{G^\wedge} < \infty$$

which implies the assertion. ■

Corollary 6.3.11 *If $X(\mu)$ is transient, then for every compact neighborhood $W \in \mathfrak{V}_{G^\wedge}(0)$ the function*

$$\operatorname{Re} \frac{1}{1 - \hat{\mu}}$$

is integrable over W.

Proof. For all $t \in [0,1[$ and $\chi \in G^\wedge$ we have that

$$\operatorname{Re} \frac{1}{1 - t\hat{\mu}} \geq 0$$

and that

$$\lim_{t \uparrow 1} \operatorname{Re} \frac{1}{1 - t\hat{\mu}} = \operatorname{Re} \frac{1}{1 - \hat{\mu}} .$$

By the Fatou Lemma together with the theorem we obtain

$$\int_W \mathrm{Re}\, \frac{1}{1-\hat{\mu}} d\omega_{G^\wedge} \leq \liminf_{t \uparrow 1} \int_W \mathrm{Re}\, \frac{1}{1-t\hat{\mu}} d\omega_{G^\wedge} < \infty$$

at first for the neighborhood W constructed in the proof of the theorem, but since

$$\mathrm{Re}\, \frac{1}{1-\hat{\mu}}$$

is continuous and $\neq 0$ on $G^\wedge \backslash \{0\}$, also for every compact neighborhood $W \in \mathfrak{V}_{G^\wedge}(0)$.

Remark 6.3.12 *The Corollary implies that every compact group G is recurrent (Compare Example 6.3.9.1).*

In fact, if G is compact, then $G^\wedge$ is discrete, so that $\omega_{G^\wedge}(\{0\}) > 0$. But this contradicts the $\omega_{G^\wedge}$-integrability of $\mathrm{Re}\, \frac{1}{1-\hat{\mu}}$ over any compact neighborhood $\in \mathfrak{V}_{G^\wedge}(0)$.

Applying the characterization theorem 6.3.10 we can now decide on recurrence or transience of random walks on a significant class of locally compact Abelian groups.

Theorem 6.3.13 *Let $X(\mu)$ be a random walk with law $\mu \in M^1(G)$, where G has the form*

$$G = \mathbf{R}^d \times \mathbf{Z}^e$$

for $d, e \geq 0$. Then

(i) if $d + e = 1$, i.e. $G = \mathbf{R}$ or $G = \mathbf{Z}$, and if

$$\int |x| \mu(dx) < \infty$$

as well as

$$\int x \mu(dx) = 0,$$

then $X(\mu)$ is recurrent.

(ii) If $d + e = 2$ and

$$\int |x|^2 \mu(dx) < \infty$$

as well as

$$\int x\mu(dx) = 0,$$

then $X(\mu)$ is recurrent.

(iii) If $d + e \geq 3$, then $X(\mu)$ is transient.

Proof. *(i)* One easily verifies the estimates

$$\mathrm{Re}\,\frac{1}{1 - t\hat{\mu}} \geq \frac{1-t}{(\mathrm{Re}\,(1 - t\hat{\mu}))^2 + t^2(\mathrm{Im}\,\hat{\mu})^2}$$

and

$$\begin{aligned}(\mathrm{Re}\,(1 - t\hat{\mu}))^2 &= ((1-t) + \mathrm{Re}\,(t(1-\hat{\mu})))^2 \\ &\leq 2(1-t)^2 + 2t^2(\mathrm{Re}\,(1-\hat{\mu}))^2\end{aligned}$$

valid for all $t \in [0, 1[$. By the assumption on the first moment of μ we see that $\hat{\mu}$ is continuously differentiable and that $\hat{\mu}'(0) = 0$. Now we apply the Taylor expansion and obtain the existence of an $\alpha > 0$ such that for $y \in \mathbf{R}^\wedge \cong \mathbf{R}$ or $y \in \mathbf{Z}^\wedge \cong \mathbf{T} \cong]-\pi, \pi] \subset \mathbf{R}$ with $|y| < \alpha$ we have

$$|\mathrm{Im}\,\hat{\mu}(y)| \leq \varepsilon |y|$$

as well as

$$\mathrm{Re}\,(1 - \hat{\mu}(y)) \leq \varepsilon |y|\,.$$

It follows that

$$\begin{aligned}\int_{-\alpha}^{\alpha} \mathrm{Re}\,\frac{1}{1 - t\hat{\mu}(y)}\lambda(dy) &\geq (1-t)\int_{-\alpha}^{\alpha} \frac{1}{2(1-t)^2 + 3t^2\varepsilon^2 y^2}\lambda(dy) \\ &\geq \frac{1}{3}\int_{-\frac{\alpha}{1-t}}^{\frac{\alpha}{1-t}} \frac{1}{1 + \varepsilon^2 y^2}\lambda(dy),\end{aligned}$$

hence that

$$\lim_{t \uparrow 1} \int_{-\alpha}^{\alpha} \mathrm{Re}\,\frac{1}{1 - t\hat{\mu}(y)}\lambda(dy) \geq \frac{\pi}{3\varepsilon}\,.$$

Now suppose that $X(\mu)$ is transient. Then there exists an $\alpha_0 > 0$ such that

$$\mathrm{Re}\,\frac{1}{1 - t\hat{\mu}(y)} > 0$$

for all $y \in [-\alpha_0, \alpha_0]$ and

$$\lim_{t \uparrow 1} \int_{-\alpha_0}^{\alpha_0} \mathrm{Re}\, \frac{1}{1 - t\hat{\mu}(y)} \lambda(dy) =: M < \infty$$

For $\alpha < \alpha_0$, however,

$$\lim_{t \uparrow 1} \int_{-\alpha}^{\alpha} \mathrm{Re}\, \frac{1}{1 - t\hat{\mu}(y)} \lambda(dy) \leq M < \infty,$$

but this contradicts the fact that ε was chosen arbitrarily.

(ii) is proved similarly to *(i)* by embedding $G^\wedge = \mathbf{R}^d \times \mathbf{T}^e$ into $\mathbf{R}^{d+e}$ with $d + e = 2$.

For the proof of

(iii) we embed $G^\wedge = \mathbf{R}^d \times \mathbf{T}^e$ into $\mathbf{R}^{d+e}$ with $d + e \geq 3$ and infer from part 1. of the proof of the transience criterion 6.2.1 that there exist a relatively compact neighborhood $U \in \mathfrak{V}_{G^\wedge}(0)$ and a constant $c > 0$ satisfying $\mathrm{Re}\, \hat{\mu}(y) \geq 0$ and

$$\mathrm{Re}\,(1 - \hat{\mu}(y)) \geq c\|y\|^2$$

for all $y \in U$ But then for $t < 1$ we obtain

$$\begin{aligned} \int_U \mathrm{Re}\, \frac{1}{1 - t\hat{\mu}(y)} \lambda^{d+e}(dy) &\leq \int_U \mathrm{Re}\, \frac{1}{1 - \hat{\mu}(y)} \lambda^{d+e}(dy) \\ &\leq \frac{1}{c} \int_U \frac{1}{\|y\|^2} \lambda^{d+e}(dy) < \infty\,, \end{aligned}$$

provided $d + e \geq 3$ which implies that all random walks in $\mathbf{R}^d \times \mathbf{Z}^e$ with $d + e \geq 3$ are transient. ■

Examples 6.3.14 *of groups in* $\mathcal{R}$

6.3.14.1 *All compact groups belong to* $\mathcal{R}$.

6.3.14.2 *All groups of the form* $G = \mathbf{R}^d \times \mathbf{Z}^e$ *with* $d + e \leq 2$ *are elements of* $\mathcal{R}$.

Properties 6.3.15 *of the class* $\mathcal{R}$

6.3.15.1 *Any open subgroup of a group in* $\mathcal{R}$ *belongs to* $\mathcal{R}$.

In particular,

6.3.15.2 *every subgroup of a discrete group in* $\mathcal{R}$ *belongs to* $\mathcal{R}$.

For the proof of Property 6.3.15.1 we start with an adapted random walk $X(\mu)$ in G and consider the Markov chain $(Z_n)_{n\in\mathbf{N}}$ in an open subgroup H of G given by

$$Z_n := X_{R_H^n}$$

where the return times R_H^n into H are defined recursively by

$$R_H^n = R_H^{n-1} + R_H \circ \theta_{R_H^{n-1}} \quad (n \geq 2)$$

and

$$R_H^1 := R_H .$$

Since $X(\mu)$ is assumed to be recurrent,

$$\mathbf{P}^0([R_H^n < \infty]) = 1$$

for every $n \in \mathbf{N}, (Z_n)_{n\in\mathbf{N}}$ is a well-defined random walk $X(\nu)$ with law

$$\nu := (\mathbf{P}^0)_{R_H} \in M^1(H) .$$

Moreover, we have

$$\limsup_{n\to\infty}[Z_n \in O] = \limsup_{n\to\infty}[X_n \in O]$$

for all open subsets O of H. Therefore, the recurrence of $X(\mu)$ yields

$$\mathbf{P}^0(\limsup_{n\to\infty}[Z_n \in O]) = 1$$

which says that $X(\nu)$ is recurrent in H.

6.3.15.3 *Let* H *be a compact subgroup of* G. *Then* $G \in \mathcal{R}$ *if and only if* $G/H \in \mathcal{R}$.

In order to see this we take $\mu \in M^1(G)$ and let $\dot{\mu}$ denote the image of μ under the canonical homomorphism $\pi : G \to G/H$. Now, for each compact neighborhood $V \in \mathfrak{V}_G(0)$ the sets VH and $\pi(VH)$ are compact neighborhoods in $\mathfrak{V}_G(0)$ and in $\mathfrak{V}_{G/H}(\dot{0})$ respectively. The identity

$$\sum_{n \to 0} \mu^n(VH) = \sum_{n \geq 0} \dot{\mu}^n(\pi(VH))$$

yields the assertion.

Theorem 6.3.16 *(Characterization of the class $\mathcal{R}$) Let G be a second countable locally compact Abelian group which by the structure theorem 4.2.20 is of the form*

$$G = \mathbf{R}^d \times G_1,$$

where G_1 contains a compact open subgroup K such that

$$G/\{0\} \times K \cong \mathbf{R}^d \times G_2$$

with a countable group G_2 of rank r. Then $G \in \mathcal{R}$ if and only if $d + r \leq 2$.

The proof of the theorem relies on

Theorem 6.3.17 *(Dudley) A countable (Abelian) group G belongs to $\mathcal{R}$ if and only if rank $(G) \leq 2$.*

Proof. 1. If rank $(G) > 2$, then there exists a subgroup of G that is isomorphic to the group $\mathbf{Z}^3$ which by Theorem 6.3.13 *(iii)* is not recurrent. On the other hand we infer from Property 6.3.15.2 that $G \notin \mathcal{R}$.

2. Conversely, we assume that rank $(G) \leq 2$. There exists a sequence $(a_n)_{n \in \mathbf{N}}$ generating G and such that a_{m+1} does not belong to the subgroup G_m generated by $\{a_1, \ldots, a_m\}$ $(m \geq 1)$. From the structure of finitely generated Abelian groups proved in the book [19] by Hewitt and Ross we infer that for every $m \geq 1$ the group G_m is of the form $\mathbf{Z}^2 \times F$ or $\mathbf{Z} \times F$ or F, where F denotes a finite group.

Applying Theorem 6.3.13 and Property 6.3.15.3 this implies that all random walks in G_m are recurrent ($m \geq 1$).

Now we define a sequence $(\mu^{(m)})_{m\geq 1}$ of measures $\mu^{(m)} \in M^1(G_m)$ by

$$\mu^{(1)}(a_1) = \mu^{(1)}(-a_1) = \frac{1}{2},$$
$$\mu^{(m)}(x) = (1 - q_m)\mu^{(m-1)}(x)$$

for $m \geq 2$ and all $x \in G_{m-1}$, and

$$\mu^{(m)}(a_m) = \mu^{(m)}(-a_m) = \frac{1}{2}q_m$$

(with a proper choice of q_m). To every $x \in G$ there exists an $m \geq 1$ such that $x \in G_m$. The number

$$\mu(x) := \prod_{i\geq m+1} (1 - q_i)\mu^{(m)}(x)$$

is independent of m. If the product $\prod_{i\geq 2}(1 - q_i)$ converges, then

$$\sum_{x\in G} \mu(g) = \lim_{m\to\infty} \sum_{x\in G_m} \mu(g)$$
$$= \lim_{m\to\infty} \prod_{i\geq m+1} (1 - q_i) = 1,$$

hence $\mu \in M^1(G)$. Plainly μ is adapted. Our aim will be to show that the number q_i can be chosen such that the random walk $X(\mu)$ with law μ is recurrent.

For every $m \geq 1$ let $X^{(m)}$ be the random walk with law μ_m defined by the canonical probability measure $\mathbf{P}^{\nu}_{(m)}$. We choose once for all a sequence $(r_n)_{n\in\mathbf{N}}$ in $]0, 1[$ such that

$$\prod_{n\geq 1}(1 - r_n) < \infty\,.$$

Next we choose a number $l_1 \in \mathbf{N}$ and a sequence $(\alpha_j^{(1)})_{j\in\mathbf{N}}$ in $]0,1[$ such that

$$\prod_{j\geq 1}(1-\alpha_j^{(1)})^{l_1} \sum_{k=1}^{l_1} \mathbf{P}^0_{(1)}([X_k^{(1)}=0]) > 1\,.$$

For $q_2 := r_1 \wedge \alpha_1^{(1)}$ the random walk $X^{(2)}$ in G_2 leads to

$$\prod_{j\geq 1}(1-\alpha_j^{(2)})^{l_2} \sum_{k=l_1+1}^{l_2} \mathbf{P}^0_{(2)}([X_k^{(2)}=0]) > 1$$

with the proper choice of l_2 and $(\alpha_j^{(2)})_{j\in\mathbf{N}}$. With $q_3 := r_2 \wedge \alpha_2^{(1)} \wedge \alpha_2^{(2)}$ we obtain the random walk $X^{(3)}$ in G_3 satisfying an analogous inequality. The inductive process continues and ends up with the inequality

$$\prod_{j\geq 1}(1-\alpha_j^{(n)})^{l_n} \sum_{k=l_{n-1}+1}^{l_n} \mathbf{P}^0_{(n-1)}([X_k^{(n-1)}=0]) > 1 \qquad (2)$$

together with the definition

$$q_n := r_{n-1} \wedge \alpha_{n-1}^{(1)} \wedge \alpha_{n-1}^{(2)} \wedge \ldots \wedge \alpha_{n-1}^{(n-1)}\,.$$

The choices taken are justified by the recurrence of the random walks $X^{(n)}$.

It is clear that
$$\prod_{i\geq 1}(1-q_i) < \infty\,.$$

For the potential kernel κ of $X(\mu)$ we have that

$$\begin{aligned}
\kappa(e) &\geq \sum_{k\geq 1} \mathbf{P}^0([X_k=0]) \\
&= \sum_{n\geq 1} \sum_{k=l_{n-1}+1}^{l_n} \mathbf{P}^0([X_k=0]) \\
&\geq \sum_{n\geq 1} \sum_{k=l_{n-1}+1}^{l_n} \prod_{i\geq n+1} (1-q_i)^k\, \mathbf{P}^0_{(n-1)}([X_k^{(n-1)}=0]),
\end{aligned}$$

where the k-th entry in the finite sum can be interpreted as the probability for $X(\mu)$ to be in $0 \in G$ at time k without having left G_{n-1}.

Moreover, by (2) we obtain that

$$\kappa(e) \geq \sum_{n\geq 1} \sum_{k=l_{n-1}+1}^{l_n} \prod_{i\geq n+1} (l - q_i)^{l_n} \ \mathbf{P}^0_{(n-1)}([X_k^{(n)} = 0])$$

$$\geq \sum_{n\geq 1} \sum_{k=l_{n-1}+1}^{l_n} \prod_{i\geq 1} (1 - \alpha_i^{(n)})^{l_n} \ \mathbf{P}^0_{(n-1)}([X_k^{(n)} = 0]) = \infty,$$

thus $X(\mu)$ is recurrent, and $G \in \mathcal{R}$. ■

The **Proof of Theorem 6.3.16** follows from Property 6.3.15.3 together with the equivalence that $\mathbf{R}^d \times G_2$ is recurrent if and only if $d + r \leq 2$, where for both implications Dudley's theorem 6.3.17 is applied. ■

Remark 6.3.18 *If in the theorem 6.3.16 $d + e \leq 2$, then G contains the dense subgroup $\mathbf{Q}^d \times G$, which belongs to $\mathcal{R}$. Consequently we have further*

Examples 6.3.19 *of groups in $\mathcal{R}$*

6.3.19.1 $\mathbf{Q}^2 \times K$ *with a compact group K belongs to $\mathcal{R}$.*

6.3.19.2 *The group $\mathbf{Q}_p$ of p-adic numbers belongs to $\mathcal{R}$.*

6.4 Classification of transient random walks

In order to study the asymptotic behavior of random walks with law on a locally compact Abelian group we need to modify their canonical construction by compactifying G in the sense of Alexandrov.

Considering a transition kernel P on $(G, \mathfrak{B}(G))$ and picking a point $\triangle$ not in G we extend P to $(G_\triangle, \mathfrak{B}(G_\triangle))$, where $G_\triangle := G \cup \{\triangle\}$ and $\mathfrak{B}(G_\triangle) := \sigma(\mathfrak{B}(G) \cup \{\triangle\})$, by

$$P(x, \{\triangle\}) := \begin{cases} 1 - P(x, G) & \text{if } x \neq \triangle \\ 1 & \text{if } x = \triangle \, . \end{cases}$$

In this situation the canonical (product) measurable space will be $(\Omega, \mathfrak{A})$ with

$$\Omega := G_\triangle^{\mathbf{Z}_+} \quad \text{and}$$

$$\mathfrak{A} := \mathfrak{B}(G_\triangle)^{\otimes \mathbf{Z}_+} \, .$$

Moreover, the sequence $(X_n)_{n \in \mathbf{Z}_+}$ of projections $X_n : G_\triangle^{\mathbf{Z}_+} \to G_\triangle$ will be supplemented by

$$X_\infty(\omega) := \triangle$$

for all $\omega \in \Omega$. So the measures $\mathbf{P}^\nu$ for $\nu \in M^1(G)$ governing the Markov chain $(X_n)_{n \in \mathbf{Z}_+}$ with transition kernel P will be interpreted as measures on the (enlarged) measurable space $(\Omega, \mathfrak{A})$.

We start the discussion by proving *some general renewal results* for transient random walks $X(\mu)$ with law $\mu \in M^1(G)$ admitting a potential kernel K.

Proposition 6.4.1 *Let $\mathcal{F}'_K$ denote the set of accumulation points of*

$$\mathcal{F}_K := \{K(x, \cdot) : x \in G\}$$

obtained as $x \to \triangle$. Then

$$\mathcal{F}'_K \subset \{c\, \omega_G : c \geq 0\} \, .$$

Proof. From Corollary 6.3.8 we infer that $\mathcal{F}_K$ is τ_v-relatively compact. Let $(x_n)_{n\in\mathbf{N}}$ be a sequence in G with $x_n \longrightarrow \triangle$ and

$$\tau_v - \lim_{n\to\infty} K(x_n, \cdot) = \nu \in M_+(G).$$

Since

$$\tau_v = \lim_{n\to\infty} \varepsilon_{x_n} = 0,$$

the equality

$$\varepsilon_{x_n} * \kappa = \varepsilon_{x_n} + \varepsilon_{x_n} * \kappa * \mu$$

valid for all $n \in \mathbf{N}$, leads as $n \to \infty$ to

$$\nu = \nu * \mu.$$

Now, the Choquet-Deny theorem 6.1.12 applies (to the adapted random walk with law $\mu \in M^1(G)$), such that ν is μ-invariant and hence a Haar measure of the form $c\,\omega_G$ with $c \geq 0$. ∎

Proposition 6.4.2 *Let $f \in C_+^c(G)$. Then*
(i) Kf is uniformly continuous.

(ii)
$$\lim_{x\to\triangle} (Kf(x+y) - Kf(x)) = 0$$

uniformly for all y from any compact subset C of G.

(iii)
$$\liminf_{x\to\triangle} Kf(x) = 0.$$

(iv)
$$\lim_{x\to\triangle} Kf(x)Kf(-x) = 0.$$

Since (i) implies that the mapping $x \mapsto K(x,\cdot)$ is continuous, hence determining the set $\mathcal{F}'_K$ is equivalent to determining the closure $\overline{\mathcal{F}}_K$ of $\mathcal{F}_K$ in $M_+(G)$ (with respect to τ_v).

Proof. *(i)* From the τ_v-relative compactness of $\mathcal{F}_K$ we infer that for every compact subset C of G there exists a constant $M_C > 0$ such that

$$K(x, C) \leq M_C$$

whenever $x \in G$. Let $H := \operatorname{supp}(f)$ and let V be a compact neighborhood in $\mathfrak{V}_G(0)$. Since f is uniformly continuous, for any $\varepsilon > 0$ we can choose a compact $V_1 \in \mathfrak{V}(0), V_1 \subset V$ such that

$$|f(x+y) - f(x)| < \varepsilon$$

for all $y \in V_1$ and all $x \in G$. For $y \in V_1$ the function

$$x \mapsto g(x) := f(x+y) - f(x)$$

vanishes outside the compact set $C := H + V$, and

$$Kg(x) = Kf(x+y) - Kf(x)$$

whenever $x \in G$. Now, from the boundedness of K we deduce that

$$|Kf(x+y) - Kf(x)| \leq \varepsilon M_C$$

for all $x \in G, y \in V$, and this implies the assertion.
(ii) Proposition 6.4.1 tells us that every sequence in G converging to $\triangle$ (as $n \to \infty$) contains a subsequence $(x_n)_{n \in \mathbf{N}}$ such that

$$\tau_v - \lim_{n \to \infty} K(x_n, \cdot) = c\,\omega_G$$

for some $c \geq 0$. But then for $y \in G$

$$\lim_{n \to \infty} (Kf(x_n + y) - Kf(x_n)) = c \left(\int f d(\omega_G * \varepsilon_y) - \int f d\omega_G \right) = 0$$

which shows that this limit exists. For every $\varepsilon > 0$ there exists a compact subset C_y of G satisfying

$$|Kf(x+y) - Kf(x)| < \frac{\varepsilon}{2}$$

whenever $x \in {C_y}^c$.

Since by *(i)* Kf is uniformly continuous, for each $\varepsilon > 0$, every $y \in G$ has a neighborhood $U_y \in \mathfrak{V}_G(y)$ such that for $z \in U_y$ we have

$$|Kf(x+z) - Kf(y+z)| < \frac{\varepsilon}{2}$$

uniformly in $x \in G$. But then

$$|Kf(x+z) - Kf(x)| < \varepsilon$$

whenever $z \in U_y$ and $x \in {C_y}^c$. Covering the given compact set C by finitely many of the sets C_y for $y \in G$ yields the assertion.

(iii) With the abbreviation

$$a := \inf_{x \in G} Kf(x)$$

we have that

$$a \leq P^n Kf(x)$$

for every $n \geq 1$ and all $x \in G$. Letting n tend to ∞ this yields $a = 0$. Now let Kf be strictly positive. Then

$$\liminf_{x \to \triangle} Kf(x) = 0\,.$$

If Kf is not strictly positive, then there exists $x_0 \in G$ such that $Kf(x_0) = 0$, hence that

$$P^n Kf(x_0) = 0$$

for every $n \in \mathbf{N}$. But for any compact subset C of G there exists an $n \in \mathbf{Z}_+$ such that

$$P^n(x_0, C^c) > 0,$$

hence there is an $x \in C^c$ with $Kf(x) = 0$. Thus, also in this case we arrive at the assertion.

(iv) Fixing $\varepsilon > 0$ we obtain from *(ii)* that there is a compact subset C_ε of G such that for $x \in {C_\varepsilon}^c$ and for every $y \in D := \operatorname{supp}(f)$ we have

$$Kf(x) \leq Kf(x+y) + \varepsilon,$$

hence

$$\frac{Kf(x)Kf(y)}{\|Kf\|} \leq Kf(x+y) + \varepsilon\,. \tag{1}$$

By the maximum principle 6.3.6.1 we obtain the validity of (1) for all $x \in {C_\varepsilon}^c$ and all $y \in G$.

Now *(iii)* provides us with a point $y_\varepsilon \in D$ such that

$$Kf(y_\varepsilon) \leq \varepsilon .$$

Replacing y by $-x + y_\varepsilon$ in (1) this yields

$$Kf(x)Kf(-x + y_\varepsilon) \leq 2\|Kf\|\varepsilon .$$

But

$$\lim_{x \to \triangle} Kf(-x) - Kf(-x + y_\varepsilon) = 0,$$

thus *(iv)* has been established. ■

Theorem 6.4.3 *$\mathcal{F}'_K \ni 0$, and there exists at most one measure $\mu \in \mathcal{F}'_K, \mu \neq 0$ which necessarily has the form*

$$\mu = c\, \omega_G$$

for some $c > 0$.

Proof. In view of Proposition 6.4.1 we have

$$\mathcal{F}'_K \subset \{c\omega_G : c \geq 0\} .$$

Let now $(x_n)_{n \in \mathbf{N}}$ be a sequence in G with $x_n \to \triangle$ such that

$$\tau_v - \lim_{n \to \infty} K(x_n, \cdot\,) = \nu \neq 0 .$$

Then, by Proposition 6.4.2 *(iv)*,

$$\tau_v - \lim_{n \to \infty} K(x_n, \cdot\,) = 0 .$$

This takes care of the first statement of the theorem.

For the proof of the second statement we assume that there exist two non-zero accumulation points in $\mathcal{F}'_K$. Then there exists a function $f \in C^c_+(G)$ with $D := \operatorname{supp}(f)$ such that the set

$$\{Kf(x) : x \in G\}$$

admits two limit points $0 < l < m$ as $x \to \Delta$. Fixing $\varepsilon > 0$ we now construct inductively a sequence $(x_n)_{n \in \mathbf{N}}$ in G such that the support of the function

$$x \mapsto g(x) := f(x) + f(x + x_1) + \ldots + f(x + x_n)$$

is contained in $\bigcup_{i=0}^{n} (-x_i + K)$.

In fact, the construction starts with $x_0 := 0$. Then, by Proposition 6.4.2 *(ii)* and *(iv)* we choose $x_1 \in G$ such that

$$Kf(x_1) < l + \frac{\varepsilon}{2^2},$$

$$Kf(-x_1) < \frac{\varepsilon}{2^2},$$

$$|Kf(x_1) - Kf(x_1 + x)| < \frac{\varepsilon}{2^2}$$

and

$$|Kf(-x_1) - Kf(-x_1 + x)| < \frac{\varepsilon}{2^2}$$

the latter two inequalities holding true for all $x \in D$. Since the sets $D \cup (-x_1 + D)$ and $D \cup (x_1 + D)$ are compact we can choose $x_2 \in G$ satisfying analogous inequalities with x_1 replaced by x_2, 2^2 by 2^3 and D by $D \cup (-x_1 + D)$ and $D \cup (x_1 + D)$ respectively, and so on until we arrive at an $x_n \in G$ satisfying the inequalities

$$Kf(x_n) < l + \frac{\varepsilon}{2^{n+1}},$$

$$Kf(-x_n) < \frac{\varepsilon}{2^{n+1}},$$

$$Kf(x_n) - Kf(x_n + x)| < \frac{\varepsilon}{2^{n+1}}$$

and

$$|Kf(-x_n) - Kf(-x_n + x)| < \frac{\varepsilon}{2^{n+1}},$$

the latter two inequalities being valid for all $x \in \bigcup_{i=0}^{n-1} (-x_i + D)$ or $x \in \bigcup_{i=0}^{n-1} (x_i + D)$ respectively. Clearly,

$$\operatorname{supp}(g) \subset \bigcup_{i=0}^{n} (-x_i + D).$$

Now, let $x \in x_p + D$ for $1 \leq p \leq n$. Then

$$Kf(x + x_p) \leq \|Kf\|,$$

$$Kf(x + x_{p+i}) = Kf(x_{p+i}) + Kf(x + x_{p+i}) - Kf(x_{p+i})$$

$$\leq l + \frac{\varepsilon}{2^{(p+i)+1}} + \frac{\varepsilon}{2^{(p+i)+1}}$$

whenever $i = 1, \ldots, n - p$, and

$$Kf(x + x_{p-i}) = Kf(z - x_p),$$

whenever $z = x_p + x + x_{p-i} \in x_{p-i} + D$ and $i = 1, \ldots, p$, so that

$$Kf(-x_p) + Kf(z - x_p) - Kf(-x_p) \leq \frac{\varepsilon}{2^{p+1}} + \frac{\varepsilon}{2^{p+1}}.$$

Forming the sum over all these inequalities we obtain that

$$Kg(x) \leq \|Kf\| + nl + \varepsilon$$

for all $x \in \bigcup_{i=0}^{n} (-x_i + D)$. The maximum principle 6.3.6.1 implies that this inequality holds for all $x \in G$. Another application of Proposition 6.4.2 *(ii)* yields

$$\limsup_{x \to \triangle} Kg(x) = (n + 1) \limsup_{x \to \triangle} Kf(x),$$

hence

$$\limsup_{x \to \triangle} Kf(x) \leq \frac{nl}{n + 1} + \frac{\|Kf\| + \varepsilon}{n}$$

for every $n \in \mathbf{N}$. Since l was the smallest of the presupposed two accumulation points > 0, the desired contradiction has been achieved. ∎

The previous theorem justifies the following

Definition 6.4.4 *A transient random walk $X(\mu)$ in G with potential kernel K is said to be* ***of type I*** *if*

$$\tau_v - \lim_{x \to \triangle} K(x, \cdot) = 0\,.$$

$X(\mu)$ is said to be ***of type II*** *if $\mathcal{F}'_K$ contains a measure $c\,\omega_G \neq 0$ (for $c > 0$).*

Remark 6.4.5 *Theorem 6.4.3 implies the dichotomy that every random walk in G is either of type I or of type II.*

Clearly, any *symmetric* random walk $X(\mu)$ in G in the sense that its determining measure $\mu \in M^1(G)$ is symmetric, is of type I.

In view of a general approach towards characterizing random walks of type I we first treat the special (classical) cases $G = \mathbf{R}$ and $G = \mathbf{Z}$. We shall apply the symbol λ for the Lebesgue measure on $\mathbf{R}$ as well as for the counting measure on $\mathbf{Z}$. The Alexandrov compactification $G_\triangle$ of G will be understood as $G \cup \{-\infty, \infty\}$.

Concerning the *renewal of the groups* $\mathbf{R}$ and $\mathbf{Z}$ we first prove

Proposition 6.4.6 *Let G be $\mathbf{R}$ or $\mathbf{Z}$, and let $X(\mu)$ be a transient random walk in G with potential kernel K. Then*

(i) $$\tau_v - \lim_{x \to \pm\infty} K(x, \cdot) = c_\pm \lambda.$$

(ii) At least one of the constants c_+ and c_- equals 0 the other one being > 0 or $= 0$.

Proof. 1. Let $G = \mathbf{Z}$, and suppose that there exist two measures in $\mathcal{F}'_K$ (as $x \to \infty$). By Theorem 6.4.3 these are the measures 0 and

$c\omega_G$ for $c > 0$. But then for any function $f \in C_+^c(G)$, any $\varepsilon > 0$ and for sufficiently large x we have either

$$Kf(x) > c\omega_G(f) - \varepsilon$$

or

$$Kf(x) < \varepsilon .$$

Consequently there exists a sequence $(x_n)_{n\in\mathbf{N}}$ in G with $x_n \to \infty$ satisfying the above inequalities with x replaced by $x_n + 1$ and x_n respectively, for all $n \in \mathbf{N}$. Hence Proposition 6.4.2 *(ii)* implies a contradiction.

2. For the case $G = \mathbf{R}$ we note that if there are two measures 0 and $c\omega_G$ for $c > 0$ in $\mathcal{F}'_K$, then for any $f \in C_+^c(G)$ the entire interval $[0, \omega_G(f)]$ is contained in $\mathcal{F}'_K$, since Kf is continuous by Proposition 6.4.2 *(i)*. The desired contradiction follows from Theorem 6.4.3. ■

A function f on $\mathbf{R}$ is called *directly Riemann integrable* (*R-integrable*) if the series $\underline{\sigma}(f,h)$ and $\overline{\sigma}(f,h)$ defined below converge and if for every $\varepsilon > 0$

$$\overline{\sigma}(f,h) - \underline{\sigma}(f,h) < \varepsilon$$

whenever $h > 0$ is sufficiently small. Here, for $h > 0$ we define $\underline{U}_n(f,h)$ and $\overline{U}_n(f,h)$ as the minimum and the maximum of f taken over $[(n-1)h, nh]$ $(n \in \mathbf{N})$ and set

$$\underline{\sigma}^{(-)}(f,h) := h \sum_{h=-\infty}^{\infty} \underline{U}_n^{(-)}(f,h)$$

respectively. Clearly, the set $R(\mathbf{R})$ of R-integrable functions on $\mathbf{R}$ contains $C^c(\mathbf{R})$. For functions f on $\mathbf{R}$ that vanish on $]-\infty, 0[$, decrease on $[0, \infty[$ and satisfy $f(\infty) = 0$ the series $\underline{\sigma}$ and $\overline{\sigma}$ either both diverge or both converge. Thus $f \in R(\mathbf{R})$ if and only if $f \in L^1(\mathbf{R}, \lambda)$. ■

Proposition 6.4.7 *If $f \in R(\mathbf{R})$, then Kf is bounded, and*

$$\lim_{x\to\pm\infty} Kf = c_\pm\, \lambda(f) .$$

Proof. 1. We first show the assertion for functions of the form

$$f_n := 1_{[(n-1)h, nh[}$$

for fixed $h \in \mathbf{R}_+^\times$ ($n \in \mathbf{N}$). By the maximum principle 6.3.6.1 there exists a constant $M > 0$ such that $Kf_n \leq M$ for all $n \in \mathbf{N}$. Let $(a_n)_{n \in \mathbf{Z}}$ denote a sequence in $\mathbf{R}_+^\times$ such that $\sum_{n=-\infty}^{\infty} a_n < \infty$. Then for

$$f := \sum_{n=-\infty}^{\infty} a_n f_n$$

we obtain that

$$\sum_{k=-m}^{m} a_k K f_k(x) \leq Kf(x) \leq \sum_{k=-m}^{m} a_k K f_k(x) + M \sum_{|k|>m} a_k$$

for all $x \in G$ and $m \in \mathbf{N}$, which implies the boundedness of Kf. The asserted convergence follows from the τ_v-convergence of the potential kernels $K(x, \cdot)$ as $x \to \infty$ ($x \to -\infty$).

2. Let now $f \in R(\mathbf{R}), f > 0$. Setting

$$\underline{f}^{(-)} = \sum_{n=-\infty}^{\infty} \underline{U}_n^{(-)} f_n$$

we immediately see that

$$K\underline{f} \leq Kf \leq K\overline{f},$$

hence that

$$\lim_{x \to \pm\infty} K\underline{f}(x) \leq \liminf_{x \to \pm\infty} Kf(x) \leq \limsup_{x \to \pm\infty} Kf(x) \leq \lim_{x \to \pm\infty} K\overline{f}(x)$$

and finally that

$$c_{\pm}\, \underline{\sigma} \leq \liminf_{x \to \pm\infty} Kf(x) \leq \limsup_{x \to \pm\infty} Kf(x) \leq c_{\pm} \overline{\sigma}$$

for all $x \in G$. This implies both assertions of the Proposition at once. ■

It follows the *computation of the constants* c_+ **and** c_-.

Proposition 6.4.8 *Let* $X(\mu)$ *admit a first moment in the sense that*

$$\int |x|\mu(dx) < \infty,$$

in which case the mean

$$m := \int x\mu(dx)$$

(of $X(\mu)$*) exists. Then* $X(\mu)$ *is of type II.*

Moreover,

(1) $m \neq 0$, *since* $X(\mu)$ *is transient.*

(2) If $m > 0$ $(m < 0)$, *then* $c_{\underset{(+)}{-}} = \frac{1}{m}$ *and* $c_{\underset{(-)}{+}} = 0$.

(3) If $\operatorname{supp}(\mu) \subset \overline{\mathbf{R}}_+ (:= \mathbf{R}_+ \cup \{\infty\})$, *then* $c_- = \frac{1}{m}$ *(and, of course* $c_+ = 0$*).*

Proof. In order to show the main statement and (2) of the Proposition we need to introduce two auxiliary functions g and k on $\mathbf{R}$ and $\mathbf{R}^2$ by

$$g(x) := \begin{cases} \mu([-x, \infty[) & \text{if } x \leq 0 \\ -\mu(]-\infty, -x[) & \text{if } x > 0 \end{cases}$$

and

$$k(x, y) := \begin{cases} 1 \text{ if } x < 0,\ y \geq -x \\ -1 \text{ if } x \geq 0,\ y < -x \\ 0 \quad \text{otherwise} \end{cases}$$

respectively. An application of the Fubini theorem to the integral

$$\int_{\mathbf{R}^2} k(x,y)\lambda(dx)\mu(dy)$$

for which

$$\int_{\mathbf{R}^2} |k(x,y)\lambda(dx)\mu(dy) = \int_{-\infty}^{\infty} |y|\mu(dy) < \infty$$

holds, yields that

$$\int_{-\infty}^{\infty} g(x)\lambda(dx) = m\,.$$

Consequently, $g \in R(\mathbf{R})$ and so Kg is bounded by Proposition 6.4.7. For $h := 1_{]-\infty,0[}$ we have

$$Ph(x) = \mu(]-\infty, -x[)$$

whenever $x \in \mathbf{R}$, hence

$$(I - P)h = g\,.$$

Applying the Choquet-Deny theorem 6.1.12 to the measures $\mu := l \cdot \omega_G \in M^{sb}(G)$ with

$$l := h - Kg$$

and $\sigma := \mu \in M^1(G)$ we obtain that l is constant $= a$ $[\omega_G]$. Now sequences $(x_n)_{\in\mathbf{N}}$ and $(y_n)_{n\in\mathbf{N}}$ in $\mathbf{R}$ can be found which converge to $-\infty$ and ∞ respectively and which satisfy $l(x_n) = l(y_n) = a$ for every $n \in \mathbf{N}$. On the other hand we have

$$h(z_n) = Kg(z_n) + l(x_n),$$

where z_n denotes either x_n or y_n. In the limit for $n \to \infty$ we obtain from Proposition 6.4.7 that

$$1 = c_{-}m + a$$

and

$$0 = c_+ m + a$$

according to the choices $(x_n)_{n\in\mathbf{N}}$ and $(y_n)_{n\in\mathbf{N}}$ respectively. But $c_+ > 0$ is impossible, since this yields $c_- = 0, a = 1$, hence $\lambda < 0$. We therefore have $c_+ = 0$, hence $a = 0$ and consequently $c_- = \frac{1}{m}$.

As to the proof of (3) we first observe that the statement is true for $m < \infty$ by what we just showed. In the case $\lambda = \infty$ we note that g vanishes on $]0, \infty[$, $Kg \geq 0$ and $Kg = h$, since h vanishes on $]0, \infty[$. Moreover, for every $n \in \mathbf{N}$ the function

$$g_n := g 1_{[-n,0]} \in R(\mathbf{R})$$

and

$$Kg_n \leq Kg = h\,.$$

In the limit for $x \to -\infty$ this inequality leads to

$$c_- \int_{-\infty}^{\infty} g_n(x)\lambda(dx) \leq 1$$

for all $n \in \mathbf{N}$. But this is a contradiction unless $c_- = 0$. ■

Given a random walk $X(\mu)$ in G with law $\mu \in M^1(G)$ and transition kernel P one introduces for any stopping time τ for $X(\mu)$ the **stopped transition kernel** P_τ by

$$P_\tau(x, B) := \mathbf{P}^x([X_\tau \in B])$$

for all $x \in G, B \in \mathfrak{B}(G)$. In the special case that τ is the **(first) entry time** (or hitting time) H_B of $X(\mu)$ into the set $B \in \mathfrak{B}(G)$ defined by

$$H_B(\omega) := \begin{cases} \inf\{n \in \mathbf{Z}_+ : X_n(\omega) \in B\} & \text{if } \{n \in \mathbf{Z}_+ : X_n(\omega) \in B\} \neq \emptyset \\ \infty & \text{otherwise,} \end{cases}$$

the strong Markov property of $X(\mu)$ implies that

$$Kf(x) = \int_0^{\infty} P_{H_{\overline{\mathbf{R}}_+}}(x, dy) Kf(y)$$

valid for all measurable functions $f \geq 0$ on $G, x \in G$.

In the following we shall write H_+ instead of $H_{\overline{\mathbf{R}}_+}$.

Properties 6.4.9

6.4.9.1 *If*

$$m^+ = \int_0^\infty x\mu(dx) = \infty,$$

then

$$\tau_v - \lim_{x \to -\infty} P_{H_+}(x, \cdot) = 0\,.$$

In fact, if $\operatorname{supp}(\mu) \subset \overline{\mathbf{R}}_+$, then by Proposition 6.4.8 (3)

$$\lim_{x \to -\infty} Kf(x) = 0,$$

and since $Kf \geq f$, this yields the assertion. Let us now suppose that μ is not carried by $\overline{\mathbf{R}}_+$. We define a sequence $(\tau_n)_{n \in \mathbf{N}}$ of stopping times τ_n for $X(\mu)$ by induction as follows:

$$\tau_1 := H_+\,,$$

$$\tau_k := \begin{cases} \inf\{n \in \mathbf{Z}_+ : X_n > X_{\tau_{k-1}}\} & \text{if } \{n \in \mathbf{Z}_+ : X_n > X_{\tau_{k-1}}\} \neq 0 \\ \infty & \text{otherwise} \end{cases}$$

for $k \geq 1$. Clearly, the random variables $X_{\tau_k} - X_{\tau_{k-1}}$ $(k \geq 1)$ are independent and equally distributed with distribution

$$\mu' := P_{H_+}(0, \cdot)\,.$$

Thus the sequence $(X_{\tau_k})_{k \in \mathbf{N}}$ forms a random walk $X(\mu')$ in $\mathbf{R}$ with $\operatorname{supp}(\mu') \subset \overline{\mathbf{R}}_+$, and since $\mu' \geq \operatorname{Res}_{\overline{\mathbf{R}}_+} \mu$, we obtain that

$$\int_0^\infty x\mu'(dx) = \infty\,.$$

Now the distributions of entry into $\overline{\mathbf{R}}_+$ with respect to $X(\mu)$ and $X(\mu')$ coincide. The first part of this proof applied to μ' instead of μ yields the assertion.

6.4.9.2 *If*

$$m^- = \int_{-\infty}^{0} (-x)\mu(dx) < \infty$$

and if $m > 0$, *then*

$$\mathbf{P}^x([H_+ < \infty]) = 1$$

for all $x \in \mathbf{R}$.

For the proof we assume that $m^+ < \infty$; the case $m^+ = \infty$ can be treated by truncating $X(\mu)$. Since $m^+ < \infty$, the strong law of large numbers implies

$$\lim_{n\to\infty} \frac{1}{n} X_n = m\,[\mathbf{P}^0]\,.$$

Thus

$$\lim_{n\to\infty} X_n = \infty\,[\mathbf{P}^0],$$

and the assertion follows.

Theorem 6.4.10 *(Renewal under non-existence of first moments)*

If the transient random walk $X(\mu)$ *in* $G = \mathbf{R}$ *or* $\mathbf{G} = \mathbf{Z}$ *admits no first moment, then it is of type I.*

Proof. We suppose that $c_- > 0$, hence $c_+ = 0$, and that $m^+ = \infty$. Then the strong Markov property of $X(\mu)$ implies that

$$Kf(x) = \int_0^\infty P_{H_+}(x, dy)Kf(y)$$

holds for each function $f \in C_+^c(G)$ vanishing on $]-\infty, 0[$ and all $x \in G$. From $c_+ = 0$ we infer that given $\varepsilon > 0$ there exists an $a > 0$ such that $Kf(y) < \varepsilon$ for all $y \in G$ with $y > a$. But then

$$Kf(x) \le \int_0^a P_{H_+}(x, dy)Kf(y) + \varepsilon$$

for all $x \in G$. Moreover,

$$\tau_v - \lim_{x \to -\infty} P_{H_+}(x, \cdot) = 0, \tag{1}$$

thus $Kf(x) \leq 2\varepsilon$ for sufficiently small $x \in G$. This, however, contradicts the assumption that $c_- > 0$. It follows that $m^+ < \infty$, hence that $m^- = \infty$. Replacing $X(\mu)$ by its *dual* in the sense that $\mu \in M^1(G)$ is replaced by $\mu^\sim$, the above assumption is equivalent to supposing that $c_- = 0$, hence $c_+ > 0$ and $m^- < \infty$, hence $m^+ = \infty$. Then

$$\mathbf{P}^x([H_+ < \infty]) = 1$$

for every $x \in \mathbf{R}$, and by (1)

$$\begin{aligned} 0 &= c_- \cdot \omega_G(f) \\ &= \lim_{x \to -\infty} Kf(x) \\ &= \lim_{x \to -\infty} \int_{\mathbf{R}} P_{H_+}(x, dy) Kf(y) \\ &= \lim_{y \to \infty} Kf(y) \\ &= c_+ \cdot \omega_G(f). \end{aligned}$$

This is the desired contradiction. ■

The previous theorem extends to random walks in groups of the form $G = \mathbf{R} \times K$ and $G = \mathbf{Z} \times K$, where K is a compact group. We shall prove the extension only for $G = \mathbf{R} \times K$, where ω_G is chosen to be $\lambda \otimes \omega_K$ with $\omega_K \in M^1(K)$. Let p denote the canonical projection from G onto $\mathbf{R}$. Then we shall employ the convention that $x \to +\infty$ $(-\infty)$ in G provided $p(x) \to +\infty$ $(-\infty)$ in $\mathbf{R}$.

Theorem 6.4.11 *Let $G \cong \mathbf{R} \times K$ or $G \cong \mathbf{Z} \times K$, and let $X(\mu)$ be a random walk with law $\mu \in M^1(G)$ which is of type II.*

Then

(*i*)
$$\int |p(x)|\mu(dx) < \infty.$$

(*ii*) If
$$m := \int_G p(x)\mu(dx) > 0,$$

then
$$\tau_v - \lim_{x\to\triangle} K(x,\cdot) = \begin{cases} 0 & \text{if } \triangle = \infty \\ \frac{1}{m}\lambda & \text{if } \triangle = -\infty. \end{cases}$$

Proof. From Proposition 6.4.1 we infer that for every sequence in G there exists a subsequence $(x_n)_{n\in\mathbf{N}}$ with $x_n := (y_n, k_n) \in \mathbf{R}\times K$ such that
$$\tau_v - \lim_{n\to\infty} K(x_n,\cdot) = c\,\omega_G$$
with $c \geq 0$. Let $f \in C_+^c(G)$ be constant on the K-cosets of G. Then *(ii)* of Proposition 6.4.2 leads to
$$\lim_{x_n\to\infty} Kf(x_n) = \lim_{y_n\to\infty} Kf(y_n, 0)\,.$$

Now we consider the random walk $X(\tilde{\mu})$ with law $\tilde{\mu} := p(\mu) \in M^1(\mathbf{R})$. With the suggestive notation for $\tilde{f}$ (with $\tilde{f}(y) = f(x)$ for all $y = K + x$) and $\tilde{K}$ we obtain
$$Kf(y_n, 0) = \tilde{K}\tilde{f}(y_n)$$
for all $n \in \mathbf{N}$, and by the discussion starting with Proposition 6.4.8 we have
$$\lim_{y_n\to\infty} \tilde{G}\tilde{f}(y_n) = c_+ \cdot \lambda(\tilde{f})$$
where $c_+ > 0$ if and only if
$$\int_{-\infty}^{+\infty} |y|\tilde{\mu}(dy) = \int_G |p(x)|\mu(dx) < \infty \text{ and } < 0\,.$$

As

$$\lambda(\tilde{f}) = \omega_G(f),$$

the theorem has been proved. ∎

The final step of our analysis will be to establish the *renewal of random walks for general locally compact Abelian groups.*

We note that the (locally compact Abelian) group G remains to be second countable. The function f appearing in the proofs is always taken from $C_+^c(G)$ and $\neq 0$.

Theorem 6.4.12 *Let G_1, be a compactly generated, non-compact open subgroup of G such that G/G_1 is infinite.*

Then all transient radom walks in G are of type I.

Proof. We suppose that the given random walk in G with potential kernel K is of type II and at the same time that there exists a sequence $(x_n)_{n \in \mathbf{N}}$ in G such that

$$\tau_v - \lim_{n \to \infty} K(x_n, \cdot) = c\,\omega_G$$

for $c > 0$. The aim is to derive a contradiction.

Given the sequence $(x_n)_{n \in \mathbf{N}}$ we may assume without loss of generality that the G_1-cosets $G_1 + x_n$ of G $(n \in \mathbf{N})$ are pairwise disjoint.

In fact, if there is no subsequence of $(x_n)_{n \in \mathbf{N}}$ with this property, then there exists a G_1-coset containing infinitely many x_n, hence $(x_n)_{n \in \mathbf{N}}$ is contained in one (and the same) G_1-coset. Now we choose a sequence $(y_k)_{k \in \mathbf{N}}$ in G such that the cosets $G_1 + y_k$ $(k \in \mathbf{N})$ are pairwise disjoint. From *(ii)* of Proposition 6.4.2 we infer that for every $k \in \mathbf{N}$ we have that

$$\lim_{n \to \infty} Kf(x_n + y_k) = c\,\omega_G(f),$$

hence that

$$|Kf(x_{n_k} + y_k) - c\,\omega_G(f)| < \frac{1}{2^k}$$

for some $n_k \in \mathbf{N}$. The sequence $(z_k)_{k \in \mathbf{N}}$ with $z_k := x_{n_k} + y_k$ fulfills the requirements.

Since G_1 is compactly generated and non-compact, there exists an $x \in G_1$ such that $nx \to \Delta$ as $n \to \infty$. Applying properties *(ii)* and *(iv)* of Proposition 6.4.2 one obtains

$$\lim_{n\to\infty} Kf(nx) = 0$$

or

$$\lim_{n\to\infty} Kf(-nx) = 0\,.$$

Without loss of generality we restrict our subsequence arguments to the first limit relationship. For every n there is a smallest integer m_n such that

$$Kf(x_n + n'x) \leq \frac{1}{2} c\,\omega_G(f)$$

whenever $n' \geq m_n$. This follows again from *(ii)* of Proposition 6.4.2. But for fixed l

$$\lim_{n\to\infty} Kf(x_n + lx) = c\,\omega_G(f),$$

hence $m_n > 1$ for sufficiently large n. We obtain, for such n,

$$Kf(x_n - m_n x - x) > \frac{1}{2} c\omega_G(f)$$

and

$$Kf(x_n + m_n x) \leq \frac{1}{2} c\omega_G(f),$$

consequently, in the limit for $n \to \infty$, $c\omega_G(f)$ and 0 respectively.

On the other hand G_1 is open, hence G/G_1 is discrete and every subset of G is contained in the union of finitely many G_1-cosets of G. Thus

$$\lim_{n\to\infty} (x_n + nx) = \Delta$$

uniformly in n, and one more application of *(ii)* of Proposition 6.4.2 yields

$$\lim_{n\to\infty} |Kf(x_n + m_n x - x) - Kf(x_n + m_n x)| = 0$$

which serves as a contradiction. ∎

Corollary 6.4.13 *If $G \cong \mathbf{R}^d \times \mathbf{Z}^e \times K$ with $d+e>1$ and a compact group K, then all transient random walks on G are of type I.*

Proof. From the proof of Theorem 6.4.11 we infer that it suffices to consider groups of the form $G = \mathbf{R}^d \times \mathbf{Z}^e$ for $d+e>1$. If $e \geq 1$, the above theorem yields the result. If, however, $G = \mathbf{R}^d$ for $d>1$, then $G = \bigcup_{n\geq 1} G_n$ for an increasing sequence $(G_n)_{n\in\mathbf{N}}$ of compact subsets G_n of G such that ${G_n}^c$ is connected ($n \in \mathbf{N}$). Suppose that for $f \in C_+^c(G), f \neq \emptyset$ we have

$$\lim_{x\to\triangle} Gf(x) = a \neq 0\,.$$

Then every point in $[0,a]$ is a limit point of $Gf(x)$ (for $x \to \triangle$). But this contradicts the statement of Theorem 6.4.3. ∎

Theorem 6.4.14 *If every element of G is compact, then all transient random walks in G are of type I.*

Proof. As in the proof of the previous theorem we assume that there is a random walk of type II in G and aim at deriving a contradiction. Let $(x_n)_{n\in\mathbf{N}}$ be a sequence in G such that

$$\tau_v - \lim_{n\to\infty} K(x_n,\cdot) = c\,\omega_G$$

with $c>0$. For every $n \in \mathbf{N}$ the set $S := \{kx_n : k \in \mathbf{N}\}$ is a compact subsemigroup, hence a subgroup of G. Indeed, if 0 is not an accumulation point of S, then S will be discrete. Since S is compact, this is a contradiction. But 0 being an accumulation point of S, $-x$ is also one. This shows that $S=-S$, hence that S is a group.

As a consequence of this we may also choose a sequence $(k_p)_{p\in\mathbf{N}}$ in $\mathbf{N}$ such that

$$\lim_{p\to\infty} k_p x_n = -x_n\,.$$

Thus, for every $\varepsilon>0$ there is $k_n \in \mathbf{N}$ satisfying

$$Kf(k_n x_n) \leq Kf(-x_n) + \frac{\varepsilon}{n}\,.$$

By *(iv)* of Proposition 6.4.2

$$\lim_{n\to\infty} Kf(-x_n) = 0,$$

hence

$$\lim_{n\to\infty} Kf(k_n x_n) = 0\,.$$

Now we choose a constant M such that $M > \|Kf\| \vee \frac{1}{4}c\,\omega_G(f)$. For sufficiently large $n \geq 1$ and the largest positive integer $m_n < k_n$ the inequalities

$$Kf(m_n x_n) \geq \frac{c^2\omega_G(f)^2}{4M}$$

and

$$Kf((m_n+1)x_n) < \frac{c^2\omega_G(f)^2}{4M}$$

imply that

$$Kf(m_n x_n)Kf(x_n) \leq M(\varepsilon + Kf((m_n+1)x_n))\,.$$

Furthermore, for sufficiently large n we have that

$$Kf(x_n) \geq \frac{1}{2}c\,\omega_G(f)$$

which leads to

$$\begin{aligned} Kf(m_n x_n) &\leq \frac{2M}{c\,\omega_G(f)}(Kf((m_n+1)x_n) + \varepsilon) \\ &\leq \frac{1}{2}c\,\omega_G(f) + \frac{2M\varepsilon}{c\,\omega_G(f)} \\ &\leq \frac{1}{2}c\,\omega_G(f) + \frac{1}{2}\varepsilon \end{aligned}$$

and

$$\frac{c^3\omega_G(f)^3}{8M^2} - \varepsilon \leq Kf((m_n+1)x_n)\,.$$

Now, by an appropriate choice of ε we arrive at the inequalities

$$0 < \alpha \leq Kf(m_n x_n) \leq \beta < c\,\omega_G(f)$$

and

$$0 < \gamma \le Kf((m_n + 1)x_n) \le \delta < c\,\omega_G(f)\,.$$

Since $x_n \to \Delta$ as $n \to \infty$, at least one of the sequences $(m_n x_n)_{n\in\mathbf{N}}$ and $((m_n+1)x_n)_{n\in\mathbf{N}}$ admits a subsequence $(y_n)_{n\in\mathbf{N}}$ with $y_n \to \Delta$. But the sequence $(Kf(y_n))_{n\in\mathbf{N}}$ has at most 0 and $c\,\omega_G(f)$ for $c > 0$ as accumulation points. This shows the desired contradiction. ■

Theorem 6.4.15 *(General renewal theorem). Let G be a second countable locally compact Abelian group, and suppose that there exists a random walk of type II in G. Then*

$$G \cong \mathbf{R} \times K \quad \text{or} \quad G \cong \mathbf{Z} \times K,$$

where K is a compact group, and the renewal results of Theorem 6.4.11 apply.

Proof. From Theorem 6.4.14 we infer that there exists a non-compact element $y \in G$. Since by Appendix A 3.5 G admits a compactly generated open subgroup G_1, the subgroup

$$G_2 := [G_1 \cup \{y\}]$$

is non-compact, compactly generated and open. Thus, by Theorem 6.4.12 the group G/G_2 must be finite, hence G itself compactly generated. But then, by the structure theorem 4.2.19

$$G \cong \mathbf{R}^d \times \mathbf{Z}^e \times K$$

with a compact group K. Corollary 6.4.13 implies the assertion. ■

Appendices

Appendices on topological groups, topological vector spaces and on commutative Banach algebras are added in order to provide the reader with the necessary prerequisites from these topics to be employed throughout the book. The exposition of basic notions and facts from functional analysis organized along specific references and presented in a unified terminology is intended to facilitate the reading of the main text.

There are excellent text books and monographs available which contain the knowledge layed out in the appendices. The citations on topological groups are justified in Chapter II of Volume I of [19] (Hewitt, Ross). The quoted material on topological vector spaces is contained in the graduate text [41] (Schaefer). A systematic treatment of the facts collected on commutative Banach algebras is given in chapters IV and V of the monograph [28] (Loomis). Only a few very special references are documented in the main text of the book.

A Topological groups

A group G (written additively with neutral element 0) which is also a topological space is said to be a **topological group** if the mapping $(x, y) \mapsto x - y$ from $G \times G$ into G is continuous.

Clearly the translates $x \mapsto a + x$ and $x \mapsto x + b$ for $a, b \in G$ as well as the inversion $x \mapsto -x$ are homeomorphisms from G onto G. As a consequence one notes that for open sets A, B of G also the sets $A + B, B + A$ and $-A$ are open. If A is closed and B is compact, then $A + B$ is compact in G.

Let $\mathfrak{V}(x)$ denote the neighborhood filter of $x \in G$ with the abbreviation $\mathfrak{V}$ for $\mathfrak{V}(0)$. Then $\mathfrak{V}(a) = a + \mathfrak{V} = \mathfrak{V} + a$ whenever $a \in G$.

Theorem A1 *(Determination of the topology of a topological group)*

(i) Let G be a topological group with neighborhood filter $\mathfrak{V}$. Then

(1) for every $U \in \mathfrak{V}$ there exists $V \in \mathfrak{V}$ such that $V + V \subset U$.

(2) If $U \in \mathfrak{V}$, then $-U \in \mathfrak{V}$.

(3) $0 \in U$ for all $U \in \mathfrak{V}$.

(4) If $U \in \mathfrak{V}, a \in G$, then $a + U - a \in \mathfrak{V}$.

(ii) Let G be a group and let $\mathfrak{V}$ be a filter in G satisfying the conditions (1) to (4) of (i). Then there exists exactly one topology in G such that G is a topological group and $\mathfrak{V}$ is the neighborhood filter of 0.

For every $a \in G$ one has

$$\mathfrak{V}(a) = a + \mathfrak{V} = \mathfrak{V} + a\,.$$

We note that for every topological group the closed symmetric neighborhoods of 0 form a fundamental system of neighborhoods of 0.

Properties A2 *of* ***subgroups, products*** *and* ***quotients*** *of a topological group*

A 2.1 *For every subgroup H of the topological group G its closure $\overline{H}$ is again a subgroup of G.*

A 2.2 *Let $(G_i)_{i\in I}$ be a family of topological groups. The product topology in $G := \Pi_{i\in I} G_i$ is compatible with the group structure in the sense that the mapping*

$$((x_i, y_i))_{i\in I} \mapsto (x_i - y_i)_{i\in I}$$

from $\Pi_{i\in I}(G_i \times G_i)$ into G is continuous.

A 2.3 *Let G be a topological group and let H be a normal subgroup of G. There exists an equivalence relation*

$$xRy \Leftrightarrow x - y \in H .$$

The group $G/H := G/R$ is a topological group with respect to the quotient topology induced by the canonical mapping $G \longrightarrow G/H$ in the sense that $\mathfrak{V}(0+H)$ satisfies the conditions (1) to (4) of Theorem A1.

One observes that an open subgroup of a topological group G is closed. Moreover, open *and* closed subgroups H of G can be generated by a symmetric neighborhood $U \in \mathfrak{V}$ in the form $H = \bigcup_{n\geq 1} nU$.

Properties A3 of locally compact groups

A 3.1 *A Hausdorff (topological) group G is locally compact if and only if* 0 *possesses a compact neighborhood.*

A 3.2 *Every closed subgroup of a locally compact group is also locally compact.*

A 3.3 *If G is a Hausdorff group and H a subgroup which is locally compact with respect to the relative topology, then H is closed.*

A 3.4 *Let G be a locally compact group and let H be a normal subgroup of G. Then G/H is a Hausdorff locally compact group if and only if H is closed.*

A 4 A topological group G is said to the **compactly generated** if it contains a compact subset F for which the subgroup generated by F coincides with G, i.e.

$$G = [F] := \{0\} \cup \bigcup_{n\geq 1} n(F \cup (-F)) .$$

For locally compact groups G this property is equivalent to the requirement that there is an open relatively compact subset U (or a neighborhood $U \in \mathfrak{V}$) such that $G = [U]$.

Further Properties A3

A 3.5 *Let G be a locally compact group and F a compact subset of G. Then there exists an open and closed compactly generated subgroup H of G with $H \supset F$.*

A 3.6 *Let G be a locally compact group, let H be a subgroup of $\mathbf{R}$ furnished with the relative topology, and let φ be a continuous homomorphism from H into G. Then either $\varphi(H)^-$ is a compact Abelian subgroup of G or φ is a topological isomorphism from H onto $\varphi(H)$.*

B Topological vector spaces

We now turn to the discussion of commutative groups, in particular to **topological vector spaces** which are vector spaces E over $\mathbf{R}$ and at the same time topological spaces such that the mappings $(x, y) \mapsto x + y$ from $E \times E$ into E and $(\lambda, x) \mapsto \lambda x$ from $\mathbf{R} \times E$ into E are continuous.

Examples B1 *of topological vector spaces are* ***semi-normed*** *vector spaces E in the sense that they admit a semi-norm p with the defining properties*

(i) $p \geq 0$.

(ii) $p(x + y) \leq p(x) + p(y)$ *for all* $x, y \in E$.

(iii) $p(\lambda x) = |\lambda| p(x)$ *for all* $\lambda \in \mathbf{R}, x \in E$.

The function ϱ on $E \times E$ given by

$$\varrho(x, y) := p(x - y)$$

for all $x, y \in E$ provides a quasi-metric on E.

B 1.1 *Every (real) Banach space is a normed vector space, hence the spaces $\mathbf{R}$ and $\mathbf{C}$ are normed vector spaces (over $\mathbf{R}$).*

Let A and B be subsets of a vector space E. A is said to **absorb** B if there exists an $\alpha \in \mathbf{R}_+^\times$ such that $B \subset \lambda A$ for all $\lambda \in \mathbf{R}$ with $|\lambda| \geq \alpha$. $A \subset E$ is called **absorbing** (radial) if A absorbs every finite subset of E, and **bounded** if it absorbs each neighborhood of 0. Finally, $A \subset E$ is called **balanced** (circled) if $\lambda A \subset A$ whenever $\lambda \in \mathbf{R}$ with $|\lambda| \geq 1$.

Theorem B2 *(Characterization of topological vector spaces by local properties)*

(i) In every topological vector space E there exists a fundamental system $\mathfrak{U}$ of closed neighborhoods of 0 such that

(1) every $U \in \mathfrak{U}$ is balanced and absorbing.

(2) For every $U \in \mathfrak{U}$ there is a $V \in \mathfrak{U}$ with $V + V \subset U$.

(ii) Let E be a vector space and let $\mathfrak{U}$ be a filter base in E satisfying properties (1) and (2) of (i). Then there exists exactly one topology in E compatible with the vector space structure of E and such that $\mathfrak{U}$ is a fundamental system of neighborhoods of 0.

For topological vector spaces one introduces linear subspaces, products and quotients in analogy to the corresponding structures for topological groups.

Theorem B 3 *(Characterization of finite dimensional vector spaces)*

Let E be a Hausdorff topological vector space with dim $E = d$ ($< \infty$). Then every linear mapping from $\mathbf{R}^d$ onto E is an isomorphism and a homeomorphism.

In particular every such topological vector space is isomorphic (as a topological vector space) to $\mathbf{R}^d$.

It is a famous result of F. Riesz that a Hausdorff topological vector space E is finite dimensional if and only if E is locally compact.

A topological vector space E is said to be a **locally convex space** if E admits a fundamental system of convex neighborhoods of 0.

Clearly every semi-normed space, hence every Banach space is locally convex.

There is a characterization of locally convex spaces by local properties analogous to Theorem B 2. In fact, the topology of a locally convex space E is determined by a fundamental system of *convex*, symmetric and absorbing closed neighborhoods of 0.

Let E be an arbitrary vector space, $(p_i)_{i \in I}$ a family Γ of semi-norms on E, and let $\mathfrak{U}$ be the system of all sets V of the form

$$\bigcap_{i \in J} \{x \in E : p_i(x) \leq \lambda_i\},$$

where J is a finite subset of $I, \lambda_i > 0$ for all $i \in J$. Then $\mathfrak{U}$ is a filter base, every set $\{x \in E : p_i(x) \leq \lambda\}$ $(i \in J, \lambda > 0)$ is convex, symmetric and absorbing, hence every $V \in \mathfrak{U}$ has these properties, and consequently there exists exactly one locally convex topology in E such that $\mathfrak{U}$ is a fundamental system of neighborhoods of 0. This topology is called the **topology defined by the set Γ of semi-norms on E** and will be denoted by τ_Γ.

Obviously the choice $\Gamma := \{p\}$ yields the topology of the space E semi-normed by p.

Theorem B 4 *Every locally convex topology τ on the vector space E is of the form τ_Γ for some family Γ of semi-norms on E.*

It is a standard procedure to introduce the inductive limit of locally convex vector spaces. A prominent example of a (strict) inductive limit of locally convex vector spaces is the space $E := C^c(X)$ for a locally compact space X. In fact,

$$C^c(X) = \varinjlim_{K \in \mathcal{K}(X)} C^c(X, K)$$

where for each $K \in \mathcal{K}(X)$ the linear subspace $C^c(X, K) := \{f \in E : \operatorname{supp}(f) \subset K\}$ of $C^c(X)$ carries the topology of uniform convergence.

Theorem B 5 *(Banach, Hahn) Let E be a vector space, p a semi-norm on E, M a linear subspace of E and f a linear functional on M.*

The following statements are equivalent:

(i) f *can be extended to a linear functional* $\overline{f}$ *on* E *satisfying*

$$|\overline{f}(x)| \leq p(x)$$

for all $x \in E$.

(ii)
$$|f(x)| \leq p(x)$$

whenever $x \in M$.

Theorem B 6 *(Closed graph) Let* E *and* F *be Banach spaces. Then any linear mapping from* E *into* F *whose graph is a closed subset of* $E \times F$ *is continuous.*

In the remaining part of this appendix we are collecting useful notions and results on the *duality of topological vector spaces.*

Let E and F be topological vector space, and let $L(E, F)$ denote the vector space of all continuous linear mappings from E into F. For a set S of a given non empty family $\mathfrak{S}$ of bounded subsets of E and for each $V \in \mathfrak{V}_F(0)$ we introduce the set

$$T(S, V) := \{u \in L(E, F) : u(S) \subset V\}.$$

$\tau_{\mathfrak{S}}$ denotes the system of all finite intersections of sets of the form $T(S, V)$. There exists exactly one topology τ on $L(E, F)$ which is compatible with the vector space structure such that $\tau_{\mathfrak{S}}$ is a fundamental system of neighborhoods of 0 for τ. The topology τ determined by $\tau_{\mathfrak{S}}$ is said to be the $\mathfrak{S}$-topology on $L(E, F)$.

Discussion B 7 *of the* $\mathfrak{S}$*-topology*

B 7.1 *If* F *is a locally convex space, then the* $\mathfrak{S}$*-topology is locally convex.*

B 7.2 *Let F be a locally convex space, and let Γ denote the set of semi-norms determining the topology of F (See Theorem B 4). For $p \in \Gamma$ and $S \in \mathfrak{S}$ we introduce the semi-norm p_S on $L(E,F)$ by*

$$p_S(u) := \sup_{x \in S} p(u(x))$$

for all $u \in L(E,F)$. With $V := \{y \in F : p(y) \leq 1\}$ we obtain

$$T(S,V) = \{u \in L(E,F) : p_S(u) \leq 1\}.$$

Therefore the family $\{p_S : S \in \mathfrak{S}, p \in \Gamma\}$ of semi-norms defines an $\mathfrak{S}$-topology on $L(E,F)$.

By $L_{\mathfrak{S}}(E,F)$ we abbreviate the vector space $L(E,F)$ equipped with the $\mathfrak{S}$-topology.

B 7.3 *Let E, F be topological vector spaces, let F be a Hausdorff space, and let $\bigcup_{S \in \mathfrak{S}} S$ be a dense subset of E. Then $L_{\mathfrak{S}}(E,F)$ is Hausdorff.*

Special cases B 8 *of $\mathfrak{S}$-topologies*

B 8.1 *If $\mathfrak{S}$ is the family $\mathcal{F}(E)$ of finite subsets of E, then $\tau_{\mathfrak{S}}$ yields the topology of* ***simple*** *(pointwise)* ***convergence****.*

B 8.2 *For $\mathfrak{S} := \mathcal{K}(E)$ $\tau_{\mathfrak{S}}$ determines the topology of* ***compact convergence****.*

B 8.3 *If $\mathfrak{S}$ is the family $\mathcal{B}(E)$ of bounded subsets of E, then $\tau_{\mathfrak{S}}$ defines the topology of* ***bounded convergence*** (which appears to be the finest of all $\mathfrak{S}$-topologies).

A set $H \subset L(E,F)$ is said to be ***equicontinuous*** if for every $V \in \mathfrak{V}_F(0)$ there exists a $U \in \mathfrak{V}_E(0)$ such that

$$u(U) \subset V$$

for all $u \in H$, or equivalently if for every $V \in \mathfrak{V}_F(0)$ we have

$$\bigcap_{u \in H} u^{-1}(V) \in \mathfrak{V}_E(0).$$

Properties B 9 *of equicontinuous sets*

B 9.1 *For any equicontinuous subset H of $L(E,F)$ the closure $\overline{H}$ of H taken with respect to the topology of simple convergence (and hence with respect to any finer topology) on $L(E,F)$ is also equicontinuous.*

B 9.2 *Every equicontinuous subset H of $L(E,F)$ is bounded with respect to each $\mathfrak{S}$-topology.*

B 9.3 *If E and F are Banach spaces, then every simply bounded subset of $L(E,F)$ is equicontinuous.*

B 9.4 *If E and F are Banach spaces, then $L(E,F)$ is itself a Banach space with respect to the topology of bounded convergence.*

Let F and G be vector spaces, and let B be a bilinear form on $F \times G$. One says that (F,G) forms a **dual pair** with respect to B if for all $x \in F, x \neq 0$ there exists a $y \in G$ such that $B(x,y) \neq 0$, and if for all $y \in G, y \neq 0$ there exists an $x \in F$ such that $B(x,y) \neq 0$.

If E is a locally convex Hausdorff space and $E' := L(E,\mathbf{R})$ its **topological dual** (space), then (E,E') forms a dual pair with respect to the bilinear form

$$(x,x') \mapsto B(x,x') = \langle x,x' \rangle := x'(x)$$

on $E \times E'$.

For a dual pair (F,G) with respect to a bilinear form B also (G,F) is a dual pair with respect to B. In the sequel we shall employ the notation $(x,y) \mapsto B(x,y) =: \langle x,y \rangle$ for the bilinear forms defining the dual pairs (F,G) and (G,F).

Given a dual pair (F,G) the **weak topology** $\sigma(F,G)$ is introduced on F by the property that all linear functionals $x \mapsto \langle x,y \rangle$ $(y \in G)$ are continuous.

Properties B 10 *of the weak topology*

B 10.1 *$\sigma(F,G)$ is determined by the set $\{p_y : y \in G\}$ of semi-norms p_y on F given by*

$$p_y(x) := |\langle x, y\rangle|$$

for all $x \in F$.

B 10.2 *$\sigma(F,G)$ is a locally convex Hausdorff topology.*

B 10.3 *F and $\sigma(F,G)$ determine G within isomorphisms, i.e. $F' \cong G$, where the prime refers to the topology $\sigma(F,G)$.*

Let E be a locally convex Hausdorff space with topology τ. Then the **weakened topology** $\sigma(E,E')$ on E is coarser than τ, and E' is also the topological dual of E with respect to $\sigma(E,E')$.

Theorem B 11 *(Alaoglu, Bourbaki) For a locally convex space E any equicontinuous subset H of E' is $\sigma(E',E)$-relatively compact.*

Let (F,G) be a dual pair of vector spaces. Without loss of generality we assume that $G \subset F^*$, where F^* denotes the algebraic dual of F. A locally convex topology τ on F is said to the **compatible with the duality** if $G = F'$, where the prime refers to the topology τ. In particular, $\sigma(F,G)$ is compatible with the duality.

One shows by employing Mazur's separation theory that all topologies on F compatible with the duality yield the same system of closed convex sets. For each convex subset of F all topologies compatible with the duality lead to the same closure.

Theorem B 12 *Let (F,G) be a dual pair.*

(i) Every locally convex Hausdorff topology τ on F is compatible with the duality.

(ii) τ is the $\mathfrak{S}$-topology for a covering $\mathfrak{S}$ of G by convex, symmetric and $\sigma(G,F)$-compact subsets.

The finest among the $\mathfrak{S}$-topologies of *(ii)* (called the Mackey topology and denoted by $\tau(F,G)$) is defined by the system $\mathfrak{S}$ of all convex, symmetric and $\sigma(G,F)$-compact subsets of G. The coarsest among those $\mathfrak{S}$-topologies is $\sigma(F,G)$.

Theorem B 13 *(Arens, Mackey) A locally convex topology τ on F is compatible with the duality if and only if*

$$\sigma(F,G) \succ \tau \succ \tau(F,G)\,.$$

Application B 14 *to the dual pair (E,E') of a Banach space E.*

Let

$$H := \{x' \in E' : \|x'\| \leq 1\}$$

be the unit ball of E'. Then

B 14.1 *H is equicontinuous, hence $\sigma(E',E)$-compact.*

In fact, H is $\sigma(E',E)$-relatively compact by Theorem B 11. Moreover, H is easily seen to be $\sigma(E',E)$ - closed.

B 14.2 *If E is separable, then H is $\sigma(E',E)$-metrizable.*

One just notes that along with E also E' is separable and that under this hypothesis every equicontinuous subset of E' is metrizable with respect to the topology of simple convergence.

B 14.3 *For every $x \in E$ one has*

$$\|x\| = \sup_{\substack{x' \in E' \\ \|x\| \leq 1}} |\langle x, x' \rangle|\,.$$

This identity follows from the fact that the unit ball of E is the polar of the unit ball of E'.

C Commutative Banach algebras

Let A be a **normed algebra** in the sense that A is an algebra and a normed vector space over $\mathbf{C}$ such that for the norm $\|\cdot\|$ in A the inequality

$$\|xy\| \leq \|x\|\|y\|$$

holds whenever $x, y \in A$. If A admits a multiplicative unit 1 then A may be renormed such that $\|\cdot\| = 1$. For the subsequent discussion we assume this renorming being done.

Theorem C 1 *(Gelfand, Mazur) Every commutative normed algebra A which at the same time is a field, is algebraically and topologically isomorphic to $\mathbf{C}$.*

Now let A be a commutative Banach algebra with unit 1. An ideal I of A is said to be maximal if $I \neq A$ and if for every ideal I of A with $I \subset J \subset A$ one has either $J = A$ or $J = I$.

Properties C 2 *of ideals I of A*

C 2.1 *$\overline{I}$ is an ideal of A.*

C 2.2 *If I is maximal then I is closed.*

C 2.3 *For each closed ideal I of A the quotient A/I is a Banach algebra.*

Theorem C 3 *(Gelfand) For each subset I of a commutative Banach algebra with unit 1 the following statements are equivalent:*

(i) I is a maximal ideal of A.

(ii) I is a closed maximal ideal of A.

(iii) There exists a continuous epimorphism $f : A \to \mathbf{C}$ such that $f^{-1}(0) = I$.

(iv) There exists an epimorphism $f : A \to \mathbf{C}$ such that $f^{-1}(0) = I$.

For the proof of the implication $(i) \Rightarrow (iii)$ one observes that by Property C 2.3 A/I is a Banach algebra and by the maximality of I that A/I is a field. But this implies the existence of an algebraic and topological isomorphism $g : A/I \to \mathbf{C}$. Considering the canonical mapping $p : A \to A/I$ and putting $f := g \circ p$ the equalities

$$f^{-1}(0) = p^{-1}(g^{-1}(0)) = p^{-1}(0) = I$$

yield the desired statement.

Preparations C 4

C 4.1 *Every continuous epimorphism* $h : A \to \mathbf{C}$ *satisfies* $\|h\| \leq 1$.

Let $\Delta(A)$ denote the set of all (continuous) epimorphism from A onto $\mathbf{C}$.

C 4.2 $\Delta(A)$ *is a* $\sigma(A', A)$*-compact subset of the unit ball of* A'.

$\Delta(A)$ is said to be the ***maximal ideal space*** or the spectrum of A.

For every $x \in A$ the mapping $\hat{x} : \Delta(A) \to \mathbf{C}$ defined by

$$\hat{x}(h) := h(x)$$

for all $h \in \Delta(A)$ is continuous, the mapping $x \mapsto \hat{x}$ from A into $C(\Delta(A))$ is a homomorphism, and

$$\|\hat{x}\| = \sup_{h \in \Delta(A)} |\hat{x}(h)| = \sup_{h \in \Delta(A)} |h(x)| \leq \|x\|$$

whenever $x \in A$.

C 4.3 $\hat{A} := \{\hat{x} : x \in A\}$ *is a subalgebra of* $C(\Delta(A))$ *which separates* $\Delta(A)$ *and contains 1.*

In general, $\hat{A}$ *is* **not** *closed in* $C(\Delta(A))$.

C 4.4 *The topology of $\Delta(A)$ is the initial topology with respect to the set $\hat{A}$.*

This topology τ_g on $\Delta(A)$ is called the ***Gelfand topology***. In this connection we also introduce the ***Gelfand transform*** $\hat{x}$ of $x \in A$ and the ***Gelfand mapping*** (representation) $x \mapsto \hat{x}$ from A into $C(\Delta(A))$.

Theorem C 5 *Let A be a commutative Banach algebra with unit 1. Then*

(i) Every epimorphism $\hat{h}$ from $\hat{A}$ onto $\mathbf{C}$ is of the form

$$\hat{h}(\hat{x}) = \hat{x}(h)$$

for some $h \in \Delta(A)$.

(ii) Every $h \in \Delta(A)$ defines an epimorphism

$$\hat{x} \mapsto \hat{x}(h)$$

from $\hat{A}$ onto $\mathbf{C}$.

A Banach algebra A is said to ***involutive*** if A admits an involution $x \mapsto x^{\sim}$ defined as a mapping $A \to A$ with the properties that

(1) $(x+y)^{\sim} = x^{\sim} + y^{\sim}$,

(2) $(\lambda x)^{\sim} = \bar{\lambda} x^{\sim}$,

(3) $(xy)^{\sim} = y^{\sim} x^{\sim}$, and

(4) $x^{\sim\sim} = x$

whenever $x, y \in A, \lambda \in \mathbf{C}$.

Moreover, an involutive Banach algebra A is called a C^*-***algebra*** if

$$\|xx^{\sim}\| = \|x\|^2$$

for all $x \in A$.

Theorem C 6 *(Gelfand's formula) Let A be a commutative Banach algebra with unit and Gelfand mapping $x \mapsto \hat{x}$. Then, for each $x \in A$ the formula*

$$\|\hat{x}\| = \lim_{n\to\infty} \|x^n\|^{\frac{1}{n}}$$

holds.

Theorem C 7 *Every commutative C^*-algebra A with 1 is* ***semisimple*** *in the sense that the Gelfand mapping $x \mapsto \hat{x}$ of A is a norm- and involution preserving isomorphism from A onto $C(\Delta(A))$.*

For the proof one starts by showing that $x \mapsto \hat{x}$ is involution-preserving. Here preparation C 4.1 is applied. The property of $x \mapsto \hat{x}$ being norm-preserving is established with the help of Gelfand's formula C 6.

In fact, we know already that $\|\hat{x}\| \leq \|x\|$ holds for all $x \in A$. For the inverse inequality we first choose $x \in A$ with $x = x^\sim$. Then

$$\|x^2\| = \|xx^\sim\| = \|x\|^2,$$

hence $\|x^2\|^{\frac{1}{n}} = \|x\|$, and by induction

$$\|x^{2^n}\|^{\frac{1}{2^n}} = \|x\|$$

whenever $n \geq 1$. Now, Theorem C 6 applies and yields

$$\begin{aligned}\|\hat{x}\| &= \lim_{n\to\infty} \|x^n\|^{\frac{1}{n}}\\ &= \lim_{n\to\infty} \|x^{2^n}\|^{\frac{1}{2^n}} = \|x\|.\end{aligned}$$

Next, for arbitrary $x \in A$ we have $(xx^\sim)^\sim = xx^\sim$, hence

$$\|x\|^2 = \|xx^\sim\|$$

and therefore

$$\begin{aligned}\|x\|^2 &= \|xx^\sim\| = \|(xx^\sim)^\wedge\|\\ &= \|\hat{x}\widehat{x^\sim}\|\\ &= \|\hat{x}\hat{\bar{x}}\|\\ &= \|\,|\hat{x}|^2\,\| = \|\hat{x}\|^2.\end{aligned}$$

Finally, $\hat{A}$ being an involutive subalgebra of $C(\Delta(A))$ with unit which separates $\Delta(A)$ is dense in $C(\Delta(A))$. But along with A also $\hat{A}$ is complete, hence closed, and $\hat{A} = C(\Delta(A))$ has been established.

Now, let A be an involutive commutative Banach algebra which does not admit a unit. By $\Delta(A)$ we again denote the set of all continuous epimorphisms from A onto $\mathbf{C}$, furnished with the initial topology with respect to the mappings $h \mapsto h(x)$ from $\Delta(A)$ into $\mathbf{C}$ (for all $x \in A$). Let $\tilde{A} := A \oplus \{1\}$ denote the involutive commutative algebra arising from adjoining a unit element 1. $\tilde{A}$ can be given a norm such that it becomes a Banach algebra with unit 1 and Gelfand mapping $x \mapsto \hat{x}$ (related to $\Delta(\tilde{A})$ and $\widehat{\tilde{A}}$). Let $h_0 \in \Delta(\tilde{A})$ be defined by

$$h_0(x) := \begin{cases} 0 & \text{if } x \in A \\ 1 & \text{if } x = 1. \end{cases}$$

Clearly, $\Delta(\tilde{A})\backslash\{h_0\}$ is a locally compact space, and the mapping $h \mapsto \mathrm{Res}_A h$ is a homeomorphism from $\Delta(\tilde{A}) \setminus \{h_0\}$ onto $\Delta(A)$. Therefore $\Delta(A)$ is locally compact, and $x \in \tilde{A}$ belongs to A if and only if $\hat{x}$ (restricted to $\Delta(A)$) vanishes at infinity.

Theorem C 8 *Let A be an involutive commutative Banach algebra. Then*

(i) $\hat{A}$ is a subalgebra of $C^0(\Delta(A))$.

(ii) If, in addition, A is a C^*-algebra, then A is **semisimple** *in the sense that $x \mapsto \hat{x}$ is a norm- and involution preserving isomorphism from A onto $C^0(\Delta(A))$.*

While (i) is clear by the remarks preceding the theorem, only (ii) requires an argument. In fact, there exists a unique norm on A which extends to $\tilde{A}$ and makes $\tilde{A}$ a C^*-algebra. From Theorem C 7 we conclude that $\tilde{A}$ is semisimple. Restricting the Gelfand mapping $x \mapsto \hat{x}$ to A and $C^0(\Delta(A))$ respectively we reach the desired conclusion.

Selected References

[1] Araujo, A. and Giné, E., *The central limit theorem for real and Banach valued random variables,* John Wiley & Sons, New York - Chichester - Brisbane - Toronto, 1980

[2] Bauer, H.. *Probability theory,* De Gruyter Studies in Mathematics, **23**. Walter de Gruyter, Berlin - New York, 1996

[3] Bendikov, A., *Potential theory on infinite-dimensional abelian groups,* De Gruyter Studies in Mathematics, **21**. Walter de Gruyter, Berlin - New York, 1995

[4] Berezansky, Yu.M. and Kalyuzhnyi, A.A., *Harmonic analysis in hypercomplex systems,* Mathematics and Its Applications, **434**. Kluwer Academic Publishers, Dordrecht - Boston - London, 1998

[5] Berg, C. and Christensen, J.P.R. and Ressel, P., *Harmonic analysis on semigroups. Theory of positive definite and related functions,* Graduate Text in Mathematics. Springer, New York - Berlin - Heidelberg - Tokyo, 1984

[6] Berg, C. and Forst, G., *Potential theory on locally compact abel ian groups,* Ergebnisse der Mathematik und ihrer Grenzgebiete, **87**. Springer, New York - Heidelberg, 1975

[7] Bloom, W. R. and Heyer, H., *Harmonic analysis of probability measures on hypergroups,* De Gruyter Studies in Mathematics, **20**. Walter de Gruyter, Berlin - New York, 1995

[8] Diaconis, P., *Group representations in probability and statistics,* Institute of Mathematical Statistics Lecture Notes - Monograph

Series, **11**. Institute of Mathematical Statistics, Hayward, California, 1988

[9] Dudley, R.M., *Random walks on abelian groups,* Proc. Am. Math. Soc. **13** (1962), 447-450

[10] Fel'dman, G.M., *Arithmetic of probability distributions, and characterization problems on abelian groups,* Translation of Mathematical Monographs, **116**. American Mathematical Society, Providence, Rhode Island, 1993

[11] Fernique, X., *Intégrabilité des vecteurs gaussiens,* C.R. Acad. Sci. Paris Ser. A 270 (1981), 1998-1699

[12] Folland, G.B., *A course in abstract harmonic analysis,* Studies in Advanced Mathematics. CRC Press, Boca Raton - Ann Arbor - London - Tokyo, 1995

[13] Forst, G., *The Lévy-Hinčin representation of negative definite functions,* Z. Wahrscheinlichkeitstheorie verw. Gebiete 34 (1976), 313-318

[14] Grenander, U., *Probabilities on algebraic structures,* Álmquist & Wiksell, Stockholm - Göteborg - Uppsala, and John Wiley & Sons, New York - London, 1963.

[15] Guichardet, A., *Analyse harmonique commutative,* Monographies Universitaires de Mathématique. Dunod, Paris, 1968

[16] Guivarc'h, Y., Keane, M. and Roynette, B., *Marches aléatoires sur les groupes de Lie,* Lecture Notes in Mathematics, **624**. Springer, Berlin - Heidelberg - New York, 1977

[17] Hannan, E.J., *Group representations and applied probability,* Methuen's Review Series in Applied Probability, **3**. Methuen & Co ltd, London, 1965

[18] Hazod, W. and Siebert, E., *Stable probability measures on Euclidean spaces and on locally compact groups. Structural properties and limit theorems,* Mathematics and Its Applications, **531**. Kluwer Academic Publishers, Dordrecht - Boston - London, 2001

[19] Hewitt, E. and Ross, K.A., *Abstract harmonic analysis I,* Die Grundlehren der Mathematischen Wissenschaften in Einzeldarstellungen, **115**. Springer, Berlin - Göttingen - Heidelberg - New York, 1970

[20] Heyer, H., *Probability measures on locally compact groups,* Ergebnisse der Mathematik und ihrer Grenzgebiete, **94**. Springer, Berlin - New York - Heidelberg, 1977

[21] Heyer, H. and Turnwald, G., *On local integrability and transience of convolution semigroups of measures,* Acta Applicandae Mathematicae 18 (1990), 283 - 296

[22] Högnäs, G. and Mukherjea, A., *Probability measures on semigroups. Convolution products, random walks, and random matrices,* The University Series in Mathematics. Plenum Press, New York - London, 1995

[23] Hoffman - Jörgensen, J., *Probability in Banach spaces,* Ecole d'Eté de Probabilités de Saint-Flour VI, 1976, pp. 1-186. Lecture Notes in Mathematics, **598**. Springer, Berlin - Heidelberg - New York, 1977

[24] Ito, M., *Transient Markov convolution semi-groups and the associated negative functions,* Nagoya Math. J. 92 (1983), 153-161. *Remarks on that paper,* Nagoya Math. J. 102 (1986), 181 - 184

[25] Ito, K. and Nishio, M., *On the convergence of sums of independent Banach space valued random variables,* Osaka J. Math. 5 (1968) , 25-48

[26] Ledoux, M. and Talagrand, M., *Probability in Banach spaces,* Ergebnisse der Mathematik und ihrer Grenzgebiete, 3. Folge, **23**. Springer, Berlin - Heidelberg - New York, 1991

[27] Linde, W., *Probability in Banach spaces. Stable and infinitely divisible distributions,* John Wiley & Sons, 1986.

[28] Loomis, L. *An introduction to abstract harmonic analysis,* D. van Nostrand, Toronto - New York - London, 1953

[29] Lukacs, E., *Characteristic functions,* Griffin's Statistical Monographs and Courses, **5**. Charles Griffin, London, 1960

[30] Meerschaert, M.M. and Scheffer, H-P., *Limit theorems for sums of independent random vectors,* John Wiley & Sons, 2001

[31] Neuenschwander, D., *Probabilities on the Heisenberg group. Limit theorems and Brownian motion,* Lecture Notes in Mathematics, **1630**. Springer, Berlin - Heidelberg, 1996

[32] Parthasarathy, K.R., *Probability measures on metric spaces,* Probability and Mathematical Statistics, **3**. Academic Press, New York - San Francisco - London, 1967

[33] Parthasarathy, K.R., *Introduction to probability and measure,* MacMillan, Delhi - Bombay - Calcutta - Madras, 1977

[34] Port, S.C. and Stone, C.J., *Infinitely divisible processes and their potential theory, I, II,* Ann. Inst. Fourier (Grenoble) 21, 2 (1971), 157-275, and 21,4 (1971), 179-265

[35] Revuz, D., *Markov chains,* North Holland Mathematical Library, **11**. North Holland Publishing Company, Amsterdam - Oxford, and American Elsevier Publishing Company, New York, 1975

[36] Rossberg, H.-J., Jesiak, B. and Siegel, G., *Analytic methods in probability theory,* Akademie-Verlag, Berlin 1985

[37] Roynette, B., *Marches aléatoires sur les groupes de Lie,* Ecole d'Eté de Probabilités de Saint-Flour VII (1977), pp. 237-379. Lecture Notes in Mathematics, **678**. Springer, Berlin - Heidelberg - New York, 1978

[38] Rudin, W., *Fourier analysis on groups,* Interscience Tracts in Pure and Applied Mathematics, **12**. Interscience Publishers, New York - London, 1962

[39] Ruzsa, I.Z. and Székely, G.J., *Algebraic probability theory,* John Wiley & Sons, Chicester - New York - Brisbane - Toronto - Singapore, 1988

[40] Sato, K., *Lévy processes and infinitely divisible distibutions,* Cambridge studies in advanced mathematics, **68**. Cambridge University Press, 1999

[41] Schaefer, H.H. (with Wolff, M.P.), *Topological vector spaces,* 2nd edition, Springer, New York - Berlin - Heidelberg, 1999

[42] Spitzer, F., *Principles of random walk,* The University Series in Higher Mathematics. D. van Nostrand, Princeton - Toronto - New York - London, 1964

[43] Stromberg, K., *Probability for analysts,* Chapman & Hall Probability Series. Chapman & Hall, New York - London, 1994

[44] Stroock, D.W., *Probability theory. An analytic view,* Cambridge University Press, 1993

[45] Tortrat, A., *Calcul des probabilités et introduction aux processus aléatoires,* Masson & Cie, Paris, 1971

[46] Vakhania, N.N., Tarieladze, V.I. and Chobanyan, S.A., *Probability distributions on Banach spaces,* Mathematics and Its Applications (Soviet Series). D. Reidel (Kluwer), Dordrecht - Boston - Tokyo, 1987

[47] Woess, W., *Random walks on infinite graphs and groups,* Cambridge Tracts in Mathematics, **138**. Cambridge University Press, 2000

[48] Yoshida, K., *Functional Analysis,* 4th edition, Die Grundlehren der mathematischen Wissenschaften in Einzeldarstellungen, **123**. Springer, Berlin - Heidelberg - New York, 1974

[49] Zygmund, A., *Trigonometric Series I,* Cambridge University Press, 1968

Symbols

A^1 . . . . . set of normed elements of A
$A^\times$ . . . . . set of elements of A different from a distinguished element
A^b . . . . . set of bounded elements of A
$A_{\mathbf{R}}$ . . . . . set of real elements of A
$A_{\mathbf{C}}$ . . . . . set of complex elements of A
$[A]$ . . . . . group generated by a set of A
$[A]^-$ . . . . . closed group generated by a set of A
$\langle A \rangle$ . . . . . semigroup generated by a set of A
$\langle A \rangle^-$ . . . . closed semigroup generated by a set of A
Res A . . . . restriction to a set A
$[\mu]$ . . . . . μ-almost everywhere
$\wedge$ . . . . . . minimum
$\vee$ . . . . . . maximum

Index